Typos
P. 56
P. 58
116
124
125
133_183
186
204
205
208
209
212
213
244
264
291
304
314 333?
338
339
356

A SECOND GENERATION OF MULTIVARIATE ANALYSIS

Volume 1

Methods

Edited by
Claes Fornell

PRAEGER SPECIAL STUDIES • PRAEGER SCIENTIFIC

Library of Congress Cataloging in Publication Data

Main entry under title:

A Second generation of multivariate analysis.
    Includes bibliographies.

    Contents: v. 1. Methods—v. 2. Measurement and
evaluation.   I. Multivariate analysis.   I. Fornell,
Claes.
QA278.S4   1982   519.5'35   82-11273
ISBN 0-03-062632-3 (set)
ISBN 0-03-061604-2 (v. 1)
ISBN 0-03-062627-7 (v. 2)

Published in 1982 by Praeger Publishers
CBS Educational and Professional Publishing
a Division of CBS Inc.
521 Fifth Avenue, New York, New York 10175, U.S.A.

© 1982 by Praeger Publishers

23456789   052   987654321

Printed in the United States of America

List of Contributors

Erling B. Andersen, University of Copenhagen

Richard P. Bagozzi, Massachusetts Institute of Technology

Peter M. Bentler, University of California, Los Angeles

Fred L. Bookstein, The University of Michigan

D. R. Cox, Imperial College, London

Daniel R. Denison, The University of Michigan

Claes Fornell, The University of Michigan

Karl G. Jöreskog, University of Uppsala

Thomas R. Knapp, University of Rochester

Harri Kiiveri, University of Western Australia

Petter Laake, University of Oslo

David F. Larcker, Northwestern University

William Love, Nova University

Willem E. Saris, Vrije Universiteit

Tore Schweder, University of Tromso

David C. Stapleton, Carnegie-Mellon University

David Stewart, University of Pittsburgh

Arnold L. van den Wollenberg, Katholieke Universiteit

Herman Wold, University of Uppsala and University of Geneva

TO MY PARENTS

Preface

In view of the many advances that have occurred in multi-variate analysis over the past few years, it seems appropriate to try to capture the essence of the major developments within a unifying framework. This volume suggests a second generation of multivariate analysis that, in many ways, represents a new way of thinking about empirical research—to describe such a framework. Its key challenge is to bring theory and data together, while at the same time recognizing their intrinsically imperfect correspondence. Its approach is to include both theoretical and observed variables in empirical analysis, specifying their hypothesized linkages and letting theory and data operate in an interactive fashion.

From the chapters included in this book, it is evident that the methods discussed have a great deal to offer the social scientist in any field, be it psychology, sociology, economics, marketing, or some other specialty. However, the offer is not without cost. It is assumed that the reader is familiar with traditional multivariate analysis and has a reasonably high level of technical sophistication. In fact, one of the objectives of this book is to demonstrate that many technical issues are too important to leave to the technical specialist. Because these issues have significant methodological and substantive implications, it is important that they be understood by the empirical researcher as well.

I wish to thank Richard P. Bagozzi, MIT; Frank J. Carmone, Jr., Drexel University; Imran S. Currim, UCLA; George John, University of Wisconsin; Robert J. Meyer, Carnegie-Mellon University; William D. Perreault, Jr., University of North Carolina; and Michael J. Ryan, University of Michigan, for reviewing the prospectus and providing many valuable comments. In addition, thanks are owed to Fred L. Bookstein, University of Michigan; Daniel R. Denison, University of Michigan; Karl G. Jöreskog, University of Uppsala; David F. Larcker, Northwestern University; Lynn W. Phillips, Stanford University; and Herman Wold, University of Uppsala and University of Geneva, for stimulating conversations that led to many of the ideas behind this book. I also appreciate the valuable interaction with the students who participated in my doctoral seminar on advanced methodology. Specifically, I would like to thank Thomas E. Buzas, now at the

University of Florida, and William T. Robinson, Gerald J. Tellis, and George M. Zinkhan, who is now at the University of Houston. Finally, I am grateful to the authors from so many different countries and research institutions who contributed chapters to this anthology. Toward integrating empirical analysis with theoretical development, their effort represents an impressive example of international scholarship.

Claes Fornell

Contents

1

A Second Generation of Multivariate Analysis— An Overview

Claes Fornell

In response to an increasing concern for measurement quality, the social sciences are beginning to draw upon a new generation of multivariate analysis methods. The common characteristic of these methods is that they are all part of a methodological recognition that scientific theory involves both abstract and empirical variables. Theoretical reasoning is, by definition, abstract. Without abstraction, there is little hope for generality, integration of concepts, or parsimony; yet in the absence of data, theory remains abstract and imaginary. It is therefore a fundamental objective of methodology to bring data and theory together; and the purpose of the second generation of multivariate analysis is to help achieve that objective.

The methods discussed in this volume all deal with abstract variables and their empirical connections in one form or another. Included are: canonical analysis, redundancy analysis, external single-set components analysis (ESSCA), the LISREL (analysis of Linear Structural Relationships) factor-analytic structural equation model, the PLS (Partial Least Squares) component structural equation model, and the CMDA (Constrained/ Confirmatory Monotone Distance Analysis) confirmatory multidimensional scaling model.[1] In order to understand what it is that these methods do and why they can be said to represent a new generation of multivariate analysis, let us briefly review the characteristics of earlier methods of multivariate analysis.

With the rapid dissemination of computers and the subsequent development of "packaged" analysis programs such as SPSS, BMD, SAS, and others, the 1960s represented the beginning of a shift in research orientation for many social sciences: from

abstract theory, often void of empirical fact, toward increasing
empiricism, now often without the justification of or interest in
making abstract claims. The next decade saw dramatic changes
in the way in which data were analyzed. Seemingly complex
phenomena could be structured into interpretable forms, multi-
dimensionality was readily portrayed, vast amounts of information
were handled simultaneously, and researchers were able to reduce
their invocation of ceteris paribus to a significant degree. Factor
analysis, cluster analysis, multidimensional scaling, MANOVA,
discriminant analysis, and principal components analysis all
became familiar terms.

It is easy to understand the rapid diffusion of multivariate
analysis. As already implied, the necessary computing resources
(both hard- and software) had become increasingly available.
It was also evident that multivariate analysis represented a considerable step forward from earlier univariate and bivariate
methods. Another factor that probably affected the adoption
rate is the fact that the attempts during the 1960s to employ
econometrics and operations research methods was often un-
successful, especially in the area of business research.[2] The
failure of some of these approaches was probably due to their
requiring more a priori knowledge than what was available at
the time.

Thus, the first generation of multivariate methods with
distinct advantages over earlier methods entered the social
sciences. Because these methods require fewer statistical
assumptions and less a priori theoretical knowledge, they
offered the promise of ridding the researcher of the often unrealistic assumptions and requirements that plagued other
quantitative approaches. The relaxation of statistical assumptions
meant that many methods were best suited for descriptive research. Similarly, a priori hypotheses were often replaced by
post hoc interpretation. Along with their empirical (as opposed
to theoretical or conceptual) precedence, virtually all social
sciences received a heavy dose of empiricism; data ruled while
theory took a back seat. Perhaps because the multivariate
methods could handle large quantities of data, relatively little
consideration was given to the quality of the data. Certainly,
the relationship between abstract conceptual schemas and empirical data received scant attention. Yet there is no question
that conceptual (theoretical) variables are necessary for scientific
progress. The exchange of inferential theory and a priori
hypotheses for fewer statistical assumptions and a process by
which data preceded conceptualization was a price paid for using
first-generation methods.

Another feature of most first-generation methods was the restriction of multivariance to one side of the equation. Although econometric models and canonical correlation (with MANOVA as a special case) were capable of handling multiple criterion and multiple predictor variables, econometric models were rarely applied outside economics and canonical correlation was used only sparingly because of its perceived interpretational difficulties. However, when multivariance is restricted to either predictor or criterion variables, we do not have a truly multivariate analysis. For example, the common practice of estimating separate multiple regressions for different criterion variables assumes that R_{yy} (the correlation matrix of y-variables) is an identity matrix (that is, the y-variables are not correlated), an assumption that is seldom justified.

Whereas first-generation methods required little in terms of a priori specification, second-generation methodology emphasizes the cumulative aspect of theory development by which a priori knowledge is incorporated into empirical analysis. Such a priori knowledge can derive from theory, previous empirical findings, or research design. The new methods are well suited to capitalize on either what is known or what is hypothesized, and thus build upon what has been accomplished by past research. Because these methods can combine as well as confront theory with empirical data, they offer a potential for scientific explanation that goes far beyond description and empirical association.

Another advantage offered by second-generation methodology is the explicit treatment of measurement error as (a) imperfect correspondence between abstract and empirical variables and/or (b) imperfect observation of empirical variables. Again, the first-generation methods have limited capabilities for assessing measurement quality or for correcting for measurement error.

In sum, then, the second generation of multivariate analysis includes methods that have a capability to

Model Measurement Error: Very few, if any, measures in the social (or natural) sciences are free from error. Variables such as earnings or sales, population growth, and age are subject to error because of inaccurate statistics, faulty record keeping, or imperfect coding. In ignoring these errors, the analyst runs the risk of obtaining biased parameter estimates.

Incorporate Abstract and Unobservable Constructs: The measurement of theoretical constructs always involves some sort of abstraction. For every

measure there is a corresponding qualitative account.
There are numerous examples of this. Attitude,
personality, and social status are constructs without
a direct linkage to observable quantitative measures;
they are abstract variables. Yet we need to include
them in empirical analysis. The second-generation
methods provide a vehicle for doing this.

Combine and Confront A Priori Knowledge and
Hypotheses with Empirical Data: If growth of knowledge is a measure of scientific progress, a priori
notions must play a dual role in empirical inquiry:
as assumptions that are given and as hypotheses
that are to be tested. The former role combines
theory and data; the latter confronts theory with
data. The second-generation models permit a good
deal of flexibility for theory to play both roles by
constraining or fixing certain parameters a priori
while estimating others. Many of the first-generation
methods, including multidimensional scaling, traditional factor analysis, principal components, and
cluster analysis, are unable to confront data for
inferential purposes. They are essentially descriptive methods void of statistical theory. As a result,
hypothesis testing is difficult. The second-
generation methods are different: They tend to be
confirmatory rather than exploratory—some more so
than others. In other words, the new methods go
beyond curve fitting; statistical inference theory
can be either incorporated in the estimation or
applied to the results.

SOME KEY CONSIDERATIONS

Before discussing the specific methods of the second generation, it will be helpful to introduce some key considerations that
provide guidelines for choosing among these methods. The
choice of a method depends on such things as the objective of
the study, the nature of the data, and the assumptions associated
with the method. These issues are covered in some detail in
the chapters in this volume. However, there are more general
considerations that, to some extent, have escaped attention in
the literature. We turn now to a discussion of these issues.
Specifically, they concern the nature of the theoretical variables
(constructs), the nature of the relationships among theoretical
variables, and the nature of the epistemic relationships.

The Nature of Theoretical Constructs

A theoretical construct is a variable (explanatory or criterion) that is of interest to the substantive context under examination. Constructs are related, via various rules of correspondence, to one or more empirical indicators (sometimes called manifest variables). In general, if measurement error is to be modeled, several indicators per construct are desirable although not always necessary. Depending on the way this error is treated, constructs can be classified in two categories: defined and indeterminate.

A defined construct is a composite (often called a component or a derived variable) of its indicators. An indeterminate construct (often called a factor or a latent variable) is a composite of its indicators plus an error term. In the psychometric literature, the controversy regarding the pros and cons of each type of construct dates back to the 1930s. Despite the long history of this debate, there are no signs that it is moving toward a settlement. On the contrary, recent psychometric literature has shown evidence of a revival of interest in the problem.[3]

Defined constructs sacrifice the theoretical desirability of allowing for imprecise measurement for the practical advantage of construct estimation and direct calculation of component scores. As the name implies, a defined construct is completely determined by its indicators and assumes that the combined effect of the indicators is free from measurement error. However, the effect of each individual indicator is allowed to be less than perfect. That is, the proportion of indicator variance that is not shared by the construct is considered to be error in measurement. In contrast, an indeterminate construct would not only allow for this type of error but also for the possibility that the combined indicator effect contains error. The drawbacks of indeterminate constructs are the interpretational difficulties that arise because an infinite number of different factor structures may have identical correlations with the indicators and because of improper or inadmissible solutions (for example, negative variances and correlations greater than unity) that often result.[4]

Since the choice between defined and indeterminate constructs involves a trade-off of characteristics, the overriding consideration must be the nature of the research. For example, if construct estimation and/or further model development based on individual scores is important, defined constructs would be preferable. On the other hand, if there is reason to believe that even the combined effect of the indicators contain significant error, indeterminate constructs would be more appropriate.

The Nature of Construct Relationships

Linear relationships can be described as orthogonal, symmetric, unidirectional, bidirectional, or causal.

Orthogonality implies a zero correlation. Orthogonality is either imposed on or assumed to exist among certain parameters, variables, and constructs. In psychometrics, orthogonality is often seen as a virtue because it facilitates interpretation by separating the various variance components into distinct parts. In econometrics, orthogonality has typically been treated in a different context—as something that may have to be imposed in order to identify a system of equations or as something suggested by substantive theory. These different perspectives on the role of orthogonality flow from the contexts in which psychometric research and econometric research traditionally are conducted. To the psychometrician, often concerned with problems of data dimensionality and post hoc interpretations, it is helpful to identify the distinct sources of variation. To the econometrician, orthogonality does not play a major role in interpretation, but it may be necessary for identifying simultaneous equations and/or for conforming with statistical assumptions.

The second generation of multivariate analysis embraces both the econometric and the psychometric tradition in its treatment of orthogonality. To the extent that the analysis is confirmatory rather than exploratory, orthogonality is invoked as suggested by the substantive theory under study, as required by statistical theory, and as required for parameter identification. To the extent that the overriding concern is optimal prediction without a priori specification of theoretical variables, orthogonality can be invoked to aid interpretation.

Whereas orthogonality implies the absence of a relationship, symmetry suggests that there is no distinction in the direction of a relationship. The correlation coefficient and the Euclidian distance are examples of measures of symmetric relationships. They are useful for indicating the strength of relationships but not how one variable affects another.

Directional relationships tell us how much a dependent variable will change given a change in an independent variable. Whereas the symmetric correlation can provide information on the scatter about the linear equation estimate (and, hence, the amount of error in equations), directional relationships reveal more of the essence of variable associations. There are two kinds of directional relationships: unidirectional (recursive) and bidirectional (nonrecursive). The regression coefficient is the primary measure of a unidirectional relationship. It implies

that x has an effect on y but says nothing about the effect of
y on x. In bidirectional relationships, there is a reciprocal
effect such that a change in x contributes to a change in y,
which in turn contributes to a further change in x, which contributes to yet a further change in y, and so on.

Although directional parameters contribute more than symmetric parameters to scientific explanation, they do not permit
inferences about causal order without highly restrictive assumptions. Causal laws cannot be proven; they are always assumed
by the researcher. When the laws appear to be violated, they
are reformulated so as to account for existing facts.[5] In experimental design, causality is traditionally assumed when there is
enough evidence to suggest that the null hypothesis is false.
This is not true for survey designs, where one cannot infer
causal ordering from a knowledge of correlations or regression
coefficients. However, it is possible to reason in the other
direction: If one knows the causal ordering, it is possible to
infer what the correlations should be.[6]

The causal parameter in some of the second-generation
models is derived from a decomposition of the correlation coefficient into direct and indirect effects. This decomposition provides a means for testing a hypothesized causal model if the
model is expressed in terms of correlations. If the model is
correctly specified, the correlation matrix should be statistically
indifferent to the sums of the direct and indirect effects specified
by the researcher.[7]

The Nature of Epistemic Relationships

An epistemic relationship describes the link between theory
and data. It is known as "rules of correspondence" or "correspondence postulates" in philosophy of science and as "auxiliary
theory" in quantitative sociology. Because abstract concepts
cannot, by definition, be directly observed, they are linked to
the empirical world via indicators or indirect observables. It is
through these "intervening" variables that theory is tested
against empirical data. There are three types of epistemic
relationships that impact upon the choice of analysis method:
reflective indicators, formative indicators, and symmetric
indicators.

Figure 1.1 depicts how abstract constructs and empirical
data are related in each of these types.[8] Abstract concepts
are drawn as circles and empirical observations as squares.
Thus, the A's in Figure 1.1 are unobserved and the x's are

FIGURE 1.1

Epistemic Relationships

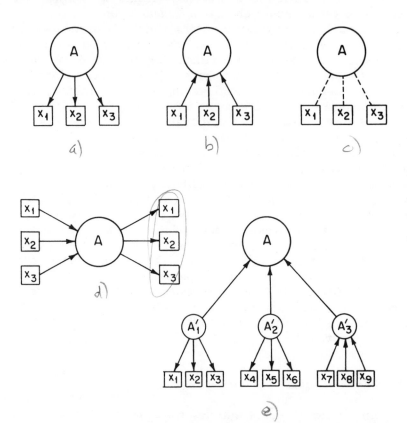

observed. In Figure 1.1a the indicators are reflective of the
unobserved theoretical construct. In other words, the unobserved construct is thought to give rise to what we observe.
This conceptualization is found in true score theory in psychometrics and is operationalized via factor analysis. Examples go
back to Spearman and Thurstone in the study of human ability.
Other constructs that are often measured empirically via reflective
indicators are personality traits and attitudes.

In Figure 1.1b, formative indicators give rise to the unobserved theoretical construct. In this case the empirical indicators produce or contribute to the construct. We may think
of "social status" as an example of such a construct. Although
we cannot observe social status directly, we may define it as
being produced by occupation, income, location of residence,
and so on.

Symmetric indicators are depicted in Figure 1.1c. In this instance we make no assumptions about the directionality or causality between constructs and empirical indicators. Instead, indicators are mapped into an abstract space of some determinate structure. We simply organize certain observed objects into a symbolic system where the functional form between construct and indicators is not necessarily specified.

Figure 1.1d illustrates a multiple-indicator multiple-cause (MIMIC) model. This representation makes use of a combination of reflective and formative indicators. For example, the unobserved social status construct may be formed by income, occupation, and so on (x-variables), and may reflect (cause) certain behaviors and attitudes (y-variables).

Figure 1.1e is an example of a higher-order model via a combination of reflective and formative indicators at two levels. Here the construct A is linked to empirical indicators via the intervening constructs A_1' to A_3', which occupy an intermediate position between theory and data. As shown in Figure 1e, A_1' and A_2' "cause" both the theoretical construct and the empirical observations. A_3', on the other hand, is formed from the empirical observations x_7 to x_9. To illustrate, consider the construct "advertising effort." Let us assume that A represents this construct, which is not directly observable but which might be defined (formed) by the quality of the media used, the quality of the copy, and the amount of advertising expenditure. For example, A_1' may be media quality measured by three empirical indicators (x_1, x_2, x_3—for example, ratings by media professionals); A_2' may be copy quality measured by three other indicators (x_4, x_5, x_6—for example, results from a copy pretest). The measures x_1 to x_3 and x_4 to x_6 are assumed to reflect media quality and copy quality, respectively. Thus, the intervening constructs A_1' and A_2' give rise to the observations and (partially) form A. As for the third intervening construct (A_3'—advertising expenditure), it is depicted as having formative indicators. Here, the allocation of the advertising budget defines expenditure, and x_7 to x_9 simply represent three different estimates of advertising expenditure (based, for example, on different ways of accounting for brands that share the same advertising budget).

Having discussed the nature of the theoretical constructs (defined, indeterminate), the nature of the relationships between constructs (orthogonal, symmetric, unidirectional, bidirectional, causal), and the nature of the epistemic relationships (reflective, formative, symmetric), let us now see how these considerations relate to each specific method.

THE METHODS

A great many methods for multivariate analysis have been developed over the past decade; only a handful, however, give explicit consideration to relationships at the abstract level. In this section six such methods are discussed.

Canonical correlation analysis, redundancy analysis, and external single-set components analysis represent a class of models that include multiple variables on both sides of the equation but, as we shall see, are somewhat limited in their capability to analyze systems of relationships. These models will be discussed first.

The LISREL formulation of factor-analytic structural equations, the PLS version of component structural equations, and the CMDA solution to confirmatory multidimensional scaling (CMDS) are not only capable of handling multiple criteria and predictors but are particularly appropriate for systems analysis. These methods are more general and also demand more sophistication on the part of the analyst. While there are many social scientists who shun technical issues, there is no way to avoid the fact that a high level of technical knowledge is necessary for dealing with the complexities of scientific inquiry. For the reader who is less familiar with the theoretical background of structural equation models, canonical correlation can be used as a bridge because it is a special case of these models. It also provides a link "backward" to such methods as multiple regression and correlation, discriminant analysis, ANOVA, MANOVA, and so on, which, in turn, are special cases of canonical correlation. In addition, both redundancy analysis and ESSCA are outgrowths of the canonical model. It seems appropriate, therefore, to introduce the second-generation methods via a discussion of canonical correlation analysis.

Let us first describe canonical correlation in terms of the key considerations discussed in the previous section. Figure 1.2 depicts a canonical model with two pairs of theoretical variables ($\hat{X}_1$, $\hat{X}_2$ and $\hat{Y}_1$, $\hat{Y}_2$); three observed x-variables (x_1, x_2, x_3); three observed y-variables (y_1, y_2, y_3); two error terms at the theoretical level (ζ_1, ζ_2); three error terms relating to the y-variables (ε_1, ε_2, ε_3); and three error terms relating to the x-variables (δ_1, δ_2, δ_3).[9] In the canonical model, all theoretical variables are defined by the corresponding empirical variables. This means that $\hat{X}_1$ and $\hat{X}_2$ can be expressed as a perfect (error free) linear combination of the x-variables and $\hat{Y}_1$ and $\hat{Y}_2$ as a perfect linear combination of the y-variables. As can be seen from the direction of the arrows, all epistemic

FIGURE 1.2

A Canonical Model

[handwritten annotation: — BOTH FORMATIVE — EXPLAINS THE DIFFERENCE BETWEEN THE PAIRED CONSTRUCTS.]

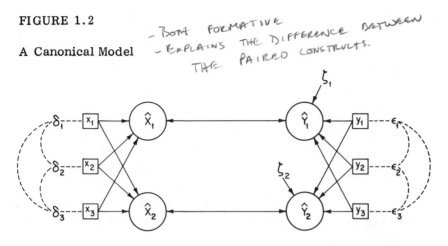

relationships are formative. The error terms, δ's and ϵ's, are not measurement error in the true score theory sense; rather they represent the variance of the empirical variables that is not a part of the theoretical variables. Although this representation of error terms is not standard, it satisfies the purpose of denoting variance unaccounted for in the relationship between a theoretical variable and an empirical observation. The dotted curved lines imply that the model does not require these error terms to be orthogonal. The nature of relationships between theoretical variables is either symmetric or orthogonal. Specifically, the relationship between corresponding pairs ($\hat{X}_1$ - $\hat{Y}_1$ and $\hat{X}_2$ - $\hat{Y}_2$) is symmetric, while the relationship between non-corresponding pairs ($\hat{X}_1$ - $\hat{Y}_2$ and $\hat{X}_2$ - $\hat{Y}_1$) and variables on the same side of the equation ($\hat{X}_1$ - $\hat{X}_2$ and $\hat{Y}_1$ - $\hat{Y}_2$) is orthogonal. In sum, then, the canonical model involves defined theoretical variables with symmetric or orthogonal relationships and formative indicators.

Chapter 2 by Knapp shows that canonical correlation analysis is a general model for a large number of other methods. Chapter 3, written by this author, argues that canonical analysis can be used in at least three different contexts, depending upon the objectives of the research. While all three approaches are basically exploratory in nature, they provide a starting point for dealing with both confirmatory canonical analysis (discussed in the introduction to the evaluation section of Volume 2 of this work) and the redundancy issue. It is also shown that the canonical model can be applied to directional relationships between theoretical variables (for example, $\hat{X}_1$) and the empirical variables relating to the other set of theoretical variables (for example y_1, y_2, y_3) without changing the properties of the model.

Chapter 4, by Stewart and Love, highlights an interpretational difficulty in canonical correlation: The canonical correlation coefficient reflects relationships at the theoretical and abstract level only. In order to provide a measure of predictive power at the level of empirical variables, these authors develop the redundancy index. Van den Wollenberg's redundancy analysis (Chapter 5) maximizes this index (instead of the canonical correlation coefficient). The result (relative to the canonical model) is higher loadings for variables to be predicted at the expense of a lower canonical correlation coefficient. This model is depicted in Figure 1.3. As can be seen from this figure, the only difference from canonical correlation is that $\hat{Y}_1$ and $\hat{Y}_2$ involve reflective indicators.

The external single-set components analysis is a MIMIC version of the redundancy model (see Chapter 6). It also maximizes redundancy, but without unobserved $\hat{Y}$-constructs. The ESSCA model forms optimal explanatory $\hat{X}$-constructs with respect to predicting the y-variables.[10] The model is shown in Figure 1.4. ESSCA is similar to the canonical model and redundancy analysis in that the constructs are defined. It differs in the sense that there are no relationships between theoretical variables; instead, the objective is for the theoretical variables to explain the observed y-variables. The error of the y-variables is thus composed of two parts (which are inseparable in all reduced-rank solutions[11]): error due to theory (the explanatory constructs account for less than the total variance in the y-variables) and error due to measurement.

FIGURE 1.3

A Redundancy Model

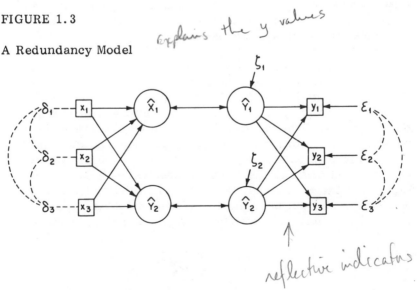

Explains the y values

reflective indicators

FIGURE 1.4

An ESSCA Model

MIMIC version of the Redundancy model

residual cannot be separated from structural error.

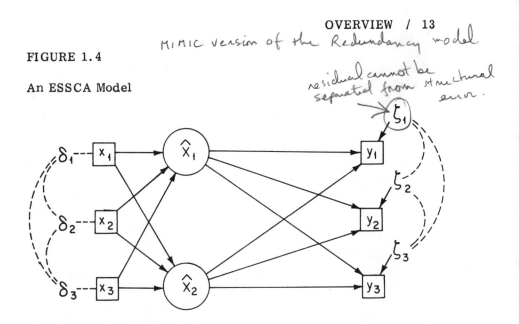

Chapter 7, by Bagozzi, Fornell, and Larcker, provides a bridge from canonical analysis to factor-analytic structural equations. Specifically, it is shown that a MIMIC model using LISREL maximum likelihood estimation, with correlated measurement errors for the y-variables and with the theoretical construct as a perfect linear combination of the x-variables, is identical to the canonical model. Even though canonical correlation analysis can be performed via factor-analytic structural equations, it should be clear that the latter method is not only more general but also represents a very different form of data analysis. The basic structure of this method was used in sociology more than a decade ago. The problem was that it was not clear how to reconcile the alternative estimates of overidentified models. This problem has now been solved. By generalizing his confirmatory factor analysis method, Karl G. Jöreskog developed the LISREL model, which is now widely used to estimate and test overidentified factor-analytic structural equations. Peter M. Bentler, another major contributor to the field,[12] provides an introduction and overview of this type of model in Chapter 8. This is followed in Chapter 9 by Willem Saris's "Linear Structural Relationships," which is a nontechnical introduction to the LISREL method. Jöreskog then describes his method in more detail in Chapter 10, "Analysis of Covariance Structures," which is followed (Chapter 11) by comments from several social scientists and statisticians. Also included is a chapter (12) in which Jöreskog replies to the comments.

The MIMIC model, which was first discussed in econometrics by Zellner and in sociology by Hauser and Goldberger,[13] is described in Chapter 13 by David Stapleton. By incorporating multiple theoretical variables, Stapleton also provides an extension of the original MIMIC formulation.

To compare and contrast the LISREL approach to canonical correlation, redundancy analysis, and ESSCA, an example of a LISREL model is shown in Figure 1.5. There are several things that should be noted. First, all indicators are reflective, and the (measurement) error terms are now drawn with arrows (pointing toward the observed variables). This is not to say that it is not possible to use formative indicators; this can be done by reformulating the model, but this is not standard practice, since it falls outside the classical test theory that underlies the measurement model in LISREL. In contrast to the methods discussed previously, the relationships between constructs in LISREL can take many forms. In Figure 1.5, for example, we have orthogonality between $\hat{X}_1$ and $\hat{Y}_2$, a symmetric relationship between $\hat{X}_1$ and $\hat{X}_2$, a directional and causal relationship between $\hat{X}_1$ and $\hat{Y}_1$, and a bidirectional and causal relationship between $\hat{Y}_1$ and $\hat{Y}_2$. Finally, the constructs in LISREL are indeterminate. It should also be mentioned that the error terms for the measure-

FIGURE 1.5

A LISREL Model

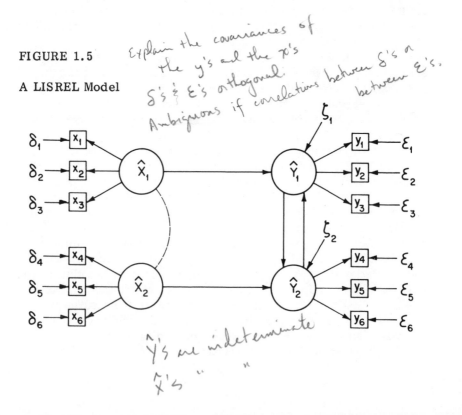

ment portion (that is, the δ's and ϵ's) can be correlated if so specified, although they are depicted as orthogonal in Figure 1.5. However, correlated measurement errors are rarely theoretically justified. They may be suggested by research design (for example, time series data or methods factors might be involved), but correlated errors should not be allowed on mere empirical grounds (regardless of how tempting it may be to improve the fit between the hypothesized model and the data by relaxing the orthogonality constraint for measurement errors).

Chapters 14, 15, and 16 deal with the Partial Least Squares approach to component structural equations. From his seminal work on fixed-point estimation, Herman Wold developed PLS as a least squares approach to structural equations with unobservables. Although PLS and LISREL have some characteristics in common—that is, the "fitting function" in both procedures is nonlinear, both are based on an iterative procedure, and both apply to a class of problems involving systems analysis of theoretical variables with fallible empirical indicators—they represent two fundamentally different approaches to analysis. These differences are related not only to statistical assumptions but to philosophical issues as well. One might speculate, therefore, that there will be, in the near future, much discussion on the relative superiority of LISREL versus PLS types of models; a debate similar to that on factor analysis versus principal components analysis. One issue of particular importance is the indeterminant versus defined nature of theoretical variables. LISREL subscribes to the first, PLS to the latter. This issue is discussed in the Fornell-Bookstein chapter (14), which compares LISREL and PLS results in a case where factor analysis assumptions do not hold. To facilitate the transition from LISREL to PLS, LISREL notation is used in this chapter. Herman Wold's "Systems under Indirect Observation Using PLS" (Chapter 15) then presents the background of the method, its formal specification, and its development to date. Fred L. Bookstein concludes the presentation of PLS in Chapter 16. By showing fixed points in diagrams of perpendiculars and linear combinations, he provides a geometrical interpretation that translates into simple command diagrams for estimating PLS models.

Figure 1.6 represents an example of a PLS model, for which it is common to use a combination of formative and reflective indicators. Here the exogenous theoretical constructs (X_1, X_2) have formative indicators and their endogenous counterparts ($\hat{Y}_1$, $\hat{Y}_2$) have reflective indicators. As mentioned, and in contrast to LISREL, all theoretical constructs are defined by their indicators. As is the case with LISREL, a number of different relationships can be portrayed. For example, the relationship between errors in measurement and corresponding manifest variables is orthogonal (otherwise PLS does not put any restrictions on the residuals in the measurement portion):[14] the relationship between $\hat{X}_1$ and $\hat{X}_2$ is symmetric; and the relationship between $\hat{X}_1$ and $\hat{Y}_1$ is undirectional and causal. Figure 1.6 is drawn without bidirectional relationships because very few applications of PLS have estimated nonrecursive equations. Recently, however, progress has been made in this direction, and it seems possible to include bidirectional relations in PLS by using techniques from econometrics.[15]

The final model in this volume is confirmatory multidimensional scaling. Although it is a very different approach compared to LISREL and PLS, it also applies to the analysis of systems of relationships of theoretical variables with measurement error. Whereas PLS requires fewer assumptions than LISREL, CMDS makes even fewer assumptions. Despite its capability for hypothesis testing, CMDS, as described in Chapter 17 by Fornell and Denison, does not specify a functional form and is not bound by traditional statistical assumptions. An example of CMDS is found in Figure 1.7, which describes a plot of nine variables (x's and y's) in a two-dimensional configuration. As in Figure 1.5 (of LISREL) and Figure 1.6 (of PLS), these observed varia-

FIGURE 1.6

A PLS Model

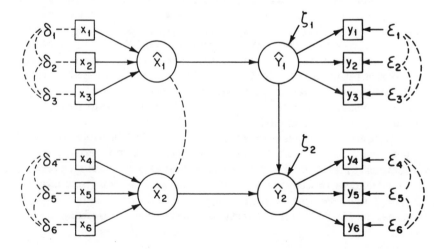

FIGURE 1.7

A Confirmatory Multidimensional Scaling Model

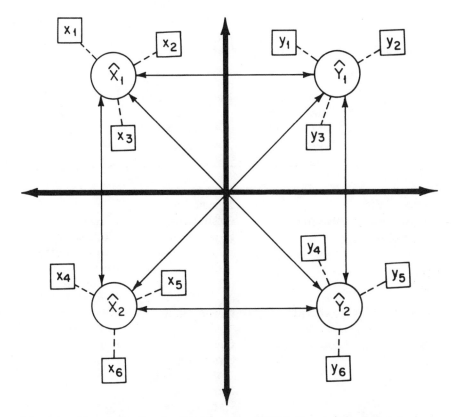

bles represent four constructs. In CMDS the constructs are
defined as the geometrical centroid of their respective indicators.
The epistemic relationships are given by the distance between
indicators and the corresponding centroid (for example, between
x_1 and $\hat{X}_1$); the greater this distance, the larger the measurement error. The relationships between theoretical variables
are measured by the distance between centroids (for example,
between $\hat{X}_1$ and $\hat{Y}_2$). If the Euclidian distance function is used,
both epistemic and construct relationships are symmetric.
Fornell and Denison discuss how a priori theory is incorporated
into the CMDS method and how results can be statistically evaluated.

Table 1.1 presents a classification of the various methods
according to the nature of theoretical constructs, the relationships between theoretical constructs, and epistemic relationships.

TABLE 1.1

A Classification of Second-Generation Methods

Methods	Nature of Theoretical Constructs	Relationships among Theoretical Constructs	Epistemic Relationships
Canonical correlation	Defined	Orthogonal Symmetric	Formative
Redundancy analysis	Defined	Orthogonal Symmetric	Formative (exogenous constructs) Reflective (endogenous constructs)
ESSCA	Defined	Orthogonal	Formative Reflective
LISREL	Indeterminate	Orthogonal Symmetric Unidirectional Bidirectional Causal	Reflective (Formative)
PLS	Defined	Orthogonal Symmetric Unidirectional (Bidirectional) Causal	Formative Reflective
CMDS	Defined	Symmetric	Symmetric

As is readily observed, LISREL and PLS are the most general and flexible methods; both allowing a variety of different types of relationships. LISREL has the capability of constraining more parameters (for example, in terms of orthogonality) than PLS, which, for example, does not generally impose constraints on residual covariances. Both types of epistemic relationships can be modeled in either model, although the theory underlying the measurement equations in LISREL refers to reflective indicators only.

NOTES

1. There are other methods available that meet the criteria (which will be discussed shortly) for inclusion among the second-generation methods. Some of these, such as latent class analysis, were excluded because of a shortage of suitable papers. The interested reader is referred to A. Mooijaart, "Latent Structure Analysis for Categorical Variables," in Systems under Indirect Observation: Causality, Structure, Prediction, ed. Karl G. Jöreskog and Herman Wold (Amsterdam: North Holland, 1982).

2. See Jagdish N. Sheth, "Multivariate Revolution in Marketing Research," Journal of Marketing 35 (January 1971): 13-14.

3. For reviews, see James H. Steiger, "Factor Independency in the 1930's and the 1970's—Some Interesting Parallels," Psychometrika 44 (1979):157-67; and James H. Steiger and Peter H. Schönemann, "A History of Factor Indeterminacy," in Theory Construction and Data Analysis, ed. S. Shye (San Francisco: Jossey-Bass, 1978), pp. 136-78.

4. For a discussion of this, see the chapters in this volume by Fornell and Bookstein, in Volume 2 by Jackson and Chan, and also Andersen's comment on Jöreskog and Jöreskog's reply in this volume.

5. For a discussion, see H. M. Blalock, Causal Inferences in Nonexperimental Research (Chapel Hill: The University of North Carolina Press, 1964).

6. Classic discussions here include the work of Herbert A. Simon, "Spurious Correlation: A Causal Interpretation," Journal of the American Statistical Association 44 (1954):464-79, and Blalock, see ibid. See also O. D. Duncan, Introduction to Structural Equation Models (New York: Academic Press, 1975).

7. Sewall Wright, "The Method of Path Coefficients," Annals of Mathematical Statistics 5 (September 1934):161-215, was the first to lay out simple path analytic rules for transform-

ing a causal model into a set of derived characteristics. Today's structural equations methodology relies heavily upon this idea. If the statistical equality between model-derived correlations and empirical correlations does not hold, the model is misspecified and the hypothesized causal ordering rejected.

8. A more general discussion of empirical meaning is found in the Bagozzi-Fornell chapter in the measurement section of Volume 2 and in Burt's chapter in the evaluation section of Volume 2.

9. All methods discussed in this chapter assume that the observed variables are correlated. For simplicity, these correlations are not drawn in the figures.

10. ESSCA thereby avoids the problem that only one of the sets of abstract variables is optimal (that is, either the $\hat{Y}$'s or the $\hat{X}$'s are optimal, given the specification of the other). For an elaboration on this and a solution within the context of redundancy analysis, see J. K. Johansson "An Extension of Wollenberg's Redundancy Analysis," Psychometrika 46, no. 1 (March 1981): 93-103. For a generalization of both canonical correlation and redundancy analysis, see Wayne S. DeSarbo, "Canonical/Redundancy Factoring Analysis," Psychometrika 46, no. 3 (September 1981): 307-29.

11. Full rank solutions of redundancy analysis and ESSCA cannot, of course, provide a higher redundancy than a full-rank canonical model. It is therefore not very meaningful to discuss full-rank models. In most analyses the researcher would consider only the dimensions that are significant in some sense, which usually implies a reduced-rank solution.

12. Bentler's structural equations program (EQS) is similar to LISREL but includes several different estimation procedures and somewhat more flexibility in the manner in which parameters can be constrained. It also facilitates higher-order factor analysis. The program is built upon the model presented by P. M. Bentler and David G. Weeks in "Linear Structural Equations with Latent Variables," Psychometrika 45, no. 3 (September 1980): 284-308.

13. A. Zellner, "Estimation of Regression Relationships Containing Unobservable Variables," International Economic Review 11 (1970): 441-54; and R. M. Hauser and A. S. Goldberger, "The Treatment of Unobservable Variables in Path Analysis," Sociological Methodology, ed. H. L. Costner (San Francisco: Jossey-Bass, 1971), pp. 81-117.

14. If residuals for reflective indicators are significantly correlated, the model is misspecified (the theoretical variable is not unidimensional). If the residuals for formative indicators

are significantly correlated, this is not necessarily due to mis-specification because the model is then not designed to account for these variances. However, if indicators have large residuals in the formative case, these variables should probably be excluded from the model.

15. See, for example, B. S. Hui, "On Building Partial Least Squares Models with Interdependent Inner Relations," in Systems under Indirect Observation: Causality, Structure, Prediction, ed. K. G. Jöreskog and H. Wold (Amsterdam: North-Holland, 1982).

2

Canonical Correlation Analysis: A General Parametric Significance–Testing System

Thomas R. Knapp

Significance tests for nine of the most common statistical procedures (simple correlation, t test for independent samples, multiple regression analysis, one-way analysis of variance, factorial analysis of variance, analysis of covariance, t test for correlated samples, discriminant analysis, and chi-square test of independence) can all be treated as special cases of the test of the null hypothesis in canonical correlation analysis for two sets of variables.

Ten years ago Cohen (1968) explained in language familiar to social scientists that the analysis of variance is a special case of multiple regression analysis and that both are subsumed under what the mathematical statisticians call the general linear model. The key concepts that link these two otherwise quite different techniques are the notion of a dummy variable (one for each of the "between" degrees of freedom) and the fact that differences between means and correlations between variables are analogous

Reprinted from Psychological Bulletin, 1978, Vol. 85, No. 2, pages 410-416. Copyright 1978 by the American Psychological Association. Reprinted by permission.

This chapter is concerned solely with traditional hypothesis testing, which continues to dominate empirical methodology (rightly or wrongly). No attempt is made to compare the relative merits of tests of significance and other modes of inference.

methodological concepts (a large difference between two sample means on a single variable conveys essentially the same information as a high correlation between that variable and the dummy variable of sample membership). Some authors of recent statistics textbooks for the social sciences, for example, Kerlinger and Pedhazur (1973), Cohen and Cohen (1975), and Roscoe (1975), have devoted full chapters or sections within chapters to similar explanations of this equivalence.

The purpose of this chapter is to extend Cohen's arguments even further by showing how virtually all of the commonly encountered parametric tests of significance can be treated as special cases of canonical correlation analysis, which is the general procedure for investigating the relationships between two sets of variables. Much of what follows has been compiled from various scattered sources. Originality is claimed only for the sections on correlated-sample t tests and chi-square tests. Consideration of chi-square tests for contingency tables as a type of canonical correlation analysis was mentioned by Darlington, Weinberg, and Walberg (1973) but was not further elaborated upon in that article.

The following section contains a formulation of the basic canonical problem in matrix notation (familiarity with matrix algebra, including a knowledge of eigenvalues and eigenvectors, is assumed)[1] and a specification of the associated F test of the significance of the correlation between linear composites of two sets of variables for sample data. The next sections describe how canonical correlation analysis can be used for simple (two-variable) correlation, the t test for independent sample means, multiple regression analysis, one-way analysis of variance, factorial analysis of variance, analysis of covariance, the t test for correlated sample means, discriminant analysis, and the chi-square test of independence of two variables. Some of these techniques are illustrated with numerical examples, using the Project Talent Test Battery data contained in the appendix of the multivariate text by Cooley and Lohnes (1971). The final section consists of a brief summary and a few remarks regarding assumptions.

THE GENERAL PROBLEM

As stated in a number of standard textbooks in multivariate analysis, for example, Anderson (1958), Morrison (1976), Cooley and Lohnes (1971), and Tatsuoka (1971), if there is one set of p variables and another set of q variables (where q is usually taken to be less than or equal to p), the principal objective of canonical correlation analysis is to find a linear combination of the p variables that correlates maximally with a linear combination of the q variables and, for sample data, to test the statistical significance of that correlation. The weights for the q variables in the second set are obtained by finding the elements of the eigenvector v_1 associated with the largest eigenvalue λ_1 of the matrix $M = R_{yy}^{-1}R_{yx}R_{xx}^{-1}R_{xy}$, where R_{yy}^{-1} is the inverse of the q × q matrix of intercorrelations among the q variables, R_{yx} is the q × p matrix of cross-correlations between the variables of the two sets, R_{xx}^{-1} is the inverse of the p × p matrix of intercorrelations among the p variables, and R_{xy} is the p × q transpose of R_{yx}. The weights for the p variables in the first set are obtained by finding the elements of the vector

$$v_2 = \lambda_1^{-\frac{1}{2}}R_{xx}^{-1}R_{xy}v_1.$$

The maximal canonical correlation r_c is the square root of λ_1. Its significance is tested by referring to a table of the F sampling distribution the following statistic for pq and ms - pq/2 + 1 degrees of freedom (the latter need not be an integer):

$$F = (1 - \Lambda^{1/3})/pq \div (\Lambda^{1/3})/(ms - pq/2 + 1),$$

where

$\Lambda = \prod\limits_{i=1}^{q} (1 - \lambda_i)$, that is, the product of all the $1 - \lambda_i$s for $i = 1, 2, \ldots, q$ (Wilks, 1932),

λ_i = ith eigenvalue of M for $i = 1, 2, \ldots, q$,

$m = N - 3/2 - (p + q)/2$, where N is the sample size,

$s = [(p^2q^2 - 4) \div (p^2 + q^2 - 5)]^{\frac{1}{2}}$.

The test (Rao, 1952) is exact if either p or q is less than or equal to two and is approximate otherwise.

Table 2.1 contains a summary of the results of a typical canonical correlation analysis, with p = q = 3. The variables in one set are Variables 17 (Sociability Inventory), 18 (Physical Science Interest Inventory), and 19 (Office Work Interest Inventory) of the Project Talent Test Battery (Cooley & Lohnes, 1971). The variables in the other set are Variables 13 (Creativity Test), 14 (Mechanical Reasoning Test), and 15 (Abstract Reasoning Test) of the same battery.

TABLE 2.1

Example of Standard Canonical Correlation Analysis

ID	X_1	X_2	X_3	Y_1	Y_2	Y_3
1	10	20	18	9	12	9
2	4	15	13	7	10	10
.	.	.	.	.	.	.
.	.	.	.	.	.	.
.	.	.	.	.	.	.
504	8	12	37	5	6	8
505	1	11	27	14	11	11

$$p = q = 3$$
$$pq = 9$$
$$m = 500.5$$
$$s = (77/13)^{\frac{1}{2}} = 2.4337$$
$$ms - pq/2 + 1 = 1214.5668^{[a]}$$

Results

$$\lambda_i = .3509, .0150, .0003$$
$$(r_c = \sqrt{.3509} = .59)$$
$$\Lambda = .6393$$
$$F(9, 1215) = 27.24$$

Note: Data taken from Cooley and Lohnes (1971). Variables are from the Project Talent Test Battery: X_1 = Variable 17 (Sociability Inventory), X_2 = Variable 18 (Physical Science Interest Inventory), X_3 = Variable 19 (Office Work Interest Inventory), Y_1 = Variable 13 (Creativity Test), Y_2 = Variable 14 (Mechanical Reasoning Test), Y_3 = Variable 15 (Abstract Reasoning Test); ID = identification number.

[a]Rounded to 1215 for use with the F sampling distribution.

SIMPLE CORRELATION

For $p = 1$ and $q = 1$ (i.e., one variable in each set), $R_{yy}^{-1}R_{yx}R_{xx}^{-1}R_{xy}$ reduces to $R_{yx}R_{xy} = r^2$, where r is the correlation between the two variables. The largest eigenvalue of the scalar r^2 is r^2 itself.

The formula for F reduces to

$$(r^2/1) \div (1 - r^2)/(N - 2),$$

since $p = q = 1$, $s = (-3/-3)^{\frac{1}{2}} = \sqrt{1} = 1$, $\Lambda^{1/3} = \Lambda = 1 - \lambda_1 = 1 - r^2$, $1 - \Lambda^{1/3} = r^2$, and $m = N - 5/2$. This is, of course, the square of the well-known t ratio for testing the significance of a correlation coefficient. Since $F(1, N-2) = t^2(N - 2)$, all is well.

do these imply that correlation = t test?

t TEST FOR INDEPENDENT SAMPLE MEANS

Several authors, for example, McNemar (1969), Welkowitz, Ewen, and Cohen (1976), and Roscoe (1975), have pointed out the equivalence of the test of the significance of the difference between two independent sample means and the test of the significance of a point-biserial correlation coefficient. The trick is to create a dichotomous dummy variable for which a score of one is indicative of membership in one of the samples and a score of zero is indicative of membership in the other sample (and therefore nonmembership in the first sample). The same trick is used when canonical correlation analysis is applied to an independent samples test on the means. The formula for F is identical to that for simple correlation in the previous section. One variable is the dummy dichotomy of group membership, and the other variable is the continuous criterion variable.

MULTIPLE REGRESSION ANALYSIS

The major difference between multiple regression analysis and canonical analysis is that the former employs just one variable in the second set, that is, $q = 1$. Therefore, $R_{yy}^{-1}R_{yx}R_{xx}^{-1}R_{xy}$ reduces to $R_{yx}R_{xx}^{-1}R_{xy}$, which is recognizable as $R_{yx}b$, where b is the column vector of beta weights (standardized partial regression coefficients), and which in

turn reduces to the scalar $r_{y \cdot x_1, x_2, \ldots, x_p}^2$ (since R_{yx} is
a row vector of the correlations of each of the variables in the
first set with the variable in the second set), that is, the
square of the multiple correlation coefficient. The largest
eigenvalue of r^2 is again r^2 itself. The formula for F reduces
to $(r^2/p) \div (1 - r^2)/(N - p - 1)$, the traditional formula for testing the significance of a multiple r, since $q = 1$, $s = [(p^2 - 4) \div$
$(p^2 - 4)]^{\frac{1}{2}} = 1$, $\Lambda^{1/3} = \Lambda = 1 - \lambda_1 = 1 - r^2$, $1 - \Lambda^{1/3} = .2$, and
$m = N - 3/2 - (p + 1)/2$.

The case of $p = 2$ and $q = 1$ presents a special difficulty,
since $p^2 + q^2 - 5 = 0$, and s is undefined. (F is still the same
multiple regression F, however.) There is a test of the significance of a canonical correlation coefficient, due to Bartlett
(1941), that is not subject to this constraint. One calculates
$\chi^2 = -[N - \frac{1}{2}(p + q + 1)]\log_e \Lambda$ and refers that value to a table
of the chi-square sampling distribution for pq degrees of freedom.

ONE-WAY ANALYSIS OF VARIANCE

The equivalence of a one-way analysis of variance for k
independent samples to a multiple regression analysis with $p = k - 1$ is well documented in Cohen's (1968) article, so a very
brief treatment should suffice here. Membership in Sample 1
is denoted by variable X_1 (which equals one if an observation
is in Sample 1 and equals zero otherwise), membership in Sample
2 is denoted by variable X_2, . . . , membership in Sample k - 1
is denoted by X_{k-1}. Members of Sample k receive scores of zero
on variables X_1 through X_{k-1}, so that a kth variable is not only
unnecessary but would produce a linear dependency that would
render R_{xx} noninvertible. The F ratio is exactly the same as
the F ratio for multiple regression analysis, with the numerator
equal to the mean square between and with the denominator
equal to the mean square within. (There is actually a total sum
of squares term in the numerator and in the denominator, but
they cancel each other.)

FACTORIAL ANALYSIS OF VARIANCE

Cohen (1968) and others have shown how the equivalence
of the analysis of variance and multiple regression analysis also
extends to factorial analysis of variance and to the analysis of
covariance (see next section). For the simplest case of two

independent variables, one with a categories and the other with b categories, one needs a - 1 dummy variables to represent group membership for the first variable, b - 1 dummy variables to represent group membership for the second variable, and (a - 1)(b - 1) = ab - a - b + 1 dummy variables to represent group membership for their interaction. This produces a total of p = (a - 1) + (b - 1) + (ab - a - b + 1) = ab - 1 new independent variables. Membership in the interaction groups can be defined by the following ordinary multiplication rules (1 and 0 are the codes for the main effect groups): $1 \times 1 = 1$, $1 \times 0 = 0$, $0 \times 1 = 0$, and $0 \times 0 = 0$. Three canonical analyses are required (one for each main effect and one for the effect of the entire set of variables; the magnitude of the interaction effect is determined by subtraction). Three F ratios are generated. The dummy dichotomies for the main effect of one independent variable are uncorrelated with the dummy dichotomies for the main effect of the other independent variable, but they are correlated with the dummy dichotomies for their interaction, even if the cell frequencies are proportional. There are other coding schemes, for example, orthogonal coding and effect coding, for which all main effect and interaction variables are uncorrelated, but all schemes produce the same desired F ratios (see Kerlinger & Pedhazur, 1973).

This procedure can be extended to take care of factorial designs for any number of independent variables, with the interaction coding determined in a similar manner, that is, by multiplying the main effect codes. For example, for four independent variables, the score on the first second-order interaction variable for an observation with main effect codes of 0, 1, and 0 would be $0 \times 1 \times 0 = 0$.

ANALYSIS OF COVARIANCE

Using canonical correlation analysis (or multiple regression analysis, for that matter) to do an analysis of covariance is cumbersome though straightforward, so only the simplest case of one principal independent variable of k categories, one continuous dependent variable, and one continuous covariable is mentioned here.

It helps to start by stating the general research question that the analysis of covariance seeks to answer, namely, What is the effect (in the most liberal sense of the word) of the principal independent variable on the dependent variable that is over and above the effect of the covariable itself? This suggests (requires) that two analyses be performed: (a) the so-called full-model test that includes the principal independent variable and the covariable and (b) the reduced-model test that includes the covariable alone. An r^2 is obtained for each test, and a subsequent F ratio that is a function of the difference between the two r^2s is determined. Cohen's (1968) article provides the necessary details.

t TEST FOR CORRELATED SAMPLE MEANS

A canonical analysis of the difference between two correlated sample means also requires full-model and reduced-model considerations. Each member of each of the two samples gets a score on the principal treatment variable (1 if in Sample 1, 0 if in Sample 2) and a score on each of N - 1 pairing variables (1 if a member of the given pair, 0 if not), where N is the number of pairs. An r^2 is obtained for the full model, which includes all of these variables, and another r^2 is obtained for the reduced model, which includes only the pairing variables. The F test of the difference between the two r^2s provides an indication of whether or not the principal independent variable has an effect that is significantly greater than the effect of the pairing variables themselves.

DISCRIMINANT ANALYSIS

This increasingly popular procedure is actually a multivariate analysis of variance in reverse; that is there is one categorical dependent variable[2] of group membership and there are two or more continuous independent variables. The p independent variables are treated as though one were carrying out a multiple regression analysis, and the dependent variable of k categories is coded in the same way that an independent variable is treated in one-way analysis of variance, that is, by creating q = k - 1 dummy dichotomies. A standard p × q canonical analysis is then applied to the resulting system, and the F test of the largest canonical correlation coefficient determines whether or not the k groups are significantly separable on the p variables.

Table 2.2 contains a summary of the results of a canonical correlation analysis of the separability of Project Talent Test Battery Variables 11 (English Test), 12 (Reading Comprehension Test), and 16 (Mathematics Test) for five groups of students defined by Variable 8 (College Plans).

TABLE 2.2

Example of Application of Canonical Analysis to a
Discriminant Problem

				Data			
ID	X_1	X_2	X_3	Y_1	Y_2	Y_3	Y_4
1	87	39	20	0	0	0	1
2	76	15	15	0	0	0	0
.	.	.	.	.	.	.	.
.	.	.	.	.	.	.	.
.	.	.	.	.	.	.	.
504	87	22	18	0	0	0	0
505	105	45	26	1	0	0	0

$$p = 3$$
$$q = 4$$
$$pq = 12$$
$$m = 497.5$$
$$s = (140/20)^{\frac{1}{2}} = 2.6458$$
$$ms - pq/2 + 1 = 1311.2606^a$$

Results

$$\lambda_i = .2706, .0163, .0033, .0000$$
$$\Lambda = .7152$$
$$F(12, 1311) = 14.83$$

Note: Data taken from Cooley and Lohnes (1971). Variables
are from the Project Talent Test Battery: X_1 = Variable 11
(English Test), X_2 = Variable 12 (Reading Comprehension Test),
X_3 = Variable 16 (Mathematics Test); Y_1, Y_2, Y_3, Y_4 = dummy
variables for five groups of students defined by Variable 8
(College Plans); ID = identification number.
[a]Rounded to 1311 for use with the F sampling distribution.

CHI-SQUARE TEST OF INDEPENDENCE
OF TWO VARIABLES

An extreme case of canonical correlation analysis is the situation in which all p variables and all q variables are dummy dichotomies. This is exactly what happens in a $j \times k$ contingency table where X_1 (equal to one or zero) indicates membership or nonmembership in row 1, X_2 indicates membership or nonmembership in row 2, . . ., and X_{j-1} indicates membership or nonmembership in row $j - 1$; Y_1 indicates membership or nonmembership in column 1, Y_2 indicates membership or nonmembership in column 2, . . ., and Y_{k-1} indicates membership or nonmembership in column $k - 1$. Since each observation can be represented by a pattern of ones and zeros on one set of $p = j - 1$ variables and another set of $q = k - 1$ variables, a test of the independence of the two original categorical variables can be regarded as just one more special instance of a $p \times q$ canonical analysis. Since $\chi^2(pq) = pq \cdot F(pq, \infty)$, the actual χ^2 value can be explicitly calculated, if desired, by multiplying the canonical F by pq(ms - pq/2 + 1 is close enough to ∞ for most data).

Table 2.3 contains a summary of the canonical approach to a chi-square test of the independence of Project Talent Test Battery Variables 1 (School Size) and 8 (College Plans).[3]

SO WHAT?

The fact that each of several popular statistical techniques can be regarded as a special case of canonical correlation analysis raises two very interesting questions:

1. Should they be taught that way?
2. Should they be run that way?

The answer to both questions is, probably not. Students who study canonical correlation analysis should surely be told that a wide variety of statistical procedures is subsumed under the canonical model, and some time should be spent in showing them why this is, using arguments and examples similar to those contained in this chapter. This is not to say, however, that all students should start with canonical correlation analysis and then go on to study its special cases (although the idea is indeed tempting), since there is so much matrix algebra that one must know before one can study canonical analysis. Nor does it necessarily follow that students should study canonical correlation analysis only and forget about all of the special jargon and formulas that are associated with t tests, analyses of variance and covariance, chi-square, and so on. They will still have to read the research literature in which such things abound and will be at a distinct disadvantage if they have not been exposed to these matters in their statistics courses.

The same sorts of considerations are relevant regarding the actual carrying out of the analyses themselves. Although a suitably written canonical correlation program is capable of handling a simple correlation or t test problem, for example, it would be computationally inefficient to do so unless it was the only program one happened to have around. (In his article, Cohen, 1968, pointed out that virtually all computer installations have a good working multiple regression program but may not have a collection of analysis of variance programs.) The t test for correlated samples is a case in point. Setting up all of those pairing variables and doing full-model and reduced-model analyses may consume an inordinate amount of time. On the other hand it could be argued that a front-end routine that is part of the canonical package can do all of the dummy coding internally for the users, that is, they would submit jobs thinking that they were doing one thing and the computer would actually do another. (Some analyses of variance are actually run as regression analyses anyhow.) Another argument in favor of a single supercanonical program is that certain computers are so fast that the difference between running an analysis of variance as an analysis of variance, for example, and running it as a special case of a general canonical problem is only a matter of milliseconds, and the latter approach would therefore not be cost-ineffective.

The pedagogical and computational implications of the generality of the canonical approach therefore depend on the kinds of students (and faculty!) one has (good or poor math backgrounds, how much statistics they will ultimately be studying, etc.) and the kinds of computing facilities (hardware and software) that are available. It is also a matter of philosophy. The deductive approach would argue for teaching the canonical approach first, then its subcases, and running all analyses as canonical analyses. The more popular and probably more psychologically defensible inductive approach would argue for working up toward canonical analyses (pointing out matters such as $F = t^2$ along the way) and having a variety of special-purpose computing routines.

One final note about the assumptions underlying Rao's F test, especially the question of what happens to continuity in general, and normality in particular, whenever one introduces

TABLE 2.3

Example of Application of Canonical Analysis to Chi-Square Test

Data

ID	X_1	X_2	Y_1	Y_2	Y_3	Y_4
1	0	1	0	0	0	1
2	1	0	0	0	0	0
.	.	.	.	.	.	.
.	.	.	.	.	.	.
.	.	.	.	.	.	.
495	0	0	0	0	0	0
496	0	0	1	0	0	0

$$p = 2$$
$$q = 4$$
$$pq = 8$$
$$m = 491.5$$
$$s = (60/15)^{\frac{1}{2}} = 2$$
$$ms - pq/2 + 1 = 980$$

Results

$$\lambda_i = .0156, .0007, .0000, .0000$$
$$\Lambda = .9837$$
$$F(8, 980) = 1.01$$
$$\chi^2 = pq \cdot F = 8(1.01) = 8.08$$

Note: Data taken from Cooley and Lohnes (1971). Variables are from the Project Talent Test Battery: X_1, X_2 = Variable 1 (School Size); Y_1, Y_2, Y_3, Y_4 = Variable 8 (College Plans); ID = identification number.

all of those dummy dichotomies: <u>Contrary to popular misconception, the significance test does not assume multivariate normality for the total system of variables.</u> What is assumed is <u>a normal distribution of one set of variables for each combination of values for the other set of variables.</u>[4] Also assumed is homoscedasticity, that is, equal dispersion for such distributions, and homogeneity of regression for covariance-type analyses. The technique that would seem to suffer most from assumption violation is the application of canonical analyses to chi-square, with no apparent hope for normality at all. But there is hidden normality whenever the expected frequencies for the cells are large enough to permit approximately normal distributions of obtained frequencies around them, which is why the authors of most statistics textbooks include so many admonitions against small expected frequencies.

NOTES

1. For those who are not familiar with eigenvalues and eigenvectors, especially the former, there are excellent sections on these topics in Cooley and Lohnes (1971), Tatsuoka (1971), and Press (1972).

2. It is of course possible to have a factorial discriminant analysis with group membership defined along two or more dependent variables, but such designs are rarely encountered in social science research.

3. Since the number of observations for one of the school sizes was so small (9), it was eliminated from the analysis, and therefore the contingency table is 3 × 5 with 496 observations rather than 4 × 5 with 505 observations.

4. As Cohen (1968) pointed out in his article, the equivalence between the analysis of variance and regression analysis holds for fixed effects.

REFERENCES

Anderson, T. W. <u>An introduction to multivariate statistical analysis</u>. New York: Wiley, 1958.

Bartlett, M. S. The statistical significance of canonical correlations. <u>Biometrika</u>, 1941, <u>32</u>, 29-38.

Cohen, J. Multiple regression as a general data-analytic system. <u>Psychological Bulletin</u>, 1968, <u>70</u>, 426-443.

Cohen, J., & Cohen, P. Applied multiple regression/correlation analysis for the behavioral sciences. Hillsdale, N.J.: Erlbaum, 1975.

Cooley, W. W., & Lohnes, P. R. Multivariate data analysis. New York: Wiley, 1971.

Darlington, R. B., Weinberg, S. L., & Walberg, H. J. Canonical variate analysis and related techniques. Review of Educational Research, 1973, 43, 433-454.

Kerlinger, F. N., & Pedhazur, E. J. Multiple regression in behavioral research. New York: Holt, Rinehart & Winston, 1973.

McNemar, Q. Psychological statistics (4th ed.). New York: Wiley, 1969.

Morrison, D. F. Multivariate statistical methods (2nd ed.). New York: McGraw-Hill, 1976.

Press, S. J. Applied multivariate analysis. New York: Holt, Rinehart & Winston, 1972.

Rao, C. F. Advanced statistical methods in biometric research. New York: Wiley, 1952.

Roscoe, J. T. Fundamental research statistics for the behavioral sciences (2nd ed.). New York: Holt, Rinehart & Winston, 1975.

Tatsuoka, M. M. Multivariate analysis: Techniques for educational and psychological research. New York: Wiley, 1971.

Welkowitz, J., Ewen, R. B., & Cohen, J. Introductory statistics for the behavioral sciences (2nd ed.). New York: Academic Press, 1976.

Wilks, S. S. Certain generalizations in the analysis of variance. Biometrika, 1932, 24, 471-474.

3

Three Approaches
to Canonical Analysis

Claes Fornell

INTRODUCTION

As multivariate statistical methods have become increasingly
popular, gaining a position as indispensable tools in the multidimensional world of the social sciences, even the most complex
methods are now finding application in marketing research.
Many of these methods come neatly packaged in standard computer
programs easily available to most researchers. Since they appear
to make the long and mazy road of analysis much shorter, the
multivariate methods are the racing machines of analysis. Like
most sophisticated racing vehicles, they are extremely sensitive
to chance variation, demanding a thorough understanding of
their delicate machinery in order to give top performance and
to avoid possible contretemps. Unfortunately, the great potential of multivariate analysis has been marred by inept usage.
The fervid rush into application has all too often prevented
sufficient analysis of the methods themselves and their appropriateness for solving the research problem at hand.

The use, and mis-use, of canonical analysis indicate that
this particular method is poorly understood (Fornell, 1978).
The purpose of this chapter is to clarify some aspects of canonical
analysis by relating it to more familiar techniques of data analysis.
Rather than focusing on its mathematical construction, it is the

---

Reprinted from Journal of the Market Research Society,
1978, Vol. 20, No. 3, pp. 166-181. Reprinted by permission.

conceptual logic and the interpretation of canonical analysis that will be discussed.

While canonical analysis might look simple and straight-forward at first glance, there are a multitude of unresolved issues associated with it. Consequently, there is a host of pitfalls, some of which have been discussed in recent marketing literature (Alpert & Peterson, 1972; Fornell, 1978; Lambert & Durand, 1975). It is clear that many issues are in need of further exploration and that both theoretical and empirical inquiries into the very structure of canonical analysis are called for. In this chapter it is suggested that the comprehension of canonical analysis will be facilitated if one understands that canonical analysis can be viewed as several related techniques, each with a somewhat different logic and a different established practice.

THE FLEXIBILITY OF CANONICAL ANALYSIS

Social phenomena may not easily lend themselves to one-dimensional conceptualization, nor to one-dimensional measurement. Indeed, many social objects may not be subject to direct measurement at all. A multivariate approach to analysis is a logical concatenation to a multidimensional approach to observing. If the objective is to describe, predict, or to explain a complex phenomenon (expressed as a set of observed variables) via its relation to other phenomena (expressed as individual variables or as sets of variables), we have multivariance in both criterion and predictor variables. Canonical analysis is a general method designed to handle this type of variable association. Through the use of dummy variables, canonical analysis can be applied in experimental settings where MANOVA or ANOVA normally are employed, or in classification studies that ordinarily belong to the domain of discriminant analysis.

The canonical solution is the maximum correlation between pairs of linear composites of variables. The objective is to find pairs of variable combinations (canonical variates), so that the correlation between them is maximized. Subject to the restriction of variate orthogonality, new pairs can be formed from residual variances with the maximum number of pairs being equal to the number of variables in the smaller of the two sets. Hence each canonical variate is a constructed, unobserved variable, regressed on the observed variables within that set.

Because of its generality, it is not surprising that canonical analysis is beginning to find increasing application in the social sciences. Although there have been several applications in marketing, they are still comparatively few in contrast to regression and correlation studies. They have also been rather limited in scope. The bulk of canonical analyses in marketing has been concerned with the relationship between personality and purchase behavior (Alpert, 1971, 1972; Baumgarten & Ring, 1971; Kernan, 1968; Sparks & Tucker, 1971; Worthing et al., 1971). As shown by Lambert and Durand (1975), one can seriously question both the validity and reliability of the results presented and the propriety of the conclusions drawn in several of these studies.

The typical analysis regarding canonical solutions begins with an examination of the size of the canonical correlations; then a decision is made on (often on the basis of a statistical test) how many variate pairs should enter the analysis; and thirdly, an interpretation of the meaning of the variates and/or an assessment of variable "importance" from the relative size of the standardized canonical weights is attempted. This pathway of analysis has been followed rather indiscriminantly and without regard to research objectives. Such a mechanical procedure is no substitute for rigorous method and serves no useful purpose; it can only add more confusion and contribute to further mis-use and misunderstanding of canonical analysis.

Writers in multivariate statistics acknowledge that canonical results can be extremely difficult to interpret, but they have been more inclined to explain the mathematical nature of the technique and less interested in providing guidance for interpretation. Tatsuoka (1971), for example, considers it quite natural that seemingly nonsense results do not hamper the potential value of canonical analysis. It only implies that the mathematical formulation may not be susceptible to "meaningful verbal descriptions of our intuitive everyday concepts."

Canonical analysis is not as rigid an approach to data analysis as its mathematics and typical application indicate. In fact, much of the confusion and misunderstanding may stem from the conception of canonical analysis as a single technique. Multiple regression, MANOVA and ANOVA, and multiple discriminant analysis can all be shown to be special cases of canonical analysis. Principal components analysis is also in the inner family circle. The difference is that in canonical analysis variates are extracted so as to maximally correlate, whereas in principal components variates are constructed from one set of variables so as to retain as much variance as possible. In both cases we obtain composite variables consisting of linear weighted averages of the observed original variables.

Consequently, canonical analysis can be depicted as consisting of several inter-related techniques. In this study, we will view canonical analysis as an approach to, rather than as a technique of, data analysis. Within this approach we will explore several techniques in an attempt to compose coherent methodologies for canonical analysis. In education and psychological research it is not uncommon to use techniques such as analysis of variance (Pruzek, 1971) and discriminant analysis (Maxwell, 1961) within the context of canonical correlations. Here we will discuss canonical analysis as: (a) a tool for examining covariation among variables (canonical correlation analysis); (b) a device for factor identification and interpretation (canonical variate analysis); (c) a regression method with several dependent variables (canonical regression). Based on this structure, we will present a framework that can be useful in tackling many of the problematical issues in canonical analysis.

CANONICAL CORRELATION ANALYSIS

Canonical correlation analysis, as defined here, applies to the examination of the overall significance and magnitude of relationships between two sets of variables. The canonical correlation is symmetric and there is no causation implied. The objectives of canonical correlation analysis are to determine the complexity of the relationship, and to provide information about the overall nature of that relationship.

Canonical correlation may be expressed as:

$$
\begin{aligned}
Y_1 &= w_{y11}y_1 + \cdots + W_{y1n}y_n & X_1 &= w_{x11}x_1 + \cdots + w_{x1m}x_m \\
&\ \ \vdots & &\ \ \vdots \\
Y_d &= w_{yd1}y_1 + \cdots + w_{ydn}y_n & X_d &= w_{xd1}x_1 + \cdots + w_{xdm}x_m
\end{aligned}
\tag{1}
$$

Y = Standardized canonical variates for variables y_1 to y_n
X = Standardized canonical variates for variables x_1 to x_m
w_y = Canonical weights in the Y-variates
w_x = Canonical weights in the X-variates
x = Standardized variables of the X-set
y = Standardized variables of the Y-set
d = The number of canonical correlations

The complexity of the relationship is reflected in the number of canonical correlations (d). The first canonical correlation is,

by definition, the largest and hence represents the most substantial relationship; the second one is the second largest and
so forth. The decision on when to stop extracting variate pairs
is, in the final analysis, up to the subjective judgment of the
researcher. One may set a criterion for the size of the canonical
correlation, below which the remaining covariation is ignored;
similarly, one may choose to ignore a variate that accounts for
less than a predetermined percentage of the variance in those
observed variables of which it is composed; one may use various
statistical tests; or one may choose to rely on the proportion of
redundant variance associated with a given canonical relationship.

These ways to determine the number of relevant dimensions
are not mutually exclusive. Instead they can often be used in
a complementary fashion. Even though one may demote the
import of statistical significance, there is seldom a good reason
for interpreting relationships that are below a reasonable statistical significance.

A commonly used test for the statistical significance of the
canonical correlations is Bartlett's (1941) chi-square approximation of Wilk's Lambda. The null hypothesis states that there is
no residual linear association between the sets following the
extraction of preceding variates. That is, according to the null
hypothesis there is no more than k canonical correlations. Normally, the testing procedure starts with k = 0 and if the hypothesis is rejected, the value of k is increased by one and the test
repeated until the hypothesis can no longer be rejected.

Since canonical correlations show relations between composites of variables and not direct associations between observed
variables the statistical significance may have little practical
significance.

Another method to determine the number of canonical correlations is to look at the proportion of redundant variance (Stewart
& Love, 1968) associated with a given relationship. The squared
k:th canonical correlation ($R_{c\,k}^2$) multiplied by the average
squared product moment correlation ($\bar{r}^2_{Y_K \cdot y_j}$) between the
canonical variate Y_k and its corresponding original variables
y_1 to y_n, divided by the total proportion of the intersection
(redundancy) between the variable sets, gives the proportion
of total redundancy (V_k) in the Y-set (given the X-set) that
is associated with variate k.

$$\bar{R}^2_{(y/x)} = R^2_{ck} \left[\sum_{j=1}^{n} \frac{(r^2_{Y_K \cdot y_j})}{n} \right] \qquad \text{Redundancy} \qquad (2)$$

$$V_{k(y/x)} = \frac{\bar{R}^2_{(y/x)}}{\sum\limits_{i=1}^{d} \bar{R}^2_{(y/x)_i}}$$

proportion of total redundancy (3)

where n is the number of variables in the Y-set and d is the number of canonical relations. We can also compute the proportion of total redundancy in the X-set, given the Y-set, by using the average squared correlation between the variate X_k and its corresponding variables x_1 to x_m, multiplied by the same squared canonical correlation R^2_{ck}.

An Illustration

A canonical correlation applied to the simultaneous analysis of a Y-set, consisting of eight variables, and its association with an X-set consisting of 14 variables, produced the following results.

In Table 3.1, there are three statistically significant canonical relations, two beyond the .001 level and a third beyond the .05 level. Thus, the following of the criterion that the relationships should be significantly different from zero implies that these three canonical correlations are deemed relevant and therefore subject to further analysis.

The measures of redundancy provide a somewhat different picture. The redundancy in the Y-set, given the X-set, is .26 for the first canonical correlation, .07 for the next, and only .03 for the third. The contribution of the third variate is a mere 8% of the total redundancy in the Y-set. As for redundancy in the X-set, given the Y-set, the third dimension accounts for a redundancy of .02 which is equivalent to a contribution to overall redundancy of 8%. So, in comparison with the proportion of total redundancy associated with the first canonical root (which is around 70%), and also relative to the second root (which has a proportion of redundancy of around 20%), the contribution of the third is meagre. Consequently, despite a statistically significant relationship, there is a strong case for ignoring the third canonical correlation in the interpretation of the results. As to the variance of the observed variables that is retained by the respective variates, the canonical loadings and the communalities will have to be inspected. We will return to this issue in the following section.

In canonical correlation we are interested in the overall association between sets of variables or in the non-complex

TABLE 3.1

Canonical Results

Canonical correlation (R_c)	.85	.72	.58
Canonical root (R_c^2)	.71	.51	.34
Significance	$p < .001$	$p < .001$	$p = .016$
Redundancy ($\bar{R}^2{}_{(y/x)}$)	.26	.07	.03
Proportion of redundancy			
($V_{(y/x)}$)	.73	.20	.08
Redundancy ($\bar{R}^2{}_{(x/y)}$)	.17	.05	.02
Proportion of redundancy			
($V_{(x/y)}$)	.71	.21	.08

Y-set	Standardized weights		
var 01	.14	-.38	-.65
var 02	-.02	-.19	.85
var 03	.07	-.41	-.08
var 04	.43	-.38	.01
var 05	.16	-.14	-.72
var 06	.00	-.02	.28
var 07	.22	.44	-.01
var 08	.43	.61	.34

X-set			
var 09	.06	-.31	.08
var 10	.08	-.30	.33
var 11	.02	.12	-.08
var 12	.11	-.07	.49
var 13	.34	-.63	.27
var 14	.17	-.17	-.88
var 15	.24	.46	.64
var 16	.28	.04	.23
var 17	.08	.21	.13
var 18	.00	-.12	.11
var 19	.25	-.03	-.14
var 20	-.04	.28	-.08
var 21	.09	.19	-.62
var 22	-.10	.18	.04

relationship between observed variables. Since the variates are left uninterpreted, this type of analysis is less complicated than canonical regression or canonical variate analysis. Through the correlation between the uninterpreted variates, the relationships between the observed variables in the sets are assessed. In the first canonical correlation in Table 3.1, variables 04 and 08 are chiefly related to variables 13 and 16, while several variables in the Y-set do not exhibit strong association with the X-variate, and most variables in the X-set show weak relationships with the corresponding variables of the other set.

On the other hand, in the second correlation we see that several variables are relatively strongly related. In the Y-set, variables 04 and 08 are not dominant to the same extent as in the first dimension, a fact which seems to make the second variate correlated to more variables in the X-set. There is no doubt that variable 13 is the one variable in the X-set that is more closely associated with the variates of the Y-set, and thus with most of the variables in the Y-set. Similarly, variables 04, 07 and 08 are closely related to most variables in the X-set.

It is a mistake to interpret pairs of variable associations as if they were independent of each other. The fact that the variate pairs are statistically unrelated does not mean that they are independent from an interpretation point of view. On the contrary, each variable relationship has to be considered simultaneously in all variates. That is, the complexity of the variable relationship has to be explored if the association is to be accurately assessed. Consider the following example from our illustration.

In the first canonical correlation we have positive relations between the major variables, but in the second, variables 07 and 08 which are significant in the first correlation, show a positive association with their composite, while the major variable coefficient 13 is strongly negatively related to its corresponding variate. So, while we may find positive relations between two or more variables in one canonical correlation, we may discover the opposite to be true in the next correlation; something that nullifies the relationship if it is to be described as a one-dimensional association.

Consider the relationship between variables 08 and 13. These two variables have a strong positive relationship in the first canonical correlation and a negative one in the second. In order to explain the nature of this relationship adequately, one-dimensional analysis does not suffice. Just as ordinary product moment correlation restricts analysis to one-dimensional relationships, canonical correlation analysis does the same if it is not extended to include variate analysis as well. That is, when the complexity of a variable relationship exceeds one, we cannot focus on individual variables only, but must also consider the structure of the composite, unobserved variables. Without reference to what the variates represent it is difficult to describe conflicting variable relationships. If more than one canonical correlation is allowed to enter the analysis, we deal with several dimensions of relationships. In cases where the directionality of variable associations differs between variate pairs, the dimensions have to be interpreted before one can turn to the examination of those associations. Thus, we are talking about two aspects of canonical dimensions. One is the problem of determining the number of canonical correlations; the other is the dimensionality of variable relationships.

Before discussing the interpretation of canonical dimensions where the complexity of variable relationships exceeds one, it should be pointed out that there is at least one alternative to the weight statistic in canonical correlation. Cross-correlations between the variate Y and the original variables x_1 to x_m are bivariate but more direct measures of the between-set associations. While weights indicate the individual contribution of each variable x to the variance of the composite variable X, cross-correlations reflect the variance shared by a composite Y and each individual variable x, without simultaneously considering the influence of the other x-variables. Consequently, cross-correlation may be particularly useful when there is high multicollinearity.

CANONICAL VARIATE ANALYSIS

Canonical variate analysis differs from canonical correlation analysis in much the same way as factor analysis differs from simple correlation analysis. To factor analyze in the most general sense means to express a variable as linear combinations of other variables. If "Canonical Factor analysis" had not been coined by Rao (1955) as a method of factoring, it would have been an appropriate label for the type of analysis we will now discuss.

Instead of moving directly to the examination of covariation between variables, canonical variate analysis begins with focusing on the within-structure of the variates. The objective of canonical variate analysis is to identify and, ultimately, label relevant dimensions in a multivariate domain, so that the original number of variables is reduced to a few meaningful constructs. This is a very different form of analysis from canonical correlation. In

lieu of assessing the covariation between observed variables via
the correlation of their composites, or through cross-correlation,
the unobserved composites are examined via their direct relationship with the observed within-set variables. The determination
of dimensionality, i.e. how many canonical correlations should
be retained, may follow any of the procedures mentioned in
canonical correlation analysis.

Because the purpose of variate analysis is different from
that of canonical correlation, the statistics for interpreting the
results are not identical either. In applications, however,
canonical weights have been used both to examine covariation
of observed variables and to identify the unobserved constructs.
This is true for canonical analysis in marketing (Baumgarten &
Ring, 1971; Kernan, 1968; Sparks & Tucker, 1971; Worthing et
al., 1971) and also in other behavioral sciences (Darlington et
al., 1973). But the heavy reliance on weights has not been
completely unchallenged. Recent articles in the marketing literature (Alpert & Peterson, 1972; Lambert & Durand, 1975) discuss
some pitfalls of canonical analysis by the indiscriminatory use
of weights as a basis for variate interpretation. Chiefly on
empirical grounds, the validity of the weight statistic has been
questioned. Meredith (1964) states flatly that when there is
even moderate multicollinearity within the sets, the possibility
of interpreting canonical variates from variable weights is practically nil.

Canonical weights are equivalent to beta weights in regression. Although beta weights are not without ambiguity, they
are usually interpreted as the independent contribution of each
explanatory variable to the variance of a criterion variable. By
analogy, the canonical weights reveal the independent contributions of each variable to the variance of the within-set composite
variable. This variable is not, as in regression, an observed
phenomenon which is operationally defined through its measurement. Instead, it is subject to interpretation following the
canonical solution since it is undefined prior to the analysis.
Thus, it is difficult to find a rationale for using weights in
canonical variate analysis, for this type of analysis resembles
factor analysis more than it does regression analysis. Rather
than relying on the independent variable contributions to the
variances of the canonical composites to identify the underlying
constructs, the focus of variate analysis is on the structure of
the composites as manifested by their product moment correlations
with the observed variables. Akin to factor analysis, these
correlations are called loadings—canonical loadings—reflecting
the degree to which a variable is represented by a canonical

variate. In canonical variate analysis the observed variables can be written:

$$z_{ij} = Y_{i1}(r_{y_j \cdot Y_1}) + \cdots + Y_{id}(r_{y_j \cdot Y_d}) + e_{ij} \qquad (4)$$

where z_{ij} is the standardized value of variable j for individual i, and e_{ij} is the residual variance. In Table 3.2, which is based on the same data as Table 3.1, we find the canonical loadings for the Y set of the first two canonical relations.

Although weights and loadings produce a rather congruous overall picture, variables 05 and 06 in the first dimension are strongly associated with the composite variable according to loadings. This is not the case if weights are used for interpretation. In the second canonical correlation these two variables are insignificant, but instead a minor variable 02 (according to weights) becomes a major one.

As in factor analysis, the sum of squared loadings divided by the number of variables is a measure of how well the composite variables retain the variance of the observed variables. The redundancy index (equation 2) is based on this measure and the size of the canonical correlation coefficient. Likewise, it is often useful, though almost never done, to compute communalities in canonical variate analysis. Communalities show the percentage

TABLE 3.2

Canonical Loadings and Communalities

	$r_{Y_1 \cdot y_j}$	$r_{Y_2 \cdot y_j}$	Communality
var 01	.48	-.41	.40
var 02	.29	-.46	.30
var 03	.60	-.41	.53
var 04	.71	-.46	.72
var 05	.52	-.07	.28
var 06	.54	-.13	.31
var 07	.71	.41	.67
var 08	.81	.39	.81
Mean squared loading and average communality	.36	.14	.50

of variance that is summarized by the canonical composites. Together, the composites in Table 3.2 retain 50% of the variance in the data.

Because canonical variate analysis is very close to factor analysis—indeed it is a type of factor analysis—there is no reason why the same interpretative devices could not be used. Should the canonical variates be difficult to comprehend in terms of general factors, the loadings can be rotated.

The possibility of rotating canonical variates has not been fully explored, however. In view of the often delicate task of interpretation, it would seem that rotation is worth further research. One may, for example, rotate to maximize the number of high or low loadings, or to minimize the number of variates with which a variable is associated. Of course, canonical variate analysis with rotation does not, unlike ordinary canonical correlation, produce a unique solution. Different rotation procedures may reveal different variate structures. These structures were always present in the data, but could perhaps not have been discovered without rotation. According to Darlington et al. (1973) canonical variates may be rotated without regard to their corresponding variate pairs. But this would mean abandoning the idea of analyzing the relationships between the sets of variables. A more useful approach would probably involve a simultaneous rotation of the sets, so that the original properties of the canonical solution (i.e. the sum of squared canonical correlations) are retained.

CANONICAL REGRESSION ANALYSIS

Thus far, we have discussed canonical analysis without partitioning the data matrix into explanatory and criterion sets. Rotation of canonical solutions may also be useful in the analysis of dependence, where the criterion canonical variates are interpreted as factors.

Canonical regression analysis differs from both canonical correlation and canonical variate analysis. It applies to the same situations as ordinary regression with the exception that the dependent variable is multidimensional. In both cases the independent variables consist of observed phenomena which, through simultaneous analysis, explain or predict the dependent variable. In canonical regression, the dependent variable is an index composed of a linear combination of observed dependent variables. Depending on the number of relevant dimensions necessary to adequately describe the relationships, there may

be more than one such index. These indices cannot be interpreted a priori. They are identified during the course of
analysis. The problem of comprehending the dependent variables
in canonical regression is identical to interpreting the variates
in canonical variate analysis.

Once the multivariate dependent variables have been interpreted and labelled, the analysis proceeds as in ordinary multiple
regression: each criterion construct is regressed on the observed
explanatory variables.

$$Y_1 = w_{y11}y_1 + \ldots + w_{y1n}y_n = (w_{x11} \cdot R_{c1})x_1 + \ldots + (w_{x1m} \cdot R_{c1})x_m + e \tag{5}$$

The standardized regression coefficients are proportional to the
standardized canonical weights. Each canonical correlation weight
is multiplied by a constant—the canonical correlation coefficient
(R_c) of that dimension. Since the dependent construct Y is a
standardized variable, the beta coefficients are equal to the unstandardized regression weights multiplied by the standard
deviation of the observed variable. R^2 is of course identical
to the canonical root (eigenvalue).

Canonical regression falls in between canonical variate
analysis and canonical correlation analysis. The interpretation
of dependent variable structure is based on loadings (equation 4),
and the assessment of individual variable contribution to the
variance of the dependent construct is based on the relative
size of the weights.

The transformation of canonical analysis into a regression
has considerable practical advantages. Not only is regression
more well-known and better understood, but it is also routinely
accompanied by a larger set of potentially helpful statistics and
extensions that are not found in standard computer packages
for canonical analysis. It might, for example, be very useful
to inspect each individual variable contribution to changes in
R^2, as well as the standard error of regression coefficients.
Various stepwise methods may also be used.

Explained Variance in Canonical Regression

If a regression program with the canonical composites as
dependent variables is used, one must keep in mind that the
measures of standard error and R^2 pertain to the unobserved
canonical variates. Whether R^2 should be used as a measure of
explanatory power or not depends on how well the criterion con-

structs reflect the phenomenon under study. It has been argued
by several (Alpert & Peterson, 1972; Darlington et al., 1973;
Lambert & Durand, 1975; Stewart & Love, 1968) that the canonical root is a highly inflated measure of explained variance, because the canonical solution operates so as to find optimal weights
for the observed variables in order to maximize the correlation
between the composites, and does not pay any attention to the
loadings. That is, the squared canonical correlation reflects
the explained variance of two maximally correlated vectors.
Naturally, if one is interested in the explanatory power of a
set of variables on the observed individual dependent variables,
and not in the explanation of the constructed dependent variables, this estimate will be overly optimistic.

Stewart and Love's (1968) redundancy index (equation 2)
has been proposed as an estimate of explained variance that
takes the observed variables into account. This index is the
mean squared loading multiplied by the corresponding canonical
root summed over the canonical dimensions. Using the first two
variates, we note that 33%* of the variance in the Y-set is accounted for by the variance in the X-set.

This is considerably lower than the measure of shared
variance between the composites.

Since the redundancy index is computed by multiplying
the mean squared loadings of a canonical vector correlation of
the corresponding vector

$$\bar{r}^2_{Y_K \cdot y_j} \cdot \bar{R}^2_{ck} = \bar{R}^2_{(y/x)} \tag{6}$$

where $\bar{r}^2_{Y_k \cdot y_j}$, $\bar{R}^2_{ck} \leq 1$

it follows that $\bar{R}^2_{(y/x)} \leq R^2_{ck}$

and in most practical situations, $\bar{R}^2_{(y/x)}$ will be much lower
than R^2_{ck}. Even though the redundancy index has some interesting features, such as being non-symmetric, it is nevertheless
a rather crude summary statistic of the average similarity between two sets of variables. There may be cases where this
statistic can be misleading. If, for example, a composite retains
a great deal of the variance in the observed variables, but is
poorly explained by a majority of the variables in the other set,

---

*(.36 × .71) + (.14 × .51) = .33

only a few of the observed variables would determine the structure and interpretation of the composites, and a single measure of average associations makes little sense.

As yet, there is no single "best" measure of explained variance in the observed dependent variables in canonical regression. A good deal can, however, be learned by inspecting communalities, mean squared loadings, canonical roots, the redundancy index and individual variable effect on R^2. One may also compare these to simple correlations and to the multiple correlation coefficients when the observed dependent variables are used in separate regressions.

Any attempt to account for the omitted variance from the original variables is bound to be extremely problematical. Actually, one may ask if it is really worthwhile or very useful to try to do this in all situations. Is it not possible that the unobserved, composite variables may better portray the phenomenon under study than the observed variables? The decision on whether to assess explained variance on R^2's and canonical roots, or on redundancy indices and simple correlations, depends on the confidence one has in the constructed variables: to what extent do they accurately reflect the phenomenon under study?

Loadings and communalities show how well the observed variables are represented by the composites: if the structure of the canonical variates is meaningful and if the variates account for a large proportion of the variance in the original data, the canonical root may be interpreted as R^2 in ordinary multiple regression. Hence, loadings and communalities are internal validation indices for canonical variates.

Confidence in canonical constructs could also, and often more effectively, be obtained (or lost) through external indices. If, for example, a variate is labelled "propensity to act" and proves to be a better correlate to that action than any of the original variables, the variate is the superior criterion variable.

Should quantitative external indices be difficult to find, there are other ways to examine the validity of canonical constructs. For example, Fornell (1976) compared non-quantitative case descriptions of respondents with their scores on the composite canonical variables. To avoid unconscious bias, it is important that such case studies are conducted independently and without knowledge of the canonical results.

SUMMARY AND CONCLUSIONS

It has been the objective of this chapter to provide a framework for the application of canonical analysis. In particular it was a concern to show the flexibility of canonical analysis and its wide applicability for different types of data analysis. The viewing of the method from various, and somewhat different, analytical perspectives demonstrated that the question of weights versus loadings is essentially a logical one.

1. In canonical correlation analysis weights or cross-correlations are used to examine variable relationships.
2. In canonical variate analysis the loadings are used to identify "meaningful" constructs from the variate structure.
3. In canonical regression analysis both weights and loadings are used: loadings for the comprehension of variate structure in the criterion variables and weights to estimate the regression for predictor variables.

Canonical correlation analysis is the simplest and the most limited type of analysis. It cannot fully handle multivariate relationships between sets of variables. Whenever we deal with several dimensions, their relationships are very difficult to describe without an understanding of what the dimensions represent. Canonical variate analysis is designed to help such understanding. It is a form of external factor analysis and, as such, the factor structure (canonical loadings) is used for interpretation. As in factor analysis, the loadings may be rotated. In canonical regression the weights are used for estimation. Naturally, the usefulness of canonical weights can be questioned on the same grounds as beta weights in regression.*

The concept of explained variance poses a special problem in canonical analysis. Since the (significant) canonical variates rarely retain all the original variance of the observed variables, there is a question of whether or not this should be taken into account by the estimate of explained variance. If it is ignored, it might be argued that the estimate (canonical root) is misleading and inflated. On the other hand, one may choose to explain the variance of the composites on the grounds that omitted variance was error variance and that the composites thus reflect the object under study more accurately. Then, of course, it is perfectly

---

*If all the variables that affect the criterion variable are either included or uncorrelated with those that are included, and if the direction of causality and the functional form of the equation are correctly specified, beta weights are the superior estimates of the true regression (Darlington, 1968).

rational to use the canonical root in the same way as R^2 in regression.

Judging from the literature, researchers seldom realize that there are several avenues to canonical analysis. Weights and canonical roots are used without regard to the objective and type of analysis. In those rare instances when there has actually been a choice between weights and loadings, the decision has been dictated by empirical considerations. While the importance of such things as multicollinearity, sample size, standard errors and stability of estimates should not be demoted, there is a significant theoretical dimension that ought to be considered as well.

REFERENCES

Alpert, Mark I. (1971) "A canonical analysis of personality and the determinants of automobile choice," Combined Proceedings. Chicago: American Marketing Association, pp. 312-16.

Alpert, Mark I. (1972) "Personality and the determinants of product choice," Journal of Marketing Research, 9, 1, pp. 89-92.

Alpert, Mark I. and Peterson, Robert A. (1972) "On the interpretation of canonical analysis," Journal of Marketing Research, 9, pp. 187-92.

Bartlett, M. S. (1941) "The statistical significance of canonical correlations," Biometrika, 32, pp. 29-38.

Baumgarten, Steven A. and Ring, L. Winston (1971) "An evaluation of media readership constructs and audience profiles by use of canonical correlation analysis," Combined Proceedings. Chicago: American Marketing Association, pp. 584-88.

Darlington, Richard B. (1968) "Multiple regression in psychological research and practice," Psychological Bulletin, 69, 3, pp. 161-82.

Darlington, Richard B., Weinberg, Sharon L., and Walberg, Herbert J. (1973) "Canonical variate analysis and related techniques," Review of Educational Research, 43, 4, pp. 433-54.

Fornell, Claes (1976) Consumer Input for Marketing Decisions: A Study of Corporate Departments for Consumer Affairs. New York: Praeger.

Fornell, Claes (1978) "Problems in the interpretation of canonical analysis: The case of power in distributive channels," Journal of Marketing Research, pp. 489-91.

Green, Paul E., Halbert, Michael H., and Robinson, Patrick J. (1966) "Canonical analysis: An exposition and illustrative application," Journal of Marketing Research, 3 1, pp. 32-39.

Green, Paul E. and Tull, Donald S. (1975) Research for Marketing Decisions. Englewood Cliffs, N.J.: Prentice-Hall.

Kernan, Jerome B. (1968) "Choice criteria, decision behaviour and personality," Journal of Marketing Research, 5, 2, pp. 155-65.

Lambert, Zarrel V. and Durand, Richard M. (1975) "Some precautions in using canonical analysis," Journal of Marketing Research, 12, 4, pp. 468-75.

Maxwell, A. E. (1961) "Canonical variate analysis when the variables are dichotomous," Educational and Psychological Measurement, XXI, 2, pp. 259-71.

Meredith, William (1964) "Canonical correlations with fallible data," Psychometrika, 29, pp. 55-56.

Pruzek, Robert M. (1971) "Methods and problems in the analysis of multivariate data," Review of Educational Research, 41, pp. 163-90.

Rao, C. Radhakrishna (1955) "Estimation and tests of significance in factor analysis," Psychometrika, 20, pp. 93-111.

Sparks, David L. and Tucker, W. T. (1971) "A multivariate analysis of personality and product use," Journal of Marketing Research, 8, 1, 67-70.

Stewart, Douglas and Love, William (1968) "A general canonical correlation index," Psychological Bulletin, 70, 3, pp. 160-63.

Tatsuoka, Maurice M. (1971) Multivariate Analysis: Techniques for Educational and Psychological Research. New York: John Wiley & Sons.

Worthing, Parker M., Venkatesan, M., and Smith, Steve (1971) "A modified approach to the exploration of personality and product use," Combined Proceedings. Chicago: American Marketing Association, pp. 363-67.

4

A General Canonical
Correlation Index

Douglas Stewart and William Love

Because a canonical correlation is the correlation
between 2 linear composites, it presents some
interpretive problems. No measure of the redundancy in 1 set of variables, given another set of
variables, has been available. A nonsymmetric
index of redundancy is proposed which represents
the amount of predicted variance in a set of variables.

The interpretation of canonical correlations presents some
problems.[1] Whereas a squared multiple correlation represents
the proportion of criterion variance predicted by the optimal
linear combination of predictors, a squared canonical correlation
represents the variance shared by linear composites of two sets
of variables, and not the shared variance of the two sets.

Unfortunately, therefore, canonical correlations cannot be
interpreted as correlations between sets of variables. It is
important to note that a relatively strong canonical correlation
may obtain between two linear functions, even though these
linear functions may not extract significant portions of variance
from their respective batteries. This is the problem of interpretation to which this chapter is addressed.

Reprinted from Psychological Bulletin, 1968, Vol. 70, No.
3, pp. 160-163. Copyright 1968 by the American Psychological
Association. Reprinted by permission.

Rozeboom (1965) suggested the relevance of information theoretic concepts in dealing with canonical correlations. Uncertainty and alienation are considered parallel, and, similarly, redundancy and correlation are treated as analogous. Given this approach, Rozeboom developed a general index which is similar to one presented by Anderson (1958, p. 244). Both measures are symmetric, that is, given two sets of variables, one number is presented which presents the magnitude of their intersection. A directional or nonsymmetric index is possible by pursuing the information theoretic analogues suggested by Rozeboom. In addition to the primitive concept of uncertainty, Shannon (Shannon & Weaver, 1949) discussed conditional uncertainty. Similarly, one may discuss the complement of conditional uncertainty as conditional redundancy. A nonsymmetric measure is considered desirable because one set of variables may be almost completely subsubsumed by a larger set; that is, redundancy can be represented as the intersection of two sets of variables, and it is desirable to represent the proportion of one set which is in the intersection (see Figure 4.1).

In the case pictured in Figure 4.1, it is clear that most of Set A is contained in Set B, whereas a relatively large portion of Set B is outside the intersection. This chapter proposes an index based on canonical correlation which is nonsymmetric and has been worthwhile in the analysis of various partitioned matrices.

If we were to component analyze two sets of variables independently and then develop weights which would rotate the two component structures to maximum correlation, we would have a canonical solution (Hotelling, 1935). In the canonical case the components are usually referred to as canonical variates. The correlation between the first variate of the left set and the first variate of the right set is the first canonical correlation (R_{c_1}). In order to take advantage of the well-developed language of component analysis, we shall call them canonical components.

FIGURE 4.1

Nonsymmetric Redundancy

Since the complete structure of a set of variables will contain as many dimensions as there are variables (this is true only where the rank of the matrix equals the order; in general this is the case and will be assumed in this study), it is obvious that if the larger set is composed of five variables and the smaller set of three variables, only three components can be extracted from the smaller set. As a result, R_c's are available between three of the components of the smaller set. The remaining two components in the larger set have no counterpart in the smaller set and do not enter into the canonical solution.

In the traditional interpretation of canonical correlations, the magnitude of the R_c's, whether or not they are significantly nonzero, and the coefficients used to obtain the R_c's are considered (Cooley & Lohnes, 1962). The interpretation of these coefficients has all the problems attendant to the beta coefficients of common multiple regression. At the suggestion of Meredith (1964), some investigators now compute the correlations between the variables in a set and the canonical variates of that set (the component loadings of component analytic parlance).

Before we consider a method of calculating an index of redundancy we should agree on vocabulary. We need one index for the redundancy in the left set given the right and another index for the reverse relation. For the sake of simplicity, we will consider one set of variables as the predictor or conditioning set and the other set as the criterion, as in multiple regression. We talk about the proportion of variance in the criterion accounted for by the predictors, but seldom if ever consider the reverse relationship. It is obvious that by reversing our definition of criterion and predictor we could develop the index in the other direction. The canonical variates of the predictor set will be FP_i and similarly FC_i for the criterion set. The variables of the predictor and criterion sets will be P_i and C_i, respectively. Since the index about to be proposed utilizes the concept of a component extracting a proportion of the variance (more appropriately proportion of trace) of a set of variables (usually a battery of tests), we will define the column sum of the squared loadings of variables within a set on a canonical variate of the set as the variance extracted by that variate. When this is divided by the number of variables in the set (M), the resulting value is the proportion of the variance of the set extracted by that canonical variate. This will be symbolized as VP_i and VC_i. The squared canonical correlations ($R_{c_i}^2$) will be written as λ_i (following Cooley & Lohnes, 1962). This is the proportion of variance in one of the ith pair of canonical variates predictable from the other member of the pair. If the VC_i is multiplied by the λ_i, the resulting figure is the proportion of the variance of the C set explained by correlation between FP_i and FC_i. If this value is calculated for each of the M_c pairs of canonical variates, the result is an index of the proportion of variance of C predictable from P, or the redundancy in C given P.

$$\overline{R_{c \cdot p}}^2 = \sum_{k=1}^{M_c} \lambda_k VC_k = \sum_{k=1}^{M_c} \lambda_k \left[\sum_{j=1}^{M_c} (L_{jk}^2 / M_c) \right]$$

(Where L_{jk} is the correlation between the jth variable and kth canonical factor.)

We have called this index $\overline{R_{c \cdot p}}^2$ because if multiple R^2 were computed between the total P set and each variable of the C set, $\overline{R_{c \cdot p}}^2 = \Sigma R^2 / M_c$. In other words $\overline{R^2}$ is the mean squared multiple correlation.

An example of the use of canonical correlation is presented by Lohnes and Marshall (1965). In this study three scores from the Pintner General Ability Test (PGAT) and 10 from the Metropolitan Achievement Test were entered into a canonical correlation with the seventh- and eighth-year course grades in English, arithmetic, social studies, and science of 230 junior high school students in a small, rural college town.[2] The first two canonical correlations were reported ($R_{c1} = .90$ and $R_{c2} = .66$). The canonical weights were reported and interpreted.

In the present analysis of the Lohnes-Marshall data, the weights were ignored and the canonical loadings and $\overline{R^2}$'s were inspected.

In the left set, loadings from .707 to .917 are found on the first variage (see Table 4.1). The loadings on the second variate drop substantially. The same condition holds in the right set. In Table 4.2, Columns 1 and 2 present the canonical correlations and their squares. Note that the upper portion of Table 4.2 considers the left set as criterion and right set as the predictor set, while the lower portion reverses these roles. The third column of Table 4.2 presents the proportions of the variance of the set extracted by each canonical variate. The fourth column is the amount of redundant variance attributed to each canonical variate. The fifth column expresses the values in the fourth column as proportions of the total redundancy.

From this we see that:

1. The eight canonical variates extract 90% of the variance of the left set.

TABLE 4.1

Canonical Loadings

Canonical Variates

Structure for Left Set

Tests							
1	-.786	.061	-.082	-.313	.054	.163	-.251	.026
2	-.828	-.163	.018	-.191	-.082	.174	-.276	.031
3	-.707	-.462	.009	-.444	.066	-.102	.018	-.152
4	-.800	-.031	.178	-.095	-.071	.451	-.026	.050
5	-.817	.061	.169	-.194	.003	.311	-.136	-.340
6	-.887	.185	-.096	.074	-.080	.005	-.081	-.005
7	-.917	.119	-.055	-.148	.205	-.016	.120	.050
8	-.836	-.066	.088	-.245	-.046	.082	-.001	.210
9	-.903	-.212	-.086	.099	.083	-.042	.069	-.182
10	-.839	-.351	.016	-.006	-.002	.008	.160	-.136
11	-.752	.048	.581	-.123	.063	.053	-.105	-.113
12	-.798	.360	.136	.011	.065	-.076	-.243	.096
13	-726	-.190	.218	-.126	.447	.321	-1.98	-.023

Structure for Right Set

Tests							
1	-.847	-.322	-.065	.094	.212	-.326	-.033	-.119
2	-.795	-.446	-.014	-.067	-.230	.255	.117	-.182
3	-.951	.140	.011	-.108	.095	.046	-.099	.206
4	-.878	.241	-.011	.025	-.194	-.055	-.057	-.354
5	-.901	.127	.315	.227	.080	.073	.093	-.002
6	-.743	.001	.540	-.134	-.189	-.021	-.180	-.263
7	-.800	.027	.088	-.222	.412	-.111	.195	-.288
8	-.727	-.079	.209	.034	.063	.335	-.361	-.416

TABLE 4.2

Components of Redundancy Measure

$$\sum \frac{(loading_i)^2}{8}$$

Root	I Canonical R R_c	II R Squared λ	III Variance Extracted VC	IV Redundancy $\lambda \cdot VC$	V Proportion of Total Redundancy
			Left Set		
1	.9021	.814	.668	.544	.927
2	.6625	.439	.049	.022	.037
3	.5015	.251	.038	.010	.016
4	.3886	.151	.039	.006	.010
5	.3098	.096	.022	.002	.004
6	.2785	.078	.038	.003	.005
7	.1500	.022	.025	.001	.001
8	.0722	.005	.020	.000	.000

Right Set

1	.9021	.814	.695	.566	.923
2	.6625	.439	.050	.022	.036
3	.5015	.251	.056	.014	.023
4	.3886	.151	.018	.003	.004
5	.3098	.096	.045	.004	.007
6	.2785	.078	.038	.003	.005
7	.1500	.022	.030	.001	.001
8	.0722	.005	.068	.000	.001

Note: Total variance extracted from left set = .899; $\bar{R}$, total redundancy for left set, given right set = .586. Total variance extracted from right set = 1.000; $\bar{R}$, total redundancy for left set, given right set = .613.

2. Fifty-nine percent of the variance of the left set is predicted by the variance in the right set (i.e., $\overline{R_{L \cdot R}}^2 = .59$).

3. Of the redundant variance, 93% is associated with the first canonical variate.

4. Despite the large value of $R_{c2} = .66$, the second canonical variates have very small amounts of variance associated (5% in both the left and right sets).

5. The eight canonical variates of the right (and smaller) set extract 100% of the variance of that set (which is always true of the smaller set in the canonical solution).

6. The redundancy of the right set (student grades) given the left set is $\overline{R_{R \cdot L}}^2 = .61$.

7. Of the redundant variance of the right set, 92% is associated with the first canonical variate.

The utility of $\overline{R}^2$ is as a summary index. In general it is not to be viewed as an analytic tool. Certain associated indices, however, have obvious analytic applications. For example, the proportion of redundant variance associated with a given root is instructive in determining whether the root deserves interpretation and further attention (in the case noted above, a canonical correlation of .66 was associated with only .05 of the variance of either side, and only 4% of the redundant variance—in short, this index instructs us differently than does the canonical correlation alone).

NOTES

1. The authors wish to express their appreciation to Paul R. Lohnes who encouraged and guided the present effort while they were Office of Education Postdoctoral Fellows (Office of Education Grant 1-6-062084) at Project TALENT, Pittsburgh, Pennsylvania.

2. Paul R. Lohnes graciously allowed the authors to use his data (Lohnes & Marshall, 1965) and modified his latest canonical program to calculate the authors' index.

REFERENCES

Anderson, T. W. Introduction to multivariate statistical analysis. New York: Wiley, 1958.

Cooley, W. W., & Lohnes, P. R. Multivariate procedures for the behavioral sciences. New York: Wiley, 1962.

Hotelling, H. The most predictable criterion. Journal of Educational Psychology, 1935, 26, 139-142.

Lohnes, P. R., & Marshall, T. O. Redundancy in student records. American Educational Research Journal, 1965, 2, 19-23.

Meredith, W. Canonical correlations with fallible data. Psychometrika, 1964, 29, 55-65.

Rozeboom, W. W. Linear correlations between sets of variables. Psychometrika, 1965, 30, 57-71.

Shannon, C. E., & Weaver, W. The mathematical theory of communication. Urbana: University of Illinois Press, 1949.

5

Redundancy Analysis:
An Alternative for
Canonical Correlation Analysis

Arnold L. van den Wollenberg

INTRODUCTION

In canonical correlation analysis components are extracted from two sets of variables simultaneously in such a way as to maximize the correlation, μ, between these components. Mathematically, the criterion to be maximized under restrictions is

$$w'R_{xy}\upsilon - \tfrac{1}{2}\mu(w'R_{xx}w - 1) - \tfrac{1}{2}\nu(\upsilon'R_{yy}\upsilon - 1), \qquad (1)$$

where μ and ν are Lagrange multipliers. Elaboration then leads to the following eigenvalue eigenvector equations [Anderson, 1958]:

$$(R_{xx}^{-1}R_{xy}R_{yy}^{-1}R_{yx} - \mu^2 I)w = 0, \qquad (2)$$

and

$$(R_{yy}^{-1}R_{yx}R_{xx}^{-1}R_{xy} - \nu^2 I)\upsilon = 0 \qquad (3)$$

The eigenvalues μ^2 and ν^2 are equal, and are also equal to the squared canonical correlation coefficient. After extraction of

---

A Fortran IV program for the method of redundancy analysis described in this chapter can be obtained from the author upon request.[1]

Reprinted from Psychometrika, June 1977, Vol. 42, No. 2, pp. 207-219. Reprinted by permission.

the first pair of canonical variates, a second pair can be determined having maximum correlation, with the restriction that the variates are uncorrelated with all other canonical variates except with their counterparts in the other set, and so forth.

Whereas in bivariate correlation and multiple correlation analysis the squared correlation coefficient is equal to the proportion of explained variance of the variables under consideration, this is not the case for the canonical correlation coefficient. Canonical correlation actually gives no information about the explained variance of the variables in one set given the other, since no attention is paid to factor loadings. Two minor components might correlate very highly, while the explained variance of the variables is very low, because of the near zero loadings of the variables on those components. As a high canonical correlation does not tell us anything about the communality of two sets of variables, it is as such an analytical tool which is hard to interpret.

As an addition to canonical correlation analysis, Stewart and Love [1968] introduced the redundancy index, which is the mean variance of the variables of one set that is explained by a canonical variate of the other set. That is, in the present notation,

$$R_{y_i} = \mu_i^2 * \frac{1}{m_y} * f_{y\hat{y}i}' f_{y\hat{y}i},$$ (4)

and

$$R_y = \sum_i R_{y_i},$$ (5)

where R_{y_i} is the redundancy of the criteria given the ith canonical variate of the predictors (R_y is the overall redundancy). The symbol m_y stands for the number of criteria; μ_i is the ith canonical correlation, while $f_{y \hat{y}_i}' = \upsilon_i' R_{yy}$ is the vector of loadings of the y-variables on their ith canonical component.

Unlike canonical correlation, redundancy is non-symmetric. Thus in general, given a canonical correlation coefficient, the associated redundancy of the Y-variables will be different from that of the X-variables. (Since the redundancy of the X-variables given the Y-variables is completely analogous to that of the Y-variables given the X-variables, we will only discuss the latter one.)

The redundancy formula can be looked upon as a two-step explained variance formula, in which μ^2 is the explained variance of the canonical variate of one set, given its counterpart in the other set whereas the second part of the formula is the mean explained variance of the variables by their ith canonical component.

In terms of this two-step explained variance approach, canonical correlation analysis only maximizes one part of the redundancy formula. It would seem reasonable, however, to try and maximize redundancy per se.

THE METHOD

To maximize redundancy it is convenient to rewrite the index. From derivation of the canonical correlation coefficient we know [Anderson, 1958, p. 291] that,

$$R_{yx}w - \mu R_{yy}\upsilon = 0,\tag{6}$$

and

$$R_{yx}w = \mu R_{yy}\upsilon.\tag{7}$$

So,

$$R_{y_i} = \frac{1}{m_y} * \mu_i * f_{y\hat{y}_i}' f_{y\hat{y}_i} * \mu_i\tag{8}$$

$$= \frac{1}{m_y} * \mu_i * \upsilon_i' R_{yy} R_{yy} \upsilon_i * \mu_i\tag{9}$$

$$= \frac{1}{m_y} w_i' R_{xy} R_{yx} w_i = \frac{1}{m_y} f_{y\hat{x}_i}' f_{y\hat{x}_i}'\tag{10}$$

where $f_{y\hat{x}_i}$ is the vector of loadings of the Y-variables on the ith canonical component of the X-variables. Therefore redundancy can also be looked upon as the mean squared loading of the variables of one set on the canonical variate under consideration of the other set.

Given two sets of variables X and Y standardized to zero mean and unit variance, we seek a variate $\xi = Xw$ with unit variance such that the sum of squared correlations of the Y variables with that variate is maximal, and a variate $\zeta = Y\upsilon$ for which the same holds in the reverse direction. The correlation of the Y-variables with the variate ξ is given by the column vector $(1/N)Y'Xw$; the sum of squared correlations is equal to the minor product moment. So we have to maximize

$$\Phi = \frac{1}{N^2} w' X'YY'Xw - \mu(\frac{1}{N} w'X'Xw - 1),$$

$$\Psi = \frac{1}{N^2} \upsilon'Y'XX'Y\upsilon - \nu(\frac{1}{N} \upsilon'Y'Y\upsilon - 1),$$

(11)

or

$$\Phi = w'R_{xy}R_{yx}w - \mu(w'R_{xx}w - 1),$$

$$\Psi = \upsilon'R_{yx}R_{xy}\upsilon - \nu(\upsilon'R_{yy}\upsilon - 1).$$

(12)

Setting the partial derivatives with respect to w and υ equal to zero we get

$$\frac{\delta\Phi}{\delta w} = R_{xy}R_{yx}w - \mu R_{xx}w = 0,$$

$$\frac{\delta\Phi}{\delta\upsilon} = R_{yx}R_{xy}\upsilon - \nu R_{yy}\upsilon = 0,$$

(13)

which can be written as two general characteristic equations:

$$(R_{xy}R_{yx} - \mu R_{xx})w = 0,$$

$$(R_{yx}R_{xy} - \nu R_{yy})\upsilon = 0.$$

(14)

The numerical solution to this problem is formally identical to that for the case of canonical correlation, since both the matrix products $R_{xy}R_{yx}$ and $R_{yx}R_{xy}$ and the matrices R_{xx} and R_{yy} are real symmetric matrices. However, the eigenvalues μ and ν are not equal, as in the case of canonical correlation analysis, so one has to compute both eigenstructures. We can interpret μ as m_y times the mean variance of the Y-variables that is explained by the first canonical variate of the X-variables. We will return to this point later.

When subsequent redundancy variates are determined we want them to be uncorrelated with the preceding variates extracted from the same set. It is not, in general, possible to have bi-orthogonal components in redundancy analysis, i.e. the components in the one set are not necessarily orthogonal to the components in the other set, since Xw and Yυ are determined separately (in canonical correlation analysis the canonical components are determined bi-orthogonally).

The functions to be maximized when the jth variates are determined are

$$\Phi_j = w_j' R_{xy} R_{yx} w_j - \mu_j (w_j' R_{xx} w_j - 1) - 2 \sum_i \alpha_i w_j' R_{xx} w_i,$$

$$(15)$$

$$\Psi_j = \upsilon_i' R_{yx} R_{xy} \upsilon_j - \nu_j (\upsilon_j' R_{yy} \upsilon_j - 1) - 2 \sum_i \beta_i \upsilon_j' R_{yy} \upsilon_i,$$

for $i = 1, \cdots, j - 1$.

Differentiating and setting equal to zero, leads to

$$\left\{ \begin{array}{l} \dfrac{\delta \Phi_j}{\delta w_j} = R_{xy} R_{yx} w_j - \mu_j R_{xx} w_j - \displaystyle\sum_i \alpha_i R_{xx} w_i = 0 \\[2em] \dfrac{\delta \Psi_j}{\delta \upsilon_j} = R_{yx} R_{xy} \upsilon_j - \nu_j R_{yy} \upsilon_j - \displaystyle\sum_i \beta_i R_{yy} \upsilon_i = 0 \end{array} \right\} \tag{16}$$

Premultiplication by w_i' and υ_i' respectively for every i gives us

$$w_i' R_{xy} R_{yx} w_j - \alpha_i = 0,$$

$$\upsilon_i' R_{yx} R_{xy} \upsilon_j - \beta_i = 0. \tag{17}$$

Because $w_i' R_{xy} R_{yx}$ is equal to $\mu_i w_i' R_{xx}$ (cf. the characteristic equation for the first variate) and $\upsilon_i' R_{yx} R_{xy}$ is equal to $\nu_i \upsilon_i' R_{yy}$, we have

$$\mu_i w_i' R_{xx} w_j - \alpha_i = 0 \rightarrow \alpha_i = 0,$$

$$\nu_i \upsilon_i' R_{yy} \upsilon_j - \beta_i = 0 \rightarrow \beta_i = 0, \tag{18}$$

which leaves us with the same characteristic equation that we found for the first variates. In other words, the vectors w_j and υ_j satisfying the above restrictions are proportional to the jth eigenvectors of the characteristic equations

$$(R_{xy} R_{yx} - \mu R_{xx}) w = 0,$$

$$(R_{yx} R_{xy} - \nu R_{yy}) \upsilon = 0. \tag{19}$$

Norming the eigenvector w_j so as to satisfy $w_j R_{xx} w_j = 1$, we have the jth factor of the predictors explaining a maximum of variance in the criteria (and vice versa).

[handwritten margin note: not $\beta'x'y$, but instead $x'y\beta'$]

MULTIPLE CORRELATION AS A SPECIAL CASE OF REDUNDANCY ANALYSIS

When we take a closer look at the matrix product $R_{xx}^{-1}R_{xy}R_{yx}$, we can see that there is some resemblance with multiple correlation. When we think of just one Y-variable, then $R_{xx}^{-1}R_{xy}$ is the column vector of β-weights in the multiple regression of that variable. Now, however, it is not the row vector of β-weights which is postmultiplied with the column vector of correlations of that variable with the predictors to give a scalar, but rather the reverse is true, to yield a matrix. In this matrix, the diagonal elements are the β-weights of the corresponding predictor in the multiple regression of the criterion times its correlation with the criterion; in other words, the partial regression coefficient of the given predictor in the multiple regression of the given criterion. The trace of that matrix is the total proportion of variance of the criterion accounted for by all predictors (the multiple correlation squared). From this it is obvious that canonical redundancy analysis includes as special cases multiple and bivariate correlation. The equations are, for only one Y-variable,

$$(R_{xx}^{-1}r_{xy}r_{yx} - \mu I)w = 0, \tag{20}$$

or

$$(\beta r_{yx} - \mu I)w = 0. \tag{21}$$

When $R_{y \cdot x}^2$ is substituted for μ, and β for w, one can show that the characteristic equation still holds:

$$\beta \, r_{yx} \, R_{xx}^{-1}r_{xy} - R_{y \cdot x}^2 \, \beta = 0; \tag{22}$$

$$\beta \, (R_{y \cdot x}^2 - R_{y \cdot x}^2) = 0. \tag{23}$$

So in the case of one criterion, μ is the multiple correlation squared and w is the vector of β-weights.

When we generalize to more Y-variables, the jth diagonal element in $R_{xx}^{-1}R_{xy}R_{yx}$ is the sum of contributions of the jth X-variable to the multiple correlation of all Y-variables with the set of X-variables. Thus it is possible to look at those diagonal elements as a kind of overall canonical partial regression coefficients. The trace of the matrix product under consideration is as expected equal to the sum of the redundancies of the variables of the other set.

So, when all redundancy components of the X-variables are determined, the explained variance of the Y-variables is equal to their respective squared multiple correlation with the X-variables. This is also true for canonical correlation analysis of which multiple correlation is a special case.

When there are residual dimensions (X and Y have a different rank) for one of the two sets, the variables of the other set will have zero loadings on them. This then implies that both analyses, when all possible components are determined, span the same space, though in a different way.

PRINCIPAL COMPONENT ANALYSIS AS A SPECIAL CASE

When the two sets upon which redundancy analysis is performed are the same, (19) leads to

$$(R_{xx}^{-1}R_{xx}R_{xx} - \mu I)w = 0, \tag{24}$$

or

$$(R_{xx} - \mu I)w = 0, \tag{25}$$

which is the characteristic equation of principal component analysis.

The characteristic equation (2) of canonical correlation analysis performed on two identical sets of variables contains an identity matrix, out of which the components are to be extracted. This is obvious in this case where all pairs of canonical variates correlate perfectly. Thus principal component analysis can be looked upon as a special case of redundancy analysis, but not, however, as a special case of canonical correlation analysis.

BI-ORTHOGONALITY

When a redundancy analysis is performed, extracting from each set p factors, for example, the explained variance of the variables in each set is a maximum, but the sets of variates spanning maximal redundant spaces are not bi-orthogonal. That is to say, the correlation matrices of variates within sets (Φ_{xx} and Φ_{yy}) are identity matrices, but the matrix of intercorrelations between sets of variates Φ_{xy} ($w'R_{xy}v$) is not a diagonal one.

For some applications bi-orthogonality could be a desirable property.

Thus we seek orthogonal rotation matrices T and S such that $\Xi^* = \Xi T$ and $Z^* = ZS$ are sets of variates that are bi-orthogonal, where Ξ is the matrix of redundancy variates ξ of the X-variables and Z is the matrix of redundancy variates ζ of the Y-variables. This now can be done by performing a canonical correlation analysis upon the redundancy variates Ξ and Z.

The characteristic equation (see [2]) becomes

$$(\Phi_{xx}^{-1}\Phi_{xy}\Phi_{yy}^{-1}\Phi_{yx} - \alpha^2 I)t = 0. \tag{26}$$

Because Φ_{xx} and Φ_{yy} are identity matrices this reduces to

$$(\Phi_{xy}\Phi_{yx} - \alpha^2 I)t = 0 \tag{27}$$

As in canonical correlation analysis, s can be found as a function of t:

$$s = \Phi_{yx}t/\alpha, \tag{28}$$

and α^2 is the squared correlation between the variates Ξt and Zs.

The whole procedure can be summarized as

$$(\Phi_{xy}\Phi_{yx} - \alpha^2 I)T = 0, \tag{29}$$

and

$$S = A^{-1}\Phi_{yx}T, \tag{30}$$

where A^{-1} is the diagonal matrix of inverses of the correlations between pairs of variates. It is possible to view T and S as orthogonal rotation matrices, containing the sines and cosines between old redundancy variates and new (bi-orthogonal) redundancy variates.

AN EXAMPLE WITH ARTIFICIAL DATA

Of four X-variables and four Y-variables the intercorrelation matrices R_{xx} and R_{yy} were constructed by means of orthogonal pattern matrices F and G; $R_{xx} = FF'$, $R_{yy} = GG'$. The matrices F and G are given in Table 5.1a. The matrix R_{xy} was constructed with the same loadings and a more or less arbitrary

TABLE 5.1

Matrices in the Construction Procedure

(a) Factor pattern

Variables	F Factors				G Factors			
	1	2	3	4	1	2	3	4
1	$+(.50^{\frac{1}{2}})$	$-(.40^{\frac{1}{2}})$	$-(.07^{\frac{1}{2}})$	$-(.03^{\frac{1}{2}})$	$+(.40^{\frac{1}{2}})$	$+(.30^{\frac{1}{2}})$	$+(.20^{\frac{1}{2}})$	$-(.10^{\frac{1}{2}})$
2	$+(.50^{\frac{1}{2}})$	$-(.40^{\frac{1}{2}})$	$+(.07^{\frac{1}{2}})$	$+(.03^{\frac{1}{2}})$	$+(.40^{\frac{1}{2}})$	$+(.30^{\frac{1}{2}})$	$-(.20^{\frac{1}{2}})$	$+(.10^{\frac{1}{2}})$
3	$+(.50^{\frac{1}{2}})$	$+(.40^{\frac{1}{2}})$	$-(.07^{\frac{1}{2}})$	$+(.03^{\frac{1}{2}})$	$+(.40^{\frac{1}{2}})$	$-(.30^{\frac{1}{2}})$	$+(.20^{\frac{1}{2}})$	$+(.10^{\frac{1}{2}})$
4	$+(.50^{\frac{1}{2}})$	$+(.40^{\frac{1}{2}})$	$+(.07^{\frac{1}{2}})$	$-(.03^{\frac{1}{2}})$	$+(.40^{\frac{1}{2}})$	$-(.30^{\frac{1}{2}})$	$-(.20^{\frac{1}{2}})$	$-(.10^{\frac{1}{2}})$

(b) Component intercorrelations

X-components	Y-components			
	1	2	3	4
1	.70	.10	-.10	.10
2	-.10	.75	.10	-.10
3	+.10	-.10	.80	.10
4	-.10	+.10	-.10	.85

(c) Resulting correlation matrix

	1	2	3	4	5	6	7
2	.800						
3	.140	.060					
4	.060	.140	.800				
5	-.003	.062	.422	.710			
6	.265	.203	.714	.440	.400		
7	.404	.709	-.142	.089	.200	.000	
8	.723	.461	-.012	-.037	.000	.200	.400

R_{xx}

R_{yy}

R_{xy}

matrix of intercorrelations between the components of the two sets. This matrix and the resulting total correlation matrix can be found in Table 5.1b and 5.1c respectively.

In Table 5.2 the matrix-products $R_{xx}^{-1}R_{xy}R_{yx}$ and $R_{yy}^{-1}R_{yx}R_{xy}$ are given. The diagonal entries of these matrices are interesting as they can be looked upon as a kind of overall canonical partial regression coefficients (above). For example, .643 is the sum of partial regression coefficients of the first X-variable in the multiple regression of each of the Y-variables on the set of X-variables. In Table 5.3 the beta-weights for the construction of the variates are given for canonical correlation analysis and redundancy analysis respectively. In Table 5.4 the loadings are given of the X- and Y-variables on both the canonical and redundancy variates. So all in all, we have eight sets of factor loadings.

Table 5.5 gives the redundancies as obtained by canonical correlation analysis and redundancy analysis respectively. As the complete set of factors is determined, the total redundancies are the same for both types of analysis (above). Differences between both methods can be easily seen by deleting, for example, the last component of both.

The redundancy (explained variance) per variable can be obtained by summating the squared loadings (as given in Tables 5.4a and 5.4b) of a variable on the components of the other set. In Table 5.6 the correlations between the sets of variates are given, illustrating the lack of bi-orthogonality for the case of redundancy analysis.

When the rotational procedure described above is performed on the complete set of redundancy factors given above, the canonical correlation solution will result, as in this case where both methods span the same space (above). However if only

TABLE 5.2

Matrix Products $R_{xx}^{-1}R_{xy}R_{yx}$ and $R_{yy}^{-1}R_{yx}R_{xy}$

$R_{xx}^{-1}R_{xy}R_{yx}$				$R_{yy}^{-1}R_{yx}R_{xy}$			
.643	.273	.054	.004	.517	.347	-.140	-.189
.146	.518	-.031	.085	.435	.657	.175	.234
-.079	-.213	.612	.111	-.024	.081	.582	.451
.128	.261	.113	.606	-.082	.097	.349	.509

TABLE 5.3

Beta Weights for Canonical Correlation- and Redundancy Analysis

	Canonical Variates				Redundancy Variates			
X-variables	1.182	-1.047	.781	-.395	.508	.184	0.842	-1.503
	-1.334	1.201	.102	-.152	.413	.210	-.530	+1.662
	-1.585	-.790	-.307	-.162	-.266	-.834	1.029	+1.197
	1.524	.619	-.330	-.667	.606	-.161	-1.157	-1.235

	Canonical Variates				Redundancy Variates			
Y-variables	.867	.185	-.485	-.530	-.101	.620	-.480	.823
	-.955	-.360	-.125	-.494	-.559	.391	.451	-.796
	-.555	.915	.337	-.208	-.473	-.416	-.728	-.613
	.644	-.558	.705	-.279	-.392	-.367	.531	.855

TABLE 5.4a

Factor Loadings in Canonical Correlation Analysis

	X-variates				Y-variates			
X-variables	-.016	-.159	.799	-.580	-.014	-.130	.615	-.415
	-.271	.403	.661	-.571	-.237	.330	.509	-.409
	-.281	-.370	-.455	-.760	-.245	-.302	-.350	-.543
	.140	.092	-.514	-.841	.122	.075	-.395	-.602
Y-variables	.327	.183	-.360	-.550	.374	.224	-.467	-.769
	-.419	-.326	-.137	-.545	-.480	-.398	-.178	-.761
	-.108	.596	.402	-.304	-.124	.729	.522	-.425
	.202	-.216	.627	-.329	.231	-.264	.815	-.460

TABLE 5.4b

Factor Loadings in Redundancy Analysis

	X-variates				Y-variates			
X-variables	.837	.226	.429	-.080	.343	-.454	-.423	-.264
	.888	.285	.043	+.359	.295	-.575	.341	.251
	.315	-.925	.189	+.099	.590	.328	-.289	.291
	.482	-.788	-.358	-.135	.538	.246	.395	-.289
Y-variables	-.622	-.332	.211	.157	-.419	.693	-.445	.382
	-.636	-.346	-.210	-.151	-.678	.565	.366	-.295
	-.370	.604	.217	-.144	-.650	-.439	-.612	-.106
	-.345	.589	-.227	.148	-.693	-.455	.330	.451

77

TABLE 5.5

Redundancies for Canonical Correlation Analysis (C.C.A.)
and Redundancy Analysis (R.A.)

	C.C.A.		R.A.	
	X-variables	Y-variables	X-variables	Y-variables
	.033	.084	.262	.210
	.056	.135	.235	.176
	.229	.176	.047	.133
	.249	.200	.023	.075
Total	.567	.595	.567	.595

TABLE 5.6

Correlation Matrices of X- and Y-Variates (ϕ_{xy})
for Both Types of Analysis

	C.C.A.				R.A.		
.873	.000	.000	.000	-.689	-.115	-.175	.146
.000	.818	.000	.000	.116	-.733	-.150	.094
.000	.000	.769	.000	-.168	-.156	.774	-.107
.000	.000	.000	.715	-.138	-.081	-.126	-.842

TABLE 5.7

Rotation Matrices T (for X-variables) and S (for Y-variables)

T		S	
.0667	.9978	-.0943	.9955
.9978	-.0667	.9955	.0943

the first two factors are retained, the solution will be definitely different from that for canonical correlation retaining just two factors. In Table 5.7 the rotation matrices T and S are given, which rotate the two pairs of variates in bi-orthogonal form.

By postmultiplication of the original sets of loadings by the appropriate rotation matrix, the loadings on the bi-orthogonal redundancy variates are found. These are presented in Table 5.8, while the diagonal matrix of intercorrelations between rotated X- and Y-variates is finally given in Table 5.9.

A more or less substantive evaluation of the above example can be given by assuming that the X- and Y-variables are intelligence tests. The canonical correlation method finds as first factors those that are maximally correlated, but unimportant in the sense of explained variance. The test batteries resemble

TABLE 5.8

Loadings after Rotation to Bi-orthogonality

	X-variates		Y-variates	
	.281	.820	-.430	.372
X-	.343	.867	-.554	.333
variables	-.902	.376	.366	.566
	-.754	.533	.281	.521
	-.272	-.651	.729	-.651
Y-	-.285	-.665	.627	-.665
variables	.636	-.311	-.376	-.311
	.619	-.288	-.388	-.288

TABLE 5.9

Correlation Matrix of the New Redundancy Variates
after Rotation

	Y-variates	
	-.7420	.0000
X-variates	.0000	-.6986

each other to a high degree, but in quite minor facets. The
first two intelligence dimensions of the X-battery yield mean
explained variances of 10.0% for the X-variables and 8.9 percent
for the Y-variables. The situation is not as bad for the first
two factors of Y. Here the results are 31.2 and 21.9 respectively.

By contrast, the first two redundancy factors of X explain
85.7 percent of the variance of the X-variables and 49.7 percent
for the Y-variables. Given the first two Y-variates, the two
figures are 68.4 and 38.6 respectively.

The danger of obtaining highly correlated, but unimportant
factors in a canonical correlation analysis is especially present
when there are two variables, one in each set, which are not
characteristic for the whole set, but yet highly correlated with
each other. Then one can find a factor pair of essentially unique
factors as the first canonical factors.

In redundancy analysis it is not necessary to extract
factors from both sets. This has an important advantage.
When we have a set of dependent and independent variables,
the predictive qualities of the independent set are found in the
redundancy factors without the complication of taking into
account the factors of the other set. It is easily seen that the
Y-variates can explain more variance of the X-variates, than
the other way around. However, the predictive power of the
X-battery with respect to the other battery is almost entirely
concentrated in the first two factors. This is less the case for
the Y-set.

Retaining the first two dimensions of each battery, the
resulting spaces are optimal in the redundancy sense. When
one wants to interpret the factors as intelligence dimensions,
bi-orthogonality could be desirable. Doing a rotation towards
bi-orthogonality does not influence the explained variances for
the total space; however, a different distribution of explained

variance over factors will result. As a result the factor pairs can be interpreted irrespective of all the other intelligence factors of both sets.

NOTE

1. Thissen, M. & Van den Wollenberg, A. L. REDANAL. A fortran IV G/H program for redundancy analysis (Research Bulletin 26). Nijmegen, the Netherlands: University of Nijmegen, Department of Mathematical Psychology, 1975.

REFERENCES

Anderson, T. W. An introduction to multivariate statistical analysis. New York: Wiley, 1958.

Stewart, D. & Love, W. A general canonical correlation index. Psychological Bulletin, 1968, 70, 160-163.

6

External Single–Set
Components Analysis of
Multiple Criterion/Multiple
Predictor Variables

Claes Fornell

INTRODUCTION

Problems in measuring and analyzing multidimensional behavioral phenomena are as old as behavioral science itself. Not
until recently, however, have survey researchers been able to
lessen their dependence on ceteris paribus protection by utilizing
techniques that simultaneously handle multiple criterion and
predictor variables.

Because of its capability to deal with multivariance on both
sides of an equation, canonical correlation has a great deal of
appeal and has also enjoyed increasing popularity in behavioral
research. But canonical correlation also stands out as perhaps
the most misused and misunderstood approach among major
multivariate techniques, and one whose interpretative properties
have caused much confusion.

Because canonical correlation coefficients pertain to unobserved linear combinations (variates) of measured variables,
there is a question of what is meant by "explained" variance. In
addition, canonical correlations is general enough to incorporate
many features of other multivariate techniques. Consequently,
canonical correlation generates a multitude of interpretative

Reprinted from Multivariate Behavioral Research, July
1979, Vol. 14, pp. 323-338. Copyright 1979 by The Society of
Multivariate Experimental Psychology, Inc. Reprinted by permission.

statistics. In applying the technique, there has been little consistency and much confusion as to the interpretative properties of the standardized regression coefficients of the variables (weights), the intra-set correlations (loadings), and the inter-set correlations (cross-loadings).

As behavioral researchers have become more familiar with sophisticated techniques of analysis, more attention is being paid to interpretative properties and the conceptual issues of the techniques in relation to both research objectives and empirical considerations. Advances have been made by a number of researchers from a variety of different academic disciplines. Barcikowski and Stevens (1975) have approached the weight versus loadings issue by studying the stability of the two statistics. Fornell (1978a) has presented a framework for determining the conditions under which one statistic is more suitable than the other.

Cross-validation procedures have been developed by Thorndike and Weiss (1973) and Wood and Erskine (1976). Thorndike (1977) has also presented a method for canonical analysis in prediction problems. Rotation of canonical solutions is seldom attempted in application, but the issue has been discussed in the literature (Cliff and Krus, 1976; Krus et al., 1976; Fornell, 1978a, 1978b; Hall, 1969; Perreault and Spiro, 1978). Horst (1961) and Carroll (1968) have extended the two-set canonical correlation to M-set analysis.

In response to the difficulties in determining "explained" variance in canonical correlation, several measures have been proposed (Rozeboom, 1965; Stewart and Love, 1968; Miller and Farr, 1971; Srikantan, 1970). Tests of significance have been discussed by Harris (1976), Knapp (1978), and Mendoza et al. (1978).

THE PROBLEM

Although these developments have been substantial and will no doubt stimulate further application of methodological procedures based on canonical correlation, there are serious problems remaining. One such problem is a direct result from the mathematical maximization principle in canonical correlation. In terms of substantive variable relationship and variate interpretation, canonical correlation provides a "compromise" solution (something in between factor analysis and regression) in the sense that a high canonical correlation coefficient can be obtained even when there are high zero-order correlations between only a few, or even between a single pair, of criterion and predictor variables. Consequently, a high canonical correlation coefficient does not necessarily mean that the variates are representative of the relationship. In spite of high shared variance between variates, it is quite possible that none of the variates extract significant variance from the original variables. If the variates are to be interpreted as factors, canonical correlation may therefore turn out to be virtually useless.

The purpose of this chapter is to offer a solution to this problem. Instead of maximizing the correlation between unobserved variates, external single-set components analysis (ESSCA) maximizes the inter-set loadings between $\hat{X}$-variates and y-variables. In order to facilitate interpretation, the resulting structure is rotated in such a manner that explanatory power is retained.

It will be empirically demonstrated that ESSCA can both increase the explanatory power relative to canonical correlation and facilitate interpretation through re-distributing intra- and inter-set loadings in an orthogonal fashion. Following a brief description of the mechanics of ESSCA, two applications will show:

1. That canonical correlation and ESSCA will yield similar results when correlations between criterion and predictor variables are fairly evenly distributed (in terms of the size of the correlation coefficients).
2. That canonical correlation and ESSCA will yield different results, and that ESSCA is superior in terms of explanatory power, if the correlations between criterion and predictor variables vary greatly in magnitude.

EXTERNAL SINGLE-SET COMPONENTS ANALYSIS (ESSCA)

If there is one set of standardized y-variables and another set of standardized x-variables, the objective of ESSCA is to find a linear combination ($\hat{X}$) of the x-variables that maximizes the common variance between the y-variables and the $\hat{X}$-variate. After the first $\hat{X}$-variate has been formed, a second one, orthogonal to the first, is found from remaining variance using the same principle. The maximum number of variates is of course equal to the number of x-variables. Thus, from one set of M equations with p predictors and q criterion variables where x and y are the two sets of measures

$$\hat{X}_1 = wx_{11} + wx_{12} + \ldots + wx_{1q} \tag{1}$$

$$\hat{X}_2 = wx_{21} + wx_{22} + \ldots + wx_{2q}$$

.
.
.

$$\hat{X}_M \quad wx_{M1} + wx_{M2} + \ldots + wx_{Mq}$$

find the vectors of weights w that maximize the summed squared correlation between $\hat{X}$ and the y-variables. In matrix notation this can be written in the characteristic equation form:

$$(R_{xy}R_{yx} - \lambda R_{xx})w = 0 \tag{2}$$

where λ is the eigenvalue and w the weight vector. The eigenvalue divided by the number of criterion variables is equal to the redundancy associated with the respective variate (Stewart and Love, 1968; Van den Wollenberg, 1977). By determining the eigenvectors from the matrix product $R_{xx}^{-1} R_{xy} R_{yx}$, a total of M or p (p = M) variates can be extracted from the x-variables. These vectors are extracted under the constraint of orthogonality.

The procedure described here differs from canonical correlation in two important regards:

1. Instead of maximizing the correlation between unobserved constructed variables (canonical variates), the weights are determined to maximize the sum of squared correlations between predictor variates and criterion variables.
2. Instead of forming pairs of variates, only one set of composite variables is constructed. There is no need to complicate the analysis by forming $\hat{Y}$-variates if the directionality of the relationship is one way and if we are not interested in the explanatory power of the y-variables on the x-variables.

The maximization principle ensures that the explanatory capability of the x-variables is optimal according to the redundancy criterion (Stewart and Love, 1968):

$$\bar{R}^2_{y/\hat{X}} = \sum_{j=1}^{M} \bar{L}^2_{y/\hat{X}_j} = \sum_{i=1}^{M} (\lambda_i) (\bar{L}^2_{y/\hat{Y}_i}) \tag{3}$$

$\bar{R}^2_{y/\hat{X}}$ is the total redundancy (the proportion of variance in the criterion variables that is explained by the predictor variates).

$\bar{L}^2_{y/\hat{X}_j}$ is the mean squared cross-loading of the y-variables on the jth $\hat{X}$-variate.

$\bar{L}^2_{y/\hat{Y}_i}$ is the mean squared loading of the y-variables on the ith $\hat{Y}$-variate.

λ_i is the ith squared canonical correlation coefficient (eigenvalue).

The fact that there is no need to extract pairs of variates simplifies interpretation considerably. In canonical correlation, the relationship between y and x is determined by the correlation between variable x and variate $\hat{X}$ because variate $\hat{X}$ is correlated with variate $\hat{Y}$, which in turn is correlated with variable y. Interpretation of external components analysis relies on a more direct association between variables.

However, if the variates are to be interpreted as factors, the results may still be difficult to comprehend. In terms of simplifying structure, the procedure described thus far offers no advantage compared to canonical correlation. But because of the orthogonality of the system, it is possible to rotate the solution without sacrificing explanatory power. If the loadings on the $\hat{X}$-variates for both criterion and predictor variables are simultaneously rotated in an orthogonal fashion, the total sum of squared loadings will be the same for the rotated solution. Only the distribution of loadings will change. For example, Kaiser's varimax rotation is likely to facilitate the task of interpretation by separating high versus low loadings on a variate.

TWO ILLUSTRATIONS

Case One

An example from organizational behavior will serve as an illustration of a case where canonical correlation and ESSCA provide similar results. The data were obtained from a survey designed to measure the involvement of corporate consumer affairs departments in management decision making. The substantive findings of this research (utilizing canonical correlation) have been published elsewhere (Fornell, 1976). For the purpose of this study, it is sufficient to briefly outline the nature of the variables and the objective of analysis.

Consumer affairs departments represent a recent addition to the organizational structure of many corporations. For the

long-term success and survival of these new organizational
units, it is important that they play an active role in the marketing decision making of their respective company. In order to
identify the characteristics of a consumer affairs department
that may promote its ability to influence marketing and other
consumer-related areas within the firm, data were obtained
from 128 consumer affairs executives. The independent variables
measured formal authority (veto power), informal authority
(education, title, previous title, and age of consumer affairs
executive), interdepartmental communication resources (representation in policy and planning committees, coordination of
consumer-related activities of other departments, education of
company personnel in consumer affairs), support and organizational independence (perceived top management support, budget
responsibility, development of own long-term goals), expertise
(consumer research activities, information publishing), and
amount of consumer criticism directed at the company (consumer
complaints relative to other unsolicited communication).

The criterion variables consisted of eight measures of
consumer affairs involvement in specific marketing decisions
and other decisions with a bearing on consumer satisfaction
and welfare.

The objective of analysis was twofold:

1. To identify the characteristics of consumer affairs departments that are related to involvement in management decision
   making.
2. To find a meaningful way to categorize consumer affairs
   departments with respect to their inter-departmental interaction.

The first objective suggests a "variable-to-variable" relationship
analysis, the second a "structure" analysis.

Table 6.1 presents the correlation matrix (where the first
14 variables are predictors). Note that most predictor variables
are moderately correlated with all or most of the criterion variables. Table 6.2 shows the results obtained from canonical
correlation and ESSCA, respectively. The variance of the y-
variables accounted for by the $\hat{X}$-variates is .33 in both solutions.
Moreover, inspection of individual variable relationships shows
that the two methods produce similar results. As typical in
canonical correlation, the intra-correlations (loadings) are used
for interpretation (due to space limitations, the cross-correlations
are not reproduced here). In ESSCA (as the name implies, there
is only one set of variates. As a result, both intra- and inter-
set correlations are used.

TABLE 6.1

Correlation Matrix—Case One

Explanatory Variables:	1	2	3	4	5	6	7	8	9	10	11	12	13	14	15	16	17	18	19	20	21	22
1. Middle-aged C.A. executive	1.00																					
2. C.A. executive's title	-.05	1.00																				
3. C.A. executive's prev. title	-.08	.49	1.00																			
4. C.A. executive's education	.10	-.15	.16	1.00																		
5. Veto power	.12	.04	.15	.02	1.00																	
6. Reps. on committees	.20	.19	.30	.21	.53	1.00																
7. Coord. consumer-related activities	.10	.18	.03	.03	.28	.43	1.00															
8. Educating personnel	.05	.19	.10	-.02	.17	.32	.59	1.00														
9. Top management support	.01	.19	.15	-.12	.21	.27	.38	.26	1.00													
10. Budget responsibility	-.10	.09	.19	.11	.12	.30	.15	.30	.24	1.00												
11. Resp. for long-term goal devlpmt.	-.00	.07	.08	.02	.08	.27	.42	.52	.40	.19	1.00											
12. Publishing factual information	-.01	.04	.01	.12	.19	.20	.40	.53	.20	.08	.45	1.00										
13. Research on consumer behavior	.06	.02	-.04	.02	.07	.19	.38	.49	.30	.06	.47	.38	1.00									
14. Consumer complaints	-.20	-.15	-.10	-.13	-.06	-.10	-.07	-.04	-.14	.04	.13	-.08	.10	1.00								
Criterion Variables—Consumer Affairs' Involvement in:																						
15. Advertising copy	.18	.17	.12	.09	.37	.48	.20	.14	.17	.19	.09	.07	.05	-.07	1.00							
16. Sales promotion	.10	.23	.15	.26	.33	.20	.13	.10	-.04	.22	-.04	.03	-.12	-.11	.54	1.00						
17. Concept generation	.20	.08	.10	.18	.45	.42	.22	.21	.26	.09	.20	.11	.08	-.26	.20	.18	1.00					
18. New product launching	.30	.22	.20	.05	.54	.44	.40	.38	.22	.15	.26	.16	.14	-.23	.23	.31	.48	1.00				
19. Market research	.06	.06	.17	.11	.26	.38	.14	.31	.08	.21	.33	.27	.40	-.04	.13	.07	.26	.18	1.00			
20. Consumer service	.28	.04	.15	.16	.30	.30	.34	.25	.25	.13	.27	.12	.18	-.03	.26	.18	.38	.42	.27	1.00		
21. New consumer programs	.08	-.04	-.01	.19	.22	.30	.56	.48	.35	.13	.51	.43	.47	-.04	.18	.09	.27	.32	.25	.43	1.00	
22. Influence on top management	-.02	.22	.17	.07	.26	.45	.63	.62	.45	.23	.50	.46	.38	-.06	.33	.18	.39	.33	.31	.34	.60	1.00

88

Significance Testing

There are several ways, both in ESSCA and in canonical correlation, to determine the number of variates to extract. One may set a criterion as to the size of the canonical correlation or amount of redundancy below which the remaining covariation is ignored. Alternatively, one may choose to discard variates that account for less than a predetermined portion of the variance of the original variables (communality).

There are also several statistical tests available. The significance of canonical correlations may be tested by Bartlett's (1941) chi-square approximation of Wilk's lambda or by partitioned chi-squares as described by Veldman (1967). The significance of redundancy in ESSCA variates (and in canonical correlation) may be tested by using Miller's (1975) analogue to the F-test in regression. As shown in Table 6.2, it makes a great deal of difference whether one is testing for the significance of canonical correlations or for the significance of redundancy. Two variates are significant beyond the .001 level according to Bartlett's test, but since one of the variates accounts for a mere 7% of the variance in the criterion variables (in both solutions), Miller's test reveals that this variate is not statistically significant in terms of redundancy.

Interpretation

The loadings in Table 6.2 are scattered across components in such a way that no apparent pattern is discernible. While this is not necessarily a drawback in variable-to-variable analysis, it certainly complicates the search for structure in the data. Examining the first ESSCA component, it is evident that "veto power," "coordination," "educating personnel," and "responsibility for long-term goal development" are strongly associated with consumer affairs' involvement and influence in both marketing and consumer-oriented matters. Due to the moderate size of the loadings and their varying signs, the relationships exhibited in the second component are more difficult to comprehend.

The rotated (Kaiser's varimax) solution presented in Table 6.3 is able to untangle the complex relationships. Note that the sum of squared loadings (and thus total explanatory power) remains unchanged after rotation. While "veto power" and "committee representation" are still cogent explanatory variables, the status of the variable measuring consumer affairs' coordination activities has been clarified. As shown in the rotated solution, it is clear that "coordination" is strongly associated with "perceived influence on top management" and involvement in

TABLE 6.2

Traditional Canonical Solution and ESSCA (Unrotated) Solution—Case One

Criterion Variables	Canonical Loadings		h^2	ESSCA (Cross) Loadings		$\sum_{i=1}^{2} L^2_y/\hat{X}_i$
15	.48	-.41	.40	-.41	-.22	= .22
16	.29	-.46	.30	-.31	-.39	.25
17	.60	-.41	.53	-.52	-.20	.31
18	.71	-.46	.72	-.65	-.23	.48
19	.52	-.07	.28	-.40	.11	.17
20	.54	-.13	.31	-.47	-.02	.22
21	.71	.41	.67	-.57	.38	.47
22	.81	.39	.81	-.65	.32	.53
Mean sum of squares	.363	.138	.503	.261	.069	.331

(handwritten annotations: $(.48)^2 + (-.41)^2$; $(-.41)^2 + (-.22)^2 =$)

Predictor Variables	Canonical Loadings		h^2	ESSCA Loadings		h^2
1	.20	-.38	.18	-.28	-.29	= .16
2	.26	-.16	.09	-.24	-.25	.12

(handwritten annotation: $(-.28)^2 + (-.29)^2 =$)

3	.25	-.16	.09	-.25	-.15	.09
4	.16	-.10	.04	-.25	-.12	.08
5	.60	-.54	.65	-.66	-.42	.61
6	.70	-.29	.57	-.73	-.13	.55
7	.75	.36	.69	-.70	.36	.62
8	.73	.33	.64	-.65	.38	.57
9	.49	.23	.29	-.47	.29	.31
10	.29	-.04	.09	-.32	-.06	.11
11	.62	.37	.52	-.57	.51	.59
12	.48	.42	.41	-.43	.45	.39
13	.52	.42	.45	-.41	.60	.53
14	-.16	.26	.09	.19	.24	.09
Mean sum of squares	.241	.102	.343	.227	.116	.344
Canonical correlation	.85	.72				
Significance (Bartlett's test)	$p < .001$	$p < .001$				
Redundancy	.26	.07	.33	.26	.07	.33
Significance (Miller's test)	$p < .001$	n.s.		$p < .001$	n.s.	
N = 128						

TABLE 6.3

Rotated ESSCA Solution—Case One

Criterion Variables	ESSCA (Cross) Loadings		$\sum\limits_{i=1}^{2} L^2\, y_j/\hat{X}_i$
15	.14	.44	.22
16	-.05	.50	.25
17	.23	.51	.31
18	.30	.62	.48
19	.36	.20	.17
20	.32	.34	.22
21	.67	.13	.47
22	.69	.23	.53

Predictor Variables	ESSCA Loadings		h^2
1	.00	.40	.16
2	.00	.35	.12
3	.07	.28	.09
4	.09	.26	.08
5	.18	.76	.61
6	.43	.60	.55
7	.75	.24	.62
8	.73	.19	.57
9	.54	.12	.31
10	.19	.27	.11
11	.77	.04	.59
12	.62	-.02	.39
13	.71	-.15	.53
14	.03	-.30	.09

the development of consumer programs and service, but only remotely related to participation in marketing.

With the rotated loadings, it is now possible to interpret the structure of the components. The first is primarily associated with high loadings on "top management influence," "consumer service," "consumer programs," and "market research" (on the criterion side), and "responsibility for long-term goal development," "coordination," "education of personnel," and "information publishing" (on the explanatory side). Consequently, this component is descriptive of a consumer affairs department that is mostly oriented toward consumer service and information, but has little impact on marketing matters such as advertising and promotion. The second component is primarily associated with specific marketing decision areas and, on the explanatory side, formal as well as informal authority, plus "committee representation."

Case Two

The structure of the initial correlation matrix (i.e., the distribution of high vs. low correlation coefficients) determines the degree of conformity between ESSCA and canonical solutions. If each predictor variable is highly to moderately correlated with one criterion variable and at the same time has a weak linear association with other predictor and criterion variables, it can be shown that canonical correlation exploits minor portions of shared variance at the expense of more substantial variable relationships. Table 6.4 presents a correlation matrix where the size of the coefficients vary greatly. It was constructed from a battery of AIO measures and preference rankings of contraceptive methods, given by a sample of male undergraduates. Six preference rankings included foam, pill, I.U.D., prophylactics, diaphragm, and rhythm. Ten AIO variables were measured on a six-point scale (agree-disagree). From Table 6.4, it can be seen that none of the preference variables is moderately correlated (in the .30 to .40 range) to more than three of the AIO variables.

Results

The standard canonical solution and the final ESSCA results are found in Table 6.5. Following Bartlett's test, three highly significant variates were extracted. However, even though canonical correlations are high (.88, .87, .76) and significant beyond the .001 level, explanatory power is extremely

TABLE 6.4

Correlation Matrix—Case Two

	1	2	3	4	5	6	7	8	9	10	11	12	13	14	15	16
1. Foam	1.00															
2. Pill	-.08	1.00														
3. I.U.D.	.04	.16	1.00													
4. Prophylactics	-.18	-.29	-.53	1.00												
5. Diaphragm	.04	-.17	.25	-.25	1.00											
6. Rhythm Method	-.30	-.48	-.48	.18	-.31	1.00										
7. I enjoy getting dressed up for parties.	.10	-.03	-.47	.15	-.17	.23	1.00									
8. Liquor adds to the party.	.42	-.33	-.15	.12	.13	.02	.11	1.00								
9. I study more than most students.	-.13	.33	.07	-.04	-.00	-.36	-.06	-.30	1.00							
10. Varsity athletes earn their scholarships.	.06	-.38	-.03	.19	.13	.18	.09	.12	-.16	1.00						
11. I am influential in my living group.	.24	-.11	-.25	-.11	-.09	.36	.33	.10	-.22	.26	1.00					
12. Good grooming is a sign of self-respect.	.18	-.24	-.35	.18	-.26	.37	.25	.13	-.10	.03	.21	1.00				
13. My days seem to follow a definite routine.	-.08	-.11	-.26	.41	.14	-.10	.13	.05	.23	-.16	-.14	.13	1.00			
14. Companies generally offer what consumers want.	.04	-.10	.02	.00	.38	-.09	-.05	-.02	-.08	.14	.10	-.25	.14	1.00		
15. Advertising results in higher prices.	.02	-.01	-.00	-.09	-.37	.26	-.07	-.13	.14	.00	.24	.18	-.22	-.21	1.00	
16. There should be a gun in every home.	.22	-.01	-.11	.29	.10	-.24	.03	.08	.00	-.01	-.02	.08	.28	.13	-.21	1.00

94

TABLE 6.5

Traditional Canonical Solution and Final ESSCA (Rotated)Solution—Case Two

Criterion Variables	Canonical Loadings			h^2	ESSCA (Cross) Loadings			$\sum_{i=1}^{3} L^2 y / \hat{X}_i$
1	-.04	-.08	.20	.05	-.04	.47	-.06	.23
2	-.13	.10	.02	.03	-.08	-.42	-.33	.29
3	-.11	-.07	-.22	.07	-.31	-.21	-.29	.22
4	.10	.11	-.03	.02	-.02	.03	.56	.32
5	-.05	.02	-.04	.01	-.48	.10	.12	.25
6	.15	-.06	.14	.05	.55	.25	.04	.37
Mean sum of squares	.011	.006	.018	.036	.106	.086	.088	.280

Predictor Variables	Canonical Loadings			h^2	ESSCA Loadings			h^2
7	.11	-.04	.00	.01	.48	.12	.29	.33
8	.03	.26	.03	.07	-.21	.78	.12	.67

(continued)

(Table 6.5 continued)

Predictor Variables	Canonical Loadings			h²	ESSCA Loadings			h²
9	.07	-.25	.02	.07	-.24	-.48	-.08	.29
10	-.04	.03	.12	.02	-.02	.36	.36	.26
11	.07	.03	.08	.01	.39	.51	-.20	.45
12	.16	.08	.08	.04	.58	.39	.17	.52
13	.15	.11	.03	.04	-.10	-.15	.77	.63
14	-.14	-.08	.02	.03	-.44	.24	.09	.26
15	.04	-.04	.01	.00	.54	.00	-.23	.35
16	.18	.07	.12	.05	-.32	.14	.35	.25
Mean sum of squares	.013	.017	.004	.034	.142	.150	.108	.401
Canonical correlation	.88	.87	.76					
Significance (Bartlett's test)	p<.001	p<.001	p<.001					
Redundancy	.009	.005	.010	.024	.106	.086	.088	.280
Significance (Miller's test)	n.s.	n.s.	n.s.		p<.001	p<.01	p<.01	p<.01

N = 340

96

low in the canonical solution. Total redundancy (for the three variate pairs) is only 2.4% and no one variate explains more than 1% of the variance in the criterion variables. Clearly, these results do not lend themselves to any form of substantive interpretation. Not surprisingly, all the relationships defined by the canonical correlation method are insignificant in terms of redundancy. The case clearly demonstrates a major caveat in canonical correlation: The results may be deceptive and easily give rise to exaggerated interpretations if communalities and redundancies are ignored.

ESSCA provides a different picture. By maximizing ex-plained variance (in the reduced rank solution), it is now possible to increase explanatory power substantially. Three rotated ESSCA variates are statistically significant and individ-ually account for between 9 to 11% of the variance in the criterion variables. Total redundancy amounts to 28%.

As in ordinary principal components or factor analysis, the loadings of the variable on the composites are examined in order to identify underlying constructs of the relationships. In the canonical solution, the loadings are so low that this is not possible. As implied by the redundancy measure, ESSCA has been more successful in this regard. For example, "con-servatism" (inferred from the loadings on variables 13, 10, 16) appears to be related to the preference of prophylactics among male college students.

CONCLUSION

While functional multivariate techniques such as regression and discriminant analysis are incapable of simultaneously handling more than one criterion variable and the structural techniques of factor analysis do not discriminate between criteria and pre-dictors, canonical correlation analysis suffers from substantial interpretative problems. Numerically large canonical correlations may lead to exaggerated interpretations since these correlations do not reflect the variance of observed variables, but rather the variance of derived canonical scores, which themselves are subject to interpretation.

The method described in this chapter addresses both the problem of explained variance and of substantive inference. By maximizing the sum of squared inter-set loadings between $\hat{X}$ and y-variables, optimal explained variance according to the redundancy formulation is ensured. As was shown, canonical correlation is not able to do this. Two-step approaches such as principal components of criterion variables followed by regressions with the components as dependent variables also fail to ensure optimal redundancy in reduced rank solutions. The principal components are defined without consideration to the predictor set and, unless all possible components are extracted, some of the variance shared by predictors and criteria is likely to be lost. Regressions using full rank principal components would yield maximum explained variance, but such an approach assumes that every component retains a sufficient amount of variance to justify its status as a separate criterion variable.

An interesting feature of ESSCA is that it employs both regression and component perspectives. The regression perspective refers to the <u>functional</u> analysis where the objective is to predict or explain phenomena via their relationship to other phenomena. The component perspective refers to the <u>structural</u> analysis where the objective is to simplify complex relationships in order to provide insights into the underlying structure of the data. As such, structural analysis typically involves data reduction and variate rotation. Since ESSCA is designed both for explanatory and simplifying purposes, it can be characterized as a joint functional/structural approach. By rotating the resulting structure in an orthogonal fashion, it was shown that interpretability could be significantly enhanced, without loss of variance.

The reliability of an ESSCA solution ultimately rests with the stability of the results. Although the stability of ESSCA has not been empirically examined, it would seem that it is less susceptible to chance variation than is canonical correlation. The reason for this is that ESSCA involves single functions, as opposed to the double (paired) functions in canonical correlation. Thus, the number of extracted variates in the canonical solution is twice the number of variates in a corresponding ESSCA.

Since only one set of components (from the predictors) is extracted in ESSCA, the method is appropriate for problems where the directionality of relationships is limited to one way (as in regression) and where the data matrix thus is clearly partitioned into one set of criterion variables and one set of predictor variables. Canonical correlation analysis does not have this restriction since its solution is symmetrical.

The proposed ESSCA method would be particularly useful in cases where correlations vary greatly, both within and between variable sets, for canonical correlation may then produce variates that are non-representative of the original variables. If, on the other hand, correlations are fairly equal across all variable relationships, canonical correlation will generate results similar to those obtained by ESSCA.

In sum, it seems that ESSCA offers most of the advantages of canonical correlation analysis, but shares few of its drawbacks. In contrast to canonical correlation, the measure of shared variance is based on both components and observed variables, and the proposed rotation makes variate interpretation much easier. However, just as the canonical correlation coefficient is a summary index of the relationship between two sets of variables, the term to be maximized in ESSCA is also a summary measure. As any measure reflecting average relationships, it is possible to obtain high values even though some of the involved variables are remotely associated. Compared to canonical correlation, however, ESSCA is at an advantage in this regard. While canonical correlation analysis limits concern to the maximization of variate pairs, without considering the loadings, ESSCA, in effect, relies on both inter- and intra-set relationships.

POSTSCRIPT

Due to a programming error, the ESSCA results in Table 6.5 are inaccurate. For the example used, the difference between canonical analysis and ESSCA in terms of redundancy is less than the values in Table 6.5 indicate.

REFERENCES

Barcikowski, R. S. and Stevens, J. P., A Monte Carlo study of the stability of canonical correlations, canonical weights, and canonical variate-variable correlations, Multivariate Behavioral Research, 1975, 10, pp. 353-364.

Carroll, J. P., Generalization of canonical correlation analysis to three or more sets of variables, Proceedings, 76th Annual Convention of the American Psychological Association, 1968, pp. 227-228.

Cliff, N. and Krus, D. J., Interpretation of canonical analysis: Rotated vs. unrotated solutions, Psychometrika, 1976, 41, pp. 35-42.

Fornell, C., Three approaches to canonical analysis, Journal of the Market Research Society, July 1978a, pp. 166-181.

Fornell, C., Problems in the interpretation of canonical analysis: The case of power in distributive channels, Journal of Marketing Research, 1978b, 15, pp. 489-491.

Fornell, C., Consumer input for marketing decisions—A study of corporate departments for consumer affairs. New York: Praeger, 1976.

Hall, C. E., Rotation of canonical variates in multivariate analysis of variance, Journal of Experimental Education, 1969, 38:2, pp. 31-38.

Harris, R. J., The invalidity of partitioned-U tests in canonical correlation and multivariate analysis of variance, Multivariate Behavioral Research, 1976, 11, pp. 353-365.

Horst, P., Relations among M sets of measures, Psychometrika, 1961, 26, pp. 129-149.

Knapp, T. J., Canonical correlation analysis: A general parametric significance-testing system, Psychological Bulletin, 1978, 85:2, pp. 410-416.

Krus, D. J., Reynolds, T. J., and Krus, P. H., Rotation in canonical variate analysis, Educational and Psychological Measurement, 1976, 36, pp. 725-730.

Mendoza, J. L., Markos, V. H., and Gonter, R., A new perspective of sequential testing procedures in canonical analysis: A Monte Carlo Evaluation, Multivariate Behavioral Research, 1978, 13, pp. 371-382.

Miller, J. K., The sampling distribution and a test for the significance of the bimultivariate redundancy statistic: A Monte Carlo study, Multivariate Behavioral Research, 1975, 10, pp. 233-244.

Miller, J. K. and Farr, S. D., Bimultivariate redundancy: A comprehensive measure of interbattery relationship, Multivariate Behavioral Research, 1971, 6, pp. 313-324.

Perreault, W. D., Jr. and Spiro, R. L., An approach for improved interpretation of multivariate analysis, Decision Sciences, 1978, 9, pp. 402-413.

Rozeboom, W. W., Linear correlations between sets of variables, Psychometrika, 1965, 30, pp. 57-71.

Srikantan, K. S., Canonical association between nominal measurements, Journal of the American Statistical Association, 1970, 65, pp. 284-292.

Stewart, D. and Love, W., A general canonical correlation index, Psychological Bulletin, 1968, 70, pp. 160-163.

Thorndike, R. M., Canonical analysis and predictor selection, Multivariate Behavioral Research, 1977, 12, pp. 75-87.

Thorndike, R. M. and Weiss, D. J., A study of the stability of canonical correlations and canonical components, Educational and Psychological Measurement, 1973, 33, pp. 123-134.

Veldman, D., Fortran Programming for the Behavioral Sciences. New York: Holt, Rinehart and Winston, 1967.

Van den Vollenberg, A. L., Redundancy Analysis: An alternative for canonical correlation analysis, Psychometrika, 1977, 42, pp. 207-219.

Wood, D. A. and Erskine, J. A., Strategies in canonical correlation with application to behavioral data, Educational and Psychological Measurement, 1976, 36, pp. 861-878.

7

Canonical Correlation Analysis as a Special Case of a Structural Relations Model

Richard P. Bagozzi, Claes Fornell,
and David F. Larcker

It has been shown by Knapp (1978) that the statistical test of the null hypothesis in canonical correlation analysis is a general model for virtually all common parametric statistical techniques. Specifically, Knapp has demonstrated that simple correlation, the t-test in independent samples, multiple regression and correlation analysis, one-way analysis of variance, factorial analysis of variance, analysis of covariance, t-test in correlated samples, discriminant analysis, and chi-square test of independence are all special cases of canonical correlation. However, the work of Jöreskog and Sörbom (1979) and Bentler and Weeks (1979) indicates that even more general models for these texts exist. This chapter will show that canonical correlation analysis is a special case of the linear structural relations model developed by Jöreskog (1970, 1973, 1977) and Jöreskog and Sörbom (1978).

There are at least two important advantages to be gained by treating canonical correlation analysis as a special case of linear structural relations. First, the problems resulting from an absence of a statistical test for canonical weights and loadings

---

Reprinted from Multivariate Behavior Research, October 1981, Vol. 16, pp. 437-454. Copyright 1981 by The Society of Multivariate Experimental Psychology, Inc. Reprinted by permission.

We wish to thank David A. Kenny, University of Connecticut, and two anonymous reviewers for their helpful comments.

are resolved. Standard errors for individual criterion and predictor weights and for cross loadings can be estimated from the inverse of the information matrix. The statistical significance of these weights and loadings can be evaluated by calculating the critical ratio (dividing the parameter estimate by its estimated standard error). Second, if one cannot reject the null hypothesis that the canonical variates are independent, the assumptions of variate orthogonality and/or component specifications might be relaxed. This analysis can be accomplished via structural relations modeling.

In order to demonstrate these advantages, formal descriptions of the canonical correlation and the structural relations models are presented, and the former is derived from the latter. Numerical illustrations are used to demonstrate (1) the usefulness of estimated standard errors for interpreting individual parameter values and (2) how the structural relations model can be used to further examine an insignificant canonical solution by altering some basic assumptions.

THE CANONICAL CORRELATION MODEL

Let q be the number of x-variables and p be the number of y variables. The objective of canonical correlation analysis is to find a linear combination of the q predictors (x_x.*) that maximally correlates with a linear combination of the p criteria (y_y.*). Following Anderson (1958), this problem is equivalent to solving the eigenstructures:

$$(R_{xx}^{-1}R_{xy}R_{yy}^{-1}R_{yx} - \lambda_j I)w = 0 \tag{1}$$

$$(R_{yy}^{-1}R_{yx}R_{xx}^{-1}R_{xy} - \lambda_j I)v = 0 \tag{2}$$

where

R_{yy} is a (p × p) criterion variable correlation matrix;
R_{xx} is a (q × q) predictor variable correlation matrix;
R_{xy} and R_{yx} are (q × p) and (p × q) cross correlation matrices;
w and v are eigenvectors (weight vectors) for the predictor and criterion variate, respectively;
λ_j is the eigenvalue (or the squared jth canonical correlation); and
I is the identity matrix.

2. The variance of the variates is constrained to equal 1, or

$$w'R_{xx}w = 1 \text{ and } v'R_{yy}v = 1. \tag{3}$$

Once the eigenvectors and the eigenvalues are determined, a second set of linear variate pairs, orthogonal to the first, can be formed from the residual variance.

The canonical loadings are obtained by premultiplying the weight vectors by their associated correlation matrix, or

$$r_{y \cdot y} = R_{yy}v \text{ and } r_{x \cdot x} = R_{xx}w. \tag{4}$$

The proportion of the variance in the y-variables accounted for by the variate x^* is given by the redundancy index (Stewart and Love, 1968):

$$\bar{R}^2_{y/x^*} = (1/p) \, (r'_{y \cdot y} r_{y \cdot y}) \, \lambda = (1/p) \, (r'_{x \cdot y} r_{x \cdot y}). \tag{5}$$

Note that the canonical variates assume:

$$r'_{x \cdot x} R_{xx}^{-1} r_{x \cdot x} = 1, \tag{6}$$

$$r'_{y \cdot y} R_{yy}^{-1} r_{y \cdot y} = 1, \tag{7}$$

$$r'_{y \cdot x} R_{xx}^{-1} r_{y \cdot x} = \lambda_j, \text{ and} \tag{8}$$

$$r'_{x \cdot y} R_{yy}^{-1} r_{x \cdot y} = \lambda_j. \tag{9}$$

The canonical correlation model with $p = 1 = 3$ is illustrated in Figure 7.1.

THE LINEAR STRUCTURAL RELATIONS MODEL

Jöreskog (1969, 1970, 1973, and 1977), Jöreskog and Goldberger (1975), and Jöreskog and Sörbom (1978 and 1979) have developed the statistical theory and algorithms for analyzing linear structural relation models which include unobservable variables (constructs), measurement error, and error in the structural equation. This formulation considers the types of structural equation models which are common in econometrics. Further, as in most psychometric studies, the variables in the structural equations may be either observable or unobservable. The unobservable constructs are conceptualized in terms of common factors and are included in the methodology via confirma-

FIGURE 7.1

The Canonical Correlation Model ($r_{x_i x_j}$ is the correlation between x_i and x_j. $r_{u_i u_j}$ is the correlation between the errors u_i and u_j. The $r_{x \cdot y_j}$ are the cross loadings between the $x \cdot$ variate and the observed y_j)

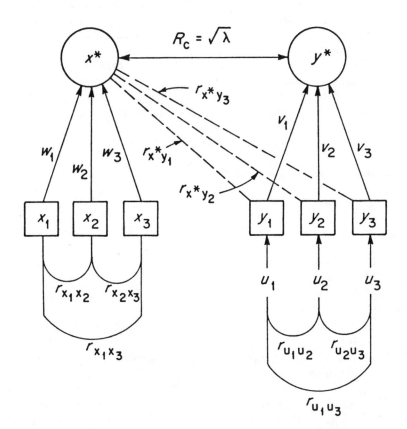

tory factor analysis. By combining structural equations with confirmatory factor analysis, Jöreskog has developed a very flexible and general model that incorporates many powerful features of both econometric and psychometric analyses.

The linear structural relations model consists of two sets of equations, the structural equations and the measurement equations. The linear structural equation is:

$$B\eta = \Gamma\xi + \zeta \tag{10}$$

where

> B is an (m × m) coefficient matrix (β_{ij} = 0 means that η_j and η_i are not related),
>
> Γ is an (m × n) coefficient matrix (γ_{ij} = 0 means that η_i is not related to ξ_j),
>
> η is an (m × 1) column vector of constructs derived from the dependent variables (y),
>
> ξ is an (n × 1) column vector of constructs derived from the independent variables (x),
>
> ζ is an (m × 1) column vector of the errors in the structural equations,
>
> m is the number of constructs (latent variables) developed from the observed dependent variables, and
>
> n is the number of constructs (latent variables) developed from the observed independent variables.

The measurement equations are:

$$y = \Lambda_y \eta + \varepsilon \tag{11}$$

and

$$x = \Lambda_x \xi + \delta \tag{12}$$

where

> y is a (p × 1) column vector of observed dependent variables,
>
> x is a (q × 1) column vector of observed independent variables,
>
> Λ_y is a (p × m) regression coefficient matrix of y on η,
>
> Λ_x is a (q × n) regression coefficient matrix of x on ξ,
>
> ε is a (p × 1) column vector of errors of measurement in y, and
>
> δ is a (q × 1) column vector of errors of measurement in x.

Given the structure imposed by (10), (11), and (12), the [(p + q) × (p + q)] covariance matrix (Σ) is:

$$\Sigma = \begin{bmatrix} \Lambda_y(B^{-1}\Gamma \Phi \Gamma'B'^{-1} + B^{-1}\Psi B^{-1})\Lambda_y' + \Xi_\varepsilon & \Lambda_y B^{-1}\Gamma \Phi \Lambda_x' \\ \Lambda_x\Phi \Gamma'B'^{-1}\Lambda_y' & \Lambda_x\Phi \Lambda_x' + \Xi_\delta \end{bmatrix} \tag{13}$$

where

Φ is defined to be an (n × n) covariance matrix of ξ,
Ψ is defined to be an (m × m) covariance matrix of ζ, and
Ξ_ϵ and Ξ_δ are the covariance matrices of ϵ and δ, respectively.

In order to estimate the unknown parameters of (13) for the identified system of equations, the log of the likelihood function is maximized, (or the fitting function [14] is minimized) via a modified Fletcher and Powell (1963) algorithm with respect to the independent unknown parameters:

$$F = \log|\Sigma| + \text{tr } (S\Sigma^{-1}) - \log|S| - (p + q) \qquad (14)$$

Assuming that the distribution of the observed y's and x's is multivariate normal, the procedure yields maximum likelihood estimates for the unknown parameters of (13). The standard errors for each of these parameters are obtained by inverting the information matrix. The computer algorithm used for these calculations is referred to as LISREL IV (Jöreskog and Sörbom, 1978).

The "goodness of fit" between the hypothesized covariance matrix (Σ) and the observed covariance matrix (S) for large samples is determined by forming the statistic:

$$N \cdot F_0 \sim \chi^2_{[1/2(p+q)(p+q+1)-Z]}, \qquad (15)$$

where

N is the sample size,
F_0 is the minimum of (14), and
Z is the number of independent parameters estimated.

The null hypothesis is S = Σ and the alternative hypothesis is that Σ is any positive definite matrix. If the null hypothesis is rejected, the theoretical structure imposed on the analysis is too restrictive and the a priori specification (number of constructs, pattern of loadings, covariances between measurement errors, etc.) is not consistent with the data.

THE RELATIONSHIP BETWEEN THE CANONICAL AND THE STRUCTURAL RELATIONS MODELS

From (5), it can be seen that redundancy for the y-variables can be calculated without knowledge of either the squared canonical correlation, λ, or the variate y$\cdot$. This means that the paired variates (y$\cdot$ and x$\cdot$) can be reduced to a model with a single variable (y$\cdot$ or x$\cdot$). As a result, the first canonical variable pair can be expressed as a Multiple Indicators/Multiple Causes (MIMIC) model of a single variate or latent construct. The single latent construct MIMIC model has been analyzed by Hauser and Goldberger (1971) and Jöreskog and Goldberger (1975), and the extension to multiple latent constructs has been discussed by Stapleton (1978).

Knowing the properties of the canonical model, we can derive the specifications necessary for obtaining a canonical solution from a MIMIC model:

(i) from (3), the variance of η is standardized to one;
(ii) from (6) and (7), η is a perfect function of ξ ($\zeta = 0$),

where $\xi = x$, i.e., ($\Lambda_x = I$, $\Xi_\delta = 0$).

The MIMIC representation of the canonical model in Figure 7.1 (p = 3, q = 3, m = 1, and n = 3) is presented in Figure 7.2. The structural equation is

$$\eta = \Gamma \xi. \tag{16}$$

The measurement equations are

$$y = \Lambda_y \eta + \varepsilon \tag{17}$$

$$x = I_\xi \tag{18}$$

where I is a (q × q) identity matrix.

Consistent with canonical correlation analysis, (16) indicates that η is an exact linear function of the x_i's = ξ_i's. Assuming that the x-variables and y-variables are standardized, the elements in Γ are identical to the canonical weights (w) for the predictor variate. However, η is not an exact function of the y_j's because Λ_y is the correlation between the predictor variate and the criterion variables (i.e., the [p × 1] matrix of cross loadings, $r_{x\cdot y}$). The determination of canonical loadings for the criterion variate requires knowledge of the canonical correlation coefficient (R_c). Recalling that λ (or the squared canonical correlation, R_c^2) is the variance of η that is explained by y and that the y-variables are assumed to be standardized, the eigenvalue can be calculated from (9) as:

$$\lambda = R_c^2 = \Lambda'_y [(y'y)^{-1}/N] \Lambda_y = r'_{x\cdot y} R_{yy}^{-1} r_{x\cdot y} \tag{19}$$

FIGURE 7.2

Canonical Correlation as a MIMIC Model—First Variate Pair
(The $r_{\varepsilon_i \varepsilon_j}$ are the correlations among the ε_i and ε_j. The correlations among the ξ_i (i.e., the φ_{ij}) are omitted from the figure
for simplicity. Note that the λ_{yj} refer to the relations between
η and the observed variables, y_j, and are <u>not</u> the eigenvalues
from the canonical model).

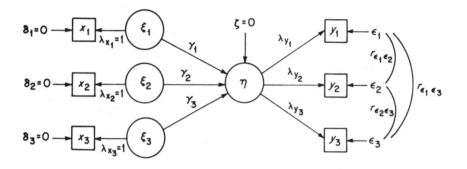

The loadings for the criterion variate in the canonical model
are

$$r_{y \cdot y} = (1/R_c) \cdot \Lambda_y. \tag{20}$$

Consistent with (5), the redundancy is

$$\bar{R}^2_{c/p} = (1/p) \Lambda'_y \Lambda_y. \tag{21}$$

The standard statistical tests (i.e., Bartlett, 1941, and Miller,
1975) associated with (19) and (21) can be used to determine
the significance of the canonical correlation and redundancy.
    By formulating canonical correlation analysis as a MIMIC
model, the goodness of fit between the theoretical structure
imposed by the canonical model and the observed data can be
examined and tested by the chi-square statistic from (15).
Standard errors for the estimated predictor weights and cross
loadings between the predictor variate and criterion variables
are also available. These standard errors can be used to calculate critical ratios and test the statistical significance of individual
coefficient weights and cross-loadings. Further, since canonical
correlation analysis is symmetric, the criterion weights, cross
loadings between the criterion variate and predictor variables,
and their associated standard errors can be estimated by simply
interchanging the x-variables and y-variables.

ILLUSTRATIONS

Example 1

As part of a larger study on university campus blood drives, a sample of 127 students, faculty, and staff were interviewed about their attitudes and intentions with respect to blood donation. Six variables were retained for the following illustration. The x-variables measured an affective attitudinal component (x_1), expectancy-value toward giving blood (x_2), and the past blood donation behavior (x_3). The y-variables consisted of three measures $(y_1, y_2, $ and $y_3)$ of the individual's behavioral intention regarding blood donation.

The correlation matrix, the canonical solution, and MIMIC solutions for the data set are provided in Tables 7.1, 7.2, and 7.3, respectively. For purposes of illustration, it is sufficient to present only the results for the first two canonical pairs with the standard errors for predictor weights and cross loadings. As shown in Table 7.3, the canonical results in Table 7.2 indicate that there are two statistically significant canonical variate pairs ($p \leq .05$ for each variate pair). The chi-square statistic associated with the first canonical pair for the MIMIC model ($\chi^2 = 10.27$, $p \leq .05$) indicates a borderline fit (the standard criterion for rejection is usually set at $p \geq .10$). In the canonical model, a high chi-square implies a significant canonical correlation. In the structural relations model, a high chi-square implies a poor fit between the model and the data. The lack of

TABLE 7.1

Correlation Matrix for Example 1 (n = 127)

	y_1	y_2	y_3	x_1	x_2	x_3
y_1	1.000					
y_2	.589	1.000				
y_3	.806	.640	1.000			
x_1	.292	.330	.372	1.000		
x_2	.354	.300	.340	.170	1.000	
x_3	.349	.494	.293	.140	.252	1.000

TABLE 7.2

Example 1: Canonical Model Parameter Estimates

	I	II
R_c	.603	.256
R_c^2 (or λ)	.363	.066
Significance of R_c^2	$p < .001$[a]	$p < .05$[a]
R_c/p^{-2}	.263	.041
Significance of $R_{c/p}^{-2}$	$p < .001$[b]	NS*

	Canonical Loadings	
Criterion Set		
y_1	.799	.280
y_2	.956	-.181
y_3	.788	.651
Predictor Set		
x_1	.593	.576
x_2	.586	.388
x_3	.811	-.585

[a]Statistical test based upon Bartlett (1941).
[b]Statistical test based upon Miller (1975).
*Not statistically significant at the .10 level.

fit for the first canonical variate pair in Table 7.3 can be attributed to the necessity of increasing the rank of a canonical solution.

To illustrate this, we incorporate a second latent construct in the MIMIC model (Figure 7.3). This was accomplished by fixing the parameter estimates (λ_{yi}'s and γ_i's) for the first canonical pair to the unstandardized values obtained from the MIMIC model in Figure 7.2, and adjusting the degrees of freedom. The chi-square statistic associated with the second canonical pair indicates a good fit for the two-construct model ($p \le .70$). Thus, one may conclude that there are two significant canonical variate pairs and that the structure imposed by the two-variate-pair canonical model is consistent with the observed data in Table 7.1.

The γ_i's from Table 7.3 are equivalent to the canonical weights for the predictor variate. From (4), the associated loadings for the first variate pair are:

TABLE 7.3

Selected MIMIC Model Parameter Estimates*

	I		II	
	Unstandardized Estimate	Standardized Estimate	Unstandardized Estimate	Standardized Estimate
λ_{y_1}	1.0 (−)[a]	.482	1.0 (−)[a]	.071
λ_{y_2}	1.196(.187)[b,c]	.576	−.658(1.485)	−.047
λ_{y_3}	.986(.114)[c]	.475	2.207(2.754)	.157
γ_1	.213(.062)[c]	.442	.043(.051)	.604
γ_2	.166(.062)[c]	.344	.035(.044)	.485
γ_3	.319(0.69)[c]	.662	−.056(.064)	−.792
	$\chi^2 = 10.27$		$\chi^2 = 1.72$	
	$df = 4$		$df = 1$	
	$p < .05$		$p < .70$	

[a] Since it is not possible to fix the variance of the η construct in LISREL, one of the λ_{y_1}'s must be fixed to 1.0 and no standard error was estimated. If one wants to assess the significance of λ_{y_1}, the analysis can be repeated with another λ_y fixed to 1.

[b] The standard error of the unstandardized estimate is in parentheses.

[c] Statistically significant at $p \leq .05$.

*This table presents cross loading between the predictor variate and the criterion variables (λ_{y_1}'s) and predictor weights (γ_1's). The cross loadings between the criterion variate and predictor variables and criterion weights can be obtained by interchanging the original x-variables and y-variables.

FIGURE 7.3

Canonical Correlation as a MIMIC Model—Second Variate Pair
(Parameters marked with a * are constrained to equal corresponding values obtained from analysis of first canonical variate pair—
see Figure 7.2. The correlations among the ξ_i [i.e., the φ_{ij}]
are omitted from the figure for simplicity.)

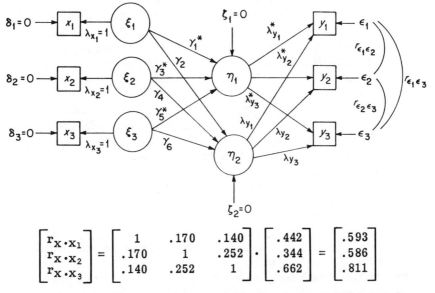

$$\begin{bmatrix} r_{x \cdot x_1} \\ r_{x \cdot x_2} \\ r_{x \cdot x_3} \end{bmatrix} = \begin{bmatrix} 1 & .170 & .140 \\ .170 & 1 & .252 \\ .140 & .252 & 1 \end{bmatrix} \cdot \begin{bmatrix} .442 \\ .344 \\ .662 \end{bmatrix} = \begin{bmatrix} .593 \\ .586 \\ .811 \end{bmatrix}$$

The first canonical correlation coefficient can be calculated from
[19] and can be shown to be equal to .603. From [21], the
criterion loadings for the first variate pair are:

$$\begin{bmatrix} r_{y \cdot y_1} \\ r_{y \cdot y_2} \\ r_{y \cdot y_3} \end{bmatrix} = \frac{1}{.603} \cdot \begin{bmatrix} .482 \\ .576 \\ .475 \end{bmatrix} = \begin{bmatrix} .799 \\ .956 \\ .788 \end{bmatrix}$$

Finally, the redundancy of the first canonical variate pair from
(21) is:

$$\bar{R}^2/_{cp} = 1/3 \ [.482 \ .576 \ .475] \begin{bmatrix} .482 \\ .576 \\ .475 \end{bmatrix} = .263.$$

As expected, the predictor and criterion loadings, canonical
correlation, and redundancy for the first variate pair as calculated from the MIMIC model are identical to those calculated
from the canonical model. It may also be verified that the load-

ings, canonical correlation, and redundancy for the second variate pair are identical for both the canonical and the MIMIC models.

An inspection of the first variate pair in Tables 7.2 and 7.3 indicates that there appears to be a relationship between each of the three predictors and intention to give blood. In particular, both the eigenvalue and the redundancy in the traditional canonical correlation analysis are significant (Table 7.2), and all the individual parameter estimates (weights and cross loadings) in the MIMIC model are significant (Table 7.3).

The second canonical variate pair illustrates a problem that has often hampered the interpretation of canonical correlation analysis. In Table 7.2, the relationship between the unobserved variates is significant. There are several high loadings, but the redundancy is insignificant. With such inconclusive results, it is important to determine which variables, if any, significantly contribute to the variate relationship. One approach to this problem involves choosing an arbitrary cutoff for interpreting a loading or weight (usually .30 to .50). The drawbacks of this procedure are that it is highly subjective and it ignores the standard error of the estimates.

The MIMIC model offers a solution to this problem. Since the standard errors of weights and cross loadings can be obtained, critical ratios can be used as criteria for determining the statistical significance of individual variables. Even though two of the three predictor loadings in Table 7.2 are large enough to pass the conservative .50 cutoff, standard errors and associated critical ratios for the unstandardized MIMIC estimates in Table 7.3 indicate that no predictor weight is statistically significant.

Example 2

The structural relations model can be helpful in cases where the analyst is uncertain about whether the theory (i.e., the relationship between constructs) should be rejected because the canonical results are insignificant or the assumptions imposed by the canonical analysis are too restrictive. Specifically, one may relax assumptions regarding variate orthogonality or components structure. Our second example illustrates a case where the canonical correlation coefficient is not significant, but when the assumptions from [6-9] are relaxed, the corresponding relationship between latent variables in a structural relations model is significant.

TABLE 7.4

Correlation Matrix for Example 2 (n = 106)

	y_1	y_2	x_1	x_2
y_1	1.000			
y_2	.420	1.000		
x_1	-.187	-.149	1.000	
x_2	-.222	-.179	.696	1.000

Consider the correlation matrix in Table 7.4. The data were obtained from a study on industrial selling (Aaker and Bagozzi, 1979). The y-variables measure dollar sales (per salesperson) and job satisfaction, respectively; the x-variables represent two measures of role ambiguity. The canonical correlation results are presented in Table 7.5. No significant relationships were found. By relaxing the requirement that the variables be exact linear combinations of their respective observed variables (6) and (9) and that the squared correlation between corresponding variates is a coefficient of determination with respect to the variables (8) and (9), we can rewrite the model as illustrated in Figure 7.4. Formally the model is:

$$\eta = \gamma \xi + \zeta \tag{22}$$

Let us also assume that all indicators (observed variables) reflect the underlying factors:

$$\begin{bmatrix} y_1 \\ y_2 \end{bmatrix} = \begin{bmatrix} \lambda_{y_1} \\ \lambda_{y_2} \end{bmatrix} [\eta] + \begin{bmatrix} \varepsilon_1 \\ \varepsilon_2 \end{bmatrix} \tag{23}$$

$$\begin{bmatrix} x_1 \\ x_2 \end{bmatrix} = \begin{bmatrix} \lambda_{x_1} \\ \lambda_{x_2} \end{bmatrix} [\xi] + \begin{bmatrix} \delta_1 \\ \delta_2 \end{bmatrix} \tag{24}$$

The results are presented in Table 7.6. Contrary to the corresponding estimate in the canonical solution, the correlation between the variates (factors η and ξ) is statistically significant. The goodness of fit ($\chi^2 \cong .0006$, $p \leq .98$) is excellent, which means that the two-construct model (Figure 7.4) is successful in accounting for the observed correlations.

TABLE 7.5

Example 2: Canonical Model Parameter Estimates

	I	II
R_{c^2}	.246	.003
R_2^2 (or λ)	.06	0
Significance of R_c^2	NS[a]	NS
$R_{c/p}^{-2}$	.042	0
Significance of $R_{c/p}^{-2}$	NS[b]	NS
	Canonical Loadings	
Criterion Set		
y_1	-.921	-.390
y_2	-.741	-.672
Predictor Set		
x_1	.822	.569
x_2	.981	.194

[a]Statistical test based upon Bartlett (1941).
[b]Statistical test based upon Miller (1975).
NS = Not statistically significant at the .10 level.

TABLE 7.6

Example 2: Structural Model Parameter Estimates

	Unstandardized Estimate	Standardized Estimate
λ_{y_1}	1.0[a] (-)	.723
λ_{y_2}	.804(.425)[b]	.581
λ_{x_1}	1.0 (-)	.764
λ_{x_2}	1.192(.421)[c]	.911
γ	-.319(.139)[c]	-.338
	$\chi^2 = .0005$	
	d.f. = 1	
	$p \leq .98$	

[a]Fixed parameter.
[b]The standard error of the unstandardized estimate is in parentheses.
[c]Statistically significant at $p \leq .05$.

FIGURE 7.4

Example 2: A Structural Relations Model

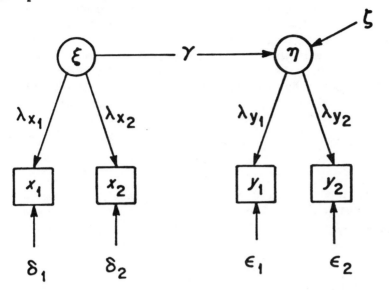

It should be emphasized that the model in Figure 7.4 (22-24) should not be regarded as an improvement over the canonical model; it is a different model with different assumptions. As such, it may, as in this example, provide different conclusions regarding the relationships between constructs.

SUMMARY

Canonical correlation analysis is often considered to be a general model for most statistical tests involving linear relationships. However, this chapter has demonstrated that the canonical model is a special case of an even more general model. Given certain specifications, a linear structural relations model with multiple causes (MIMIC) of a single construct was shown to be identical to the canonical model. As was illustrated, the extension to multiple constructs is straightforward.

By performing canonical correlation analysis via a structural relations model, two important advantages are gained. First, the problem of determining statistical significance for individual parameters in canonical correlations is overcome. For large samples, the estimated standard errors can be used to calculate critical ratios for the evaluation of individual weights and cross loadings.[1]

Second, it was shown that the structural relations model is a more flexible tool for data analysis compared to canonical correlation. It is possible that a canonical model may fail to reject the null hypothesis regarding the relationship between two variates, although this relationship is significant in a less restrictive structural relations model. Thus, if the data are inconsistent with the canonical model it does not necessarily follow that one's theory should be rejected. By altering some of the assumptions of canonical analysis, such as the requirement that the variates be exact linear combinations of the variables, it is possible to arrive at a different conclusion.

NOTE

1. It should be noted that the accuracy of the standard errors obtained from the information matrix has yet to be fully established (Lee and Jennrich, 1979) and further research is necessary on this issue.

REFERENCES

Aaker, D. A., & Bagozzi, R. P. Unobservable variables in structural equation models with an application in industrial selling. Journal of Marketing Research, 1979, 16, 147-158.

Anderson, T. W. Introduction to Multivariate Statistical Analysis. New York: Wiley, 1958.

Bartlett, M. S. The statistical significance of canonical correlations. Biometrika, 1941, 32, 29-38.

Bentler, P. M., & Weeks, D. G. Interrelations among models for the analysis of moment structures. Multivariate Behavioral Research, 1979, 14, 169-185.

Fletcher, R., & Powell, M. J. D. A rapidly convergent descent method for minimization. Computer Journal, 1963, 6, 163-168.

Hauser, R. M., & Goldberger, A. S. The treatment of unobservable variables in path analysis. In H. L. Costner (ed.), Sociological Methodology, pp. 81-177. San Francisco: Jossey-Bass, 1971.

Jöreskog, K. G. A general approach to confirmatory maximum likelihood factor analysis. Psychometrika, 1969, 34, 183-202.

Jöreskog, K. G. A general method for analysis of covariance structures. Biometrika, 1970, 57, 239-251.

Jöreskog, K. G. A general method for estimating a linear structural equation system. In A. S. Goldberger and O. D. Duncan (ed.), Structural Equation Models in the Social Sciences, pp. 85-112. New York: Seminar Press, 1973.

Jöreskog, K. G. Structural equation models in the social sciences: Specification, estimation, and testing. In P. R. Krishnaiah (ed.), Application of Statistics, pp. 265-287. New York: North Holland Publishing Company, 1977.

Jöreskog, K. G., & Goldberger, A. S. Estimation of a model with multiple indicators and multiple causes of a single latent variable. Journal of the American Statistical Association, 1975, 10, 631-639.

Jöreskog, K. G., & Sörbom, D. LISREL IV: Analysis of Linear Structural Relationships by the Method of Maximum Likelihood. Chicago: National Educational Resources, Inc., 1978.

Jöreskog, K. G., & Sörbom, D. (eds.), Advances in Factor Analysis and Structural Equation Models. Cambridge, Mass.: Abt Books, 1979.

Knapp, T. R., Canonical correlation analysis: A general parametric significance-testing system. Psychological Bulletin, 1978, 85, 410-416.

Lee, S. Y., & Jennrich, I. A study of algorithms for covariance structure analysis with specific comparisons using factor analysis. Psychometrika, 1979, 4, 99-113.

Miller, J. K. The sampling distribution and a test for the significance of the bimultivariate redundancy statistic: A Monte Carlo study. Multivariate Behavioral Research, 1975, 10, 233-244.

Stapleton, D. C. Analyzing political participation data with a MIMIC model. In K. F. Schuesster (ed.), Sociological Methodology 1978, pp. 52-74. San Francisco: Jossey-Bass, 1978.

Stewart, D., & Love, W. A general canonical correlation index
   Psychological Bulletin, 1968, 70, 160-163.

8

Multivariate Analysis with Latent Variables: Causal Modeling

Peter M. Bentler

INTRODUCTION

This chapter was commissioned by the editorial board of the Annual Review of Psychology in recognition of the increasing relevance of "multivariate analysis" to psychological research. A simple projection based on my literature search and on Anderson et al. (1972) suggests that over 10,000 articles have been published in this field. Obviously only 1-2% of these can be mentioned here. After consulting friends and colleagues—who often provided contradictory advice—I decided to inaugurate this coverage by concentrating on the topic that I believe holds the greatest promise for furthering psychological science: multivariate analysis with latent variables and, more narrowly, linear structural equation (simultaneous equation, path analysis, structural relations, covariance structure) models with latent (unobserved, unmeasured) variables. The topic is to some extent controversial, as will be outlined below, since a latent variable is a variable that an investigator has not measured and, in fact, typically cannot measure. Latent variables are hypothetical constructs invented by a scientist for the purpose of understanding a research area; generally there exists no operational method for directly measuring these constructs.

The constructs are related to each other in certain ways as specified by the investigator's theory. When the relations among all constructs and the relation of all constructs to manifest (measured) variables are specified in mathematical form—here simply a simultaneous system of highly restricted linear regression equations—one obtains a model having a certain structural form and certain unknown parameters. The model purports to explain the statistical properties of the measured variables in terms of the hypothesized latent variables. The primary statistical problem is one of optimally estimating the parameters of the model and determining the goodness-of-fit of the model to sample data on the measured variables. If the model does not acceptably fit the data, the proposed model is rejected as a possible candidate for the causal structure underlying the observed variables. If the model cannot be rejected statistically, it is a plausible representation of the causal structure. Since different models typically generate different observed data, carefully specified competing models can be compared statistically. Obviously it is not necessary to take a stand on the meaning of "cause" to see why the modeling process is colloquially called causal modeling (with latent variables). The word "cause" is meant to provide no philosophical meaning beyond a shorthand designation for a hypothesized unobserved process, so that phrases such as "process" or "system" modeling would be viable substitute labels for "causal" modeling. In such a definitional context, one need not worry about the criticism that "causal analysis does not analyze causes" (Guttman 1977, p. 103).

It has become popular to categorize statistical methods as exploratory or confirmatory (supportive) in nature (Dempster 1971, Tukey 1977). Multivariate analysis with latent variables can be similarly considered, e.g. traditional factor analysis is a data exploration method, but recent developments emphasize its hypothesis-testing nature (Jöreskog 1969). In this chapter, I emphasize the theory-testing aspects of causal modeling methods rather than procedures aimed at finding "causes" through exploratory data analysis. In the 1960s the idea of searching for causes was entertained in psychology and sociology, in part as a consequence of the pioneering ideas of Campbell & Stanley (1963) on cross-lagged panel correlations and Simon's (1954) application of simultaneous equation methodology to the three variable case (cf. Blalock 1963, 1964; Duncan 1969; Pelz & Andrews 1964; Rozelle & Campbell 1969). The recent literature, in contrast, proposes specifying competing causal models via mathematical equations and testing the models by statistical means. A compendium of early literature on causal models has been gathered by Blalock (1971a); this literature will not be documented here in detail since it has been largely superseded by recent developments. Nonetheless, even though the goal of causal modeling is explanation rather than description, an appropriate interplay between theory and data surely involves exploration as well as confirmation (Bentler 1978; Wold 1956, 1979). By emphasizing confirmatory methodology, I do not wish to suggest that the final word has been written on exploratory latent variable methods (Kaiser 1976, Yates 1979).

This chapter was commissioned to reach psychologists rather than psychometricians or statisticians; Bentler & Weeks (1980) provide a technical overview of the field. Since the basic concepts of causal modeling will be new to many readers, the chapter has been oriented to have a didactic as well as review function.

BASIC CONCEPTS

Structural Equations and Path Diagrams

The basic building block of a causal model is the linear regression equation (linear by assumption). Such an equation specifies the hypothesized effects of certain variables (here called predictors) on another variable (here called criterion). To illustrate, consider the equation $Y = b_1X_1 + b_2X_2 + b_3X_3 + e$. In such an equation, the intercept term has been dropped as irrelevant, and one considers the four variables Y, X_1, X_2, and X_3 as deviations from their means. (To be strictly accurate, one could add a subscript to the variables indicating the scores of a given entity or subject; there would be as many equations as entities. However, these equations are identical in form and are governed by the same parameters, so only one generic equation is needed.) The parameters b_1, b_2, and b_3 represent the regression weights to be used in optimally predicting Y from the Xs, and e represents an error of prediction. The variable e is not actually measured; in the population, however, one would know the weights b_i, and hence e could be calculated exactly as the residual $(Y - b_1X_1 - b_2X_2 - b_3X_3)$. I shall call such variables unmeasured, but not latent. In this equation, there are four predictor variables X_1 - X_3 and e_1, and Y is the criterion variable.

A path diagram for this equation is shown in Figure 8.1a. Squares are used to enclose variables that are measured (MVs); the unmeasured variable e has a circle inside the square. The predictors are shaded, and the criterion is light. Causal or directional influences of predictors on the criterion are indicated by unidirectional arrows; the strength of each effect is indicated by the weight for each arrow. The diagram can be read as indicating that "Y equals b_1 times X_1, plus b_2 times X_2, plus b_3 times X_3, plus 1.0 times e," thus completely summarizing the equation.

A regression equation in the context of a causal model is called a structural equation, and the parameters, structural parameters. Structural parameters presumably represent relatively invariant parameters of a causal process, and are considered to have more theoretical meaning than ordinary predictive regression weights. A problem with structural equations is that they do not adequately represent the parameters of a causal process. Implicit in each equation are parameters associated with the variances of the predictor variables (here, σ_1^2, σ_2^2, σ_3^2, σ_e) as well as their covariances (here, σ_{12}, σ_{13}, σ_{23}, since the residual e is forced to be independent of the Xs by construction). Hence, there are more parameters associated with a causal process than are represented in the structural equation. In the example, there are 10 parameters in the causal system, but only three of them are shown in the equation. Figure 8.1b presents a more complete representation of the model. Although such a representation is not standard, it mirrors the hypothesized causal process. Covariances or correlations among the predictor variables are shown by two-headed arrows, and variances are marked inside the squares.

Typically, only the form of the model is known, and the parameter values need to be estimated from the data. Often it is desirable to test hypotheses about given parameters, e.g. that $b_1 = 0$. If this null hypothesis cannot be rejected, the path diagram would need to be redrawn, and the arrow corresponding to b_1 would be removed. Such a revision of the path diagram points out an important feature: missing paths are as important to accurate representation as existing paths. In Figure 8.1, the absence of a connection between X_1 and e shows that these variables are neither causally dependent nor correlated. If X_1 and X_2 were independent, the σ_{12} two-headed arrow would be missing.

Causal models typically have more than one equation. Figure 8.2a shows a confirmatory factor analysis model that proposes that there are two common factors (X_1, X_2), which may have a nonzero covariance (σ_{12}), as well as four independent error or unique factors. Latent variables (LVs) are simply common or unique factors. The six LVs are represented in

FIGURE 8.1

Two Representations for a Structural Equation: (a) Path Diagram for the Equation; (b) The Complete Model

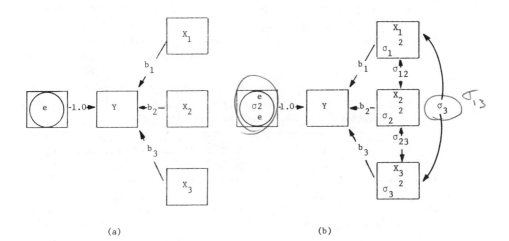

(a)   (b)

FIGURE 8.2

Latent and Manifest Variable Models: (a) A Factor Analytic Model; (b) A Path Model

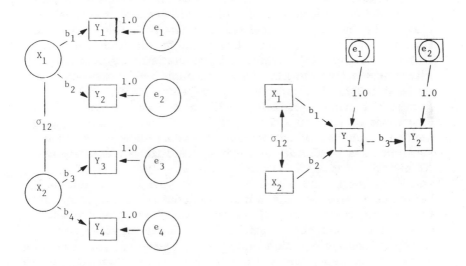

(a)   (b)

circles, the four MVs in squares. If Y_1 and Y_2 represented mathematical tasks, while Y_3 and Y_4 represented verbal tasks, X_1 and X_2 might be hypothesized quantitative and verbal factors of intelligence. In such a model, the common factors generate the correlations among the MVs, and the MVs are indicators of the LVs; note that the arrows go from LVs to MVs. This diagram follows typical practice in not showing the variances of the predictors, which are parameters of the model. The predictors X and e are independent in each of the four equations, $(Y_1 = b_1X_1 + e_1, Y_2 = b_2X_1 + e_2, Y_3 = b_3X_2 + e_3, Y_4 = b_4X_2 + e_4)$. Since there are no two-way arrows, the e's are independent across equations. In all LV models, one common factor regression weight must be fixed as known to identify each factor (see below), or else the factor variance must be fixed; thus, one can set $b_1 = b_3 = 1.0$ and let σ_1^2 and σ_2^2 be free, unknown parameters, or else one can set $\sigma_1^2 = \sigma_2^2 = 1.0$ and let b_1 - b_4 be free parameters (in this case, $\sigma_{12} = r_{12}$, a correlation). Hence, there are 9 parameters in the model (e.g. with factor variances fixed, there are 4 variances, 1 correlation, and 4 regression weights), one less than the model of Figure 8.1. An alternate theory for the MVs might be that there is only a single common factor (say, general intelligence) rather than two separate but correlated factors. Such a theory would be verified in current representation if X_1 and X_2 were perfectly correlated, i.e. if $\sigma_{12} = r_{12} = 1.0$. Then the X part of the diagram could be redrawn to show a single common factor X with an arrow pointing to each Y_i. Bentler & Peeler (1979) applied these ideas to Masters & Johnson's (1966) theory of the unidimensionality of female orgasmic responsiveness. The crucial statistical test evaluated whether coital and masturbatory responsiveness LVs could be considered as perfectly correlated; this hypothesis was statistically untenable. Alternative approaches to evaluating such hypotheses are given by Lord (1973), Kristof (1973), and Healy (1979).

In the previous models, each variable was either a predictor or criterion. In Figure 8.2a, the four MVs are always criteria, never predictors. More complicated models allow variables to serve both as predictors and criteria in different equations. An example is given in Figure 8.2b. The MVs are X_1, X_2, Y_1, and Y_2, while e_1 and e_2 are unmeasured variables. Variable Y_1 serves as a criterion, since it is predicted by X_1, X_2, and the unmeasured residual e_1. However, it also serves as a predictor of Y_2. I shall call those variables that never serve as criteria in any structural equation independent variables; the remaining variables are dependent variables. There are, then, exactly as many structural equations in a causal model as dependent varia-

bles. Independent variables are shaded in the figures, dependent variables remain light; thus there are two equations in the
last-described model. Using this convention, one can describe
the complexity of a causal model by the number of equations,
the number of independent variables, the number of unknown
regression weights, and the number of unknown, nonzero interrelations among the independent variables. (Note that the
independent, i.e. nondependent, variables need not be mutually
statistically independent or uncorrelated.)

The examples illustrate structural equations and path diagrams in the context of MV path and LV factor models. In
factor analytic models, the LVs are independent variables only;
more general LV models would have LVs as dependent variables
also. If the model in Figure 8.2b represented relations among
theoretical constructs only, one would replace the squares by
circles. In addition, one would need to provide MVs as indicators
for each of the LVs, using unidirectional arrows emanating from
each LV to several MVs, with their residuals.

Model Specification and Research Design

A theory that is to be tested via causal modeling will have
to be specified mathematically, i.e. translated into structural
equations. Such a specification should assure that all the relevant constructs are being considered simultaneously, and that
their uni- and bidirectional interrelations are made explicit.
Ideally, a competing theory should be specified at the same
time, since the only clean test of competing theories can be
made when one theory can be represented as part of another,
i.e. their path diagrams are identical except that certain paths
can be taken as known (usually zero). Such nested comparisons
can be assessed statistically. For example, in Figure 8.2a,
two-factor and one-factor theories can be compared by evaluating
whether $\sigma_{12} = r_{12} = 1.0$; a third two-factor theory might propose
that $\sigma_{12} = 0$, i.e. that the factors are independent; yet another
theory might propose that all regression weights $b_1, \ldots, b_4$
are equal.

The theoretical relations between constructs can be specified
without anchoring the constructs in measurement operations.
Such a specification must represent a reasonable translation of
theory into equations or a diagram. If the scientific community
cannot be convinced about the logic of the representation, there
is no point in relating the constructs to indicators. If the theoretical representation is adequate, care must still be taken to

provide adequate indicators of each construct. Each construct is considered to be a common factor LV that affects several MVs; each MV in turn is also affected by a residual LV (e's in Figure 8.2a) that is typically not an explicit part of a theory. Since the LVs are in practice abstractions that presumably underlie MVs, a poor choice of MVs will create doubt as to whether a theory's constructs are in fact embedded in a model. Choosing the right number of indicators for each LV is something of an art: in principle, the more the better; in practice, too many indicators make it difficult if not impossible to fit a model to data. While one might hypothesize that certain MVs represent only a given LV, MVs tend not to behave as well as expected. Since linearity of relations and continuity of variables are typical causal modeling assumptions, MVs should be chosen with these properties in mind. Statistical tests also typically are based on independence of observations and normality of MVs; these conditions should be met if possible.

The structural equations associated with a causal model imply some very specific consequences for the moment structure of the data, specifically the variances and covariances of the MVs (if the MVs are standardized, covariances are correlations). Thus causation implies correlation—but of a very specific form. If the hypothesized causal process is correct, only certain values will be observed for these variances and covariances (which is why structural equation models often are called covariance structure models). For example, in Figure 8.2a, if a model specifies that $\sigma_{12} = 0$ and $\sigma_1^2 = \sigma_2^2$ are fixed at 1.0, then the covariance of Y_1 with Y_2 must equal the product $b_1 b_2$. This may be seen as follows. Assuming that Y_1 and Y_2 are standardized, this model proposes that X_1 generates the correlation between Y_1 and Y_2. If this model is true, partialing X_1 out of Y_1 and Y_2 should leave these MVs uncorrelated. But the partial correlation formula, rearranged, states that $r_{12.3}[(1 - r_{13}^2) (1 - r_{23}^2)]^{1/2} = r_{12} - r_{13}r_{23}$ (where 1,2,3 are Y_1, Y_2, X_1). If the hypothesis $r_{12.3} = 0$ is true, it follows that $r_{12} = r_{13}r_{23}$. Then, with $r_{13} = b_1$ and $r_{23} = b_2$, the conclusion follows. Implications of this sort are drawn for every variance and covariance of the MVs, based on any assumed model. Any observed covariances other than the expected ones would be inconsistent with the proposed model, thus providing a basis for rejecting a model. However, even if a model is consistent with data, one cannot conclude that it mirrors the true causal process, since other models also might be able to reproduce the moment structure of the data.

Not all causal models can be tested. Models can only be tested if the parameters of the model can be uniquely specified or identified. In the example, it was possible to conclude that $r_{12} = r_{13}r_{23}$ only because the variance of X_1 was fixed at 1.0; it would not also be possible to treat this value as unknown, to be determined. The parameters of a model are identified if all parameters would have identical values in various causal models that generate identical observed data. In simple models it is possible to evaluate the identification problem by finding transformations that take the MV moments into parameter values, but this task is virtually impossible in large models and one typically uses certain heuristics instead (e.g. residual e variables in Figures 8.1 and 8.2 have 1.0 paths to other variables). A useful causal model must be overidentified, meaning, loosely speaking, it should have fewer parameters than data points (usually variances and covariances), because only then is the model potentially able to be rejected by data. If a model is just identified, meaning, loosely speaking, that there is a one-to-one transformation possible between parameters and data, the model is not scientifically interesting because it can never be rejected. The regression model of Figure 8.1, taken by itself, is such a model since it can be fit to any data; only if certain parameters are taken as known (e.g. $b_1 = .5$) would a potentially rejectable model exist. On the other hand, if the MVs were replaced by LVs and several factor analytic indicators were chosen for each LV in Figure 8.1, one would have a model for regression with LVs that could be tested. If the parameters of a model are underidentified, meaning, loosely speaking, that they can take on many values rather than be determined uniquely, the model is not statistically testable and thus scientifically useless. Since certain parts of a model may be overidentified and other parts underidentified, identification is an issue for every equation in a model as well as for the model as a whole. The analysis of identification, while difficult, can provide insight into possible deficiencies of a theory (Bentler 1978).

Model specification thus should be closely related to research design. Arbitrarily gathered data usually will not have the characteristics needed to provide an adequate test of a theory. It is necessary to assure that the conditions for data gathering are theoretically appropriate and statistically adequate. If theory specifies that certain influences occur only across time, with a certain causal lag, the time of measurement must reflect this lag; if various populations are expected to differ in specific ways, it is necessary to use a multiple population modeling method; if an identification problem exists, use of additional MVs may cure the problem; and so on. Enough subjects must be tested to assure that conditions for evaluating theory are appropriate.

Not only is the statistical theory applicable primarily to large samples, but one can guarantee the inability to reject a model by using a small enough sample. This is an inappropriate way to make a theory look good! Unfortunately, the researcher is to some extent in a double bind, since with extremely large samples almost any theory will be rejected by current procedures.

Model Testing

Once a model has been specified and the relevant data gathered in the context of an appropriate design, it is possible to compare the hypothesized model to data. The raw data are irrelevant to this process. Only the variances and covariances typically are utilized, though means are relevant in some contexts. The unknown parameters of the model are estimated so as to make the variances and covariances (possibly, means) that are reproduced from the model in some sense close to the observed data. Obviously, a good model would allow very close approximation to the data. If even the best choice of parameter values leads to a poor approximation to the data, the model can be rejected as a plausible representation of the causal process that generated the data.

In general it is impossible to estimate model parameters without a computer program because there is no available algebraic solution, and iterative approximations that refine a user-provided initial solution are utilized. The programs COFAMM and LISREL (Sörbom & Jöreskog 1976, Jöreskog & Sörbom 1978) use the maximum likelihood method of estimation, while the author's program additionally uses generalized least-squares and least-squares criteria; the least-squares method has no statistical test associated with it. These programs may yield meaningless results even when the user sets up the problem correctly: an identification problem may preclude obtaining a solution; the initial values may be so far off optimum that the program may not converge on a final solution; and some parameter estimates may be completely unreasonable (i.e. negative variances). An appropriate statistical solution will yield a chi-squared (χ^2) value to evaluate the goodness-of-fit of the model to data and standard errors to reflect the sampling variability of each parameter estimate. The χ^2 statistic provides a test of the proposed model against the general alternative that the MVs are simply correlated to an arbitrary extent. If the χ^2 is large compared to degrees of freedom, one concludes that the model does not appropriately mirror the causal process that generated the data. If the statistic is small compared to degrees of free-dom, one concludes that the model provides a plausible repre-sentation of the causal process. The standard error for each parameter estimate can be used to provide an indication of the importance of that parameter to the model as a whole. If the critical ratio formed by dividing the estimate by its standard error is large, the parameter is essential to the model; if it is small, the parameter is probably unnecessary to the model. These critical ratios have an approximate z distribution, so that the standard normal curve provides the index for deciding be-tween "large" or "small." (LISREL calls these ratios "T-values"; also it calls parameter estimates "LISREL estimates" rather than maximum likelihood estimates.) Parameter estimates can also be transformed into values that would be obtained if the common factor LVs have unit variances, yielding a standardized solution. (Some computer centers' versions of LISREL as yet have errors in this solution.)

In very large samples, the most trivial discrepancy between model and data will require rejection of a model by the χ^2 test. This test has drawbacks, since it is affected by the extent to which crucial assumptions such as multinormality and linearity are violated. Even if all assumptions are met, one might wish to take the significance level chosen to evaluate a model as a decreasing function of sample size (Leamer 1978). Unfortunately, no standard methodology exists for such choices. The goodness-of-fit of a model to data should certainly be evaluated by methods besides the χ^2 statistic, for example, by examining residuals or evaluating a coefficient that does not depend on sample size (Tucker & Lewis 1973, Bentler & Bonett 1979). In such evalua-tions, one frequently wants to show that a model provides a plausible representation of data. This is difficult to do with statistical hypothesis-testing procedures, since it entails accept-ing the null hypothesis that the model provides a plausible representation of the data. Within such a framework, statistical power plays a paradoxical role.

Any competing model can be estimated and evaluated as stated above. However, if one can specify an alternative model that is a subset of the initial model, the difference in χ^2 values between the two models is itself a χ^2 statistic which can be used to test the importance of the parameters that differentiate the models. When there is no alternative model, and a model does not fit, one may wish to modify the model in a heuristic manner. Bentler & Bonett (1979) propose a step-up procedure of model testing that begins with a null model that yields a modified test of independence. This procedure strengthens inference with small as well as large samples.

Model Modification and Reconfirmation

Parameters whose estimates are small compared to their standard errors can be eliminated from a model and the resulting model reestimated. This process amounts to modifying the path diagram by removing paths. Paths can also be added by examining the residuals, i.e. the specific patterns of lack of fit of model to data. Certain derivatives can be examined (Sörbom 1975). Parameters in a causal model are embedded in a matrix representation, and zero entries in various matrices correspond to missing paths in a diagram. If derivatives of the fit function are large with respect to these missing paths, it is possible that adding the associated parameters may improve the fit. Additional paths representing the correlations between errors can also be added. While procedures such as these usually produce a new model with better fit to data, it should be recognized that they can capitalize on chance associations in the data. They will also not necessarily find an alternative model that might provide a far superior fit.

Cross-validation provides an appropriate way of establishing whether empirically based model modifications represent genuinely valuable information about a model. For example, a sample may be split in two halves and one half used to develop a model and the second to provide a clean test of the developed model. It is possible to use tight, moderate, and loose replication strategies (G. J. Huba, J. A. Woodward, P. M. Bentler & J. A. Wingard, unpublished). In tight replication, one would attempt to fit the model to the second sample using the first sample's exact parameter estimates. In loose replication, the identical model and fitting procedures are used in both samples. In moderate replication, critical theoretical parameters (such as factor loadings) are held constant but others (such as error variances) can be estimated in the new sample. Research is required to differentiate these methods, but factor invariance theory (Bloxom 1972) would favor the moderate strategy.

Example: Models of Attitude-Behavior Relations

Fishbein & Ajzen (1975, Ajzen & Fishbein 1977) have advanced a major theoretical statement on the interrelations among attitudes, subjective norms, intentions, and behavior. Assuming certain research design considerations to hold, they proposed a model that can be represented as in Figure 8.2b. With the notation X_1 = attitude, X_2 = subjective norm, Y_1 = intention, and Y_2 = behavior, they proposed that attitudes and subjective norms influence future behavior only through the mediation of the intention to perform the behavior. Note that there are no direct influences of attitudes on behavior, nor of subjective norms on behavior. Such a model is theoretical, of course, and hence the MVs (squares) in the figure should be replaced by LVs (circles); to make it operational, multiple indicators of each LV would be needed. Bentler & Speckart (1979) proposed some alternative models for these relations into which the Fishbein-Ajzen model can be embedded. In one model, a significant direct effect of attitude on future behavior (i.e. a path from X_1 to Y_2) was hypothesized. In another model, past behavior was hypothesized to be a significant predictor of intentions to engage in a behavior as well as future behavior itself. Such a model is shown in Figure 8.3, which clearly differentiates between the theoretical constructs (circles) and MVs (squares). In this model, initial behavior (B_1), attitudes (A), and subjective norms (SN) are considered to be independent variables, allowed to correlate freely. Attitudes are shown to have a direct effect on future

FIGURE 8.3

A Model for Attitude-Behavior Relations

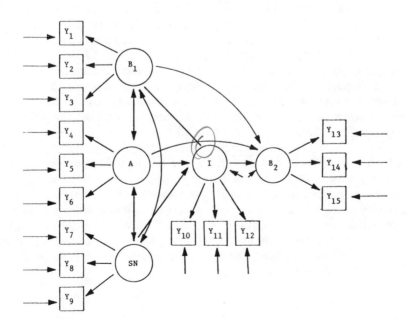

behavior (B_2), and initial behavior a direct effect on intentions (I) and future behavior; the Fishbein-Ajzen theory would predict that these three effects are not necessary to the model. In addition to the three shaded LVs, errors of measurement and prediction are independent LVs in the system. These 17 LVs are indicated by arrows without a source.

Bentler & Speckart (1979) obtained questionnaire data about alcohol, marihuana, and hard drug use from 228 subjects on two occasions. These behavior domains were chosen to allow all model tests to be replicated three times. Each domain was represented by 15 variables, and each LV by three MVs in the manner shown in Figure 8.3. The three MVs were chosen to generate some variation in type or context of substance usage, so that the LV would represent the shared content of three MVs.

The simple Fishbein-Ajzen model (in LVs, as in Figure 8.2) was tested first and compared to a model with an additional path from attitudes to behavior; thus, B_1 and its indicators and consequences (as in Figure 8.3) were excluded from these analyses. The Fishbein-Ajzen model could be rejected for two substances (p < .05), but it was marginally acceptable for one (p = .05). However, the model with a direct attitude-behavior effect was acceptable in all cases (p > .05), and a χ^2 difference test evaluating the improvement in fit due to the additional path showed that it provided a significant increment in fit of model to data (p < .001) in all cases. The complete model of Figure 8.3 showed an acceptable fit to the data in all replications, so that it provides a plausible representation of attitude behavior relations. In this model the 15 MVs have $15(16)/2 = 120$ variances and covariances; fixing one path from each LV to an MV, there are 10 unknown such weights, 15 error variances, 2 residual error of prediction variances (for I and B_2), 6 variances and covariances of the shaded LVs, and 6 direct LV effects, making 39 parameters in all. The χ^2s with 81 degrees of freedom (d.f.) ranged from 89 to 95 (p > .05 in all cases). In contrast, the Fishbein-Ajzen version of the model without B_1 - B_2, B_1 - I, and A - B_2 effects did not fit in any of the behavior domains; χ^2s were 143-163 with 84 d.f. (p < .05). The 3 d.f. χ^2 difference tests, of course, showed that the removal of the three paths yielded a significant decrement in fit of the model; thus, they are essential to an adequate understanding of the data in the context of the theories that were compared. Bentler & Speckart also provide a number of other model comparisons that need not be described here.

OVERVIEW OF THE FIELD

Historical and Scientific Context

General linear LV causal models represent the convergence of relatively independent research traditions in psychometrics, econometrics, and biometrics. The concepts of LV and errors in variables have had the longest tradition of development and use in psychometrics, beginning with Spearman (1904). These developments have come to be known as factor analysis and reliability theory. Similarly, the simultaneous directional influences of some variables on others have been studied intensively for decades in econometrics, primarily with MVs under the label of simultaneous equation models. Finally, a specialized tradition in biometrics, associated primarily with Wright (1934), has dealt with simultaneous equation models, sometimes with LVs, in the context of representation and estimation schemes known as path analysis. Although there have been occasional attempts to merge these traditions (Frisch 1934), they remained relatively independent and unaware of each other until the 1960s. Stimulated by such recent statistical sources as Simon (1954), Tukey (1954), and Wold (1956), and biometric papers by Turner & Stevens (1959) and Wright (1960a,b), sociological methodologists such as Blalock (1961, 1963), Boudon (1965), and Duncan (1966) started a movement that demonstrated the value of combining the simplicity of path analytic representations with the rigor of specifying equations simultaneously, possibly with LV models. By the early 1970s, causal modeling was a major sociological research method (Blalock 1971a), and LV models were being studied in depth. Excellent perspectives on these developments are provided by the econometrician Goldberger (1971, 1972a), who has become a major supporter and contributor to the field (see Goldberger & Duncan 1973, Aigner & Goldberger 1977). Psychologists were not major contributors to these developments. While Campbell & Stanley (1963) had begun to focus on problems of causal inference in nonexperimental data (e.g. with cross-lagged panel correlation), the main introduction of these ideas into psychology is attributable to Werts & Linn (1970a,b). However, their papers, and the related presentation by Isaac (1970), did not inspire extensive imitation. In my opinion, this occurred in large part because unifying mathematical and statistical principles, and simple procedures for their implementation, had not yet been developed for the methodology.

Many publications up to the early 1970s followed Simon (1954) in providing highly detailed analyses of artificial and small problems. It must have become painfully obvious to many researchers that such approaches were too limited in scope to have any hope of providing a broad enough framework to solve more realistic problems. Clearly, what was needed was a general method, analogous to multiple factor analysis or analysis of variance, that could in principle deal with extremely complex models in a routine way. While some existing models (Corballis & Traub 1970, Jöreskog 1970a) could have served this purpose after some modification, such a framework was finally provided by the JKW model (Jöreskog 1973a, 1977; Keesling 1972; Wiley 1973). The JKW model (widely known as LISREL) appears to have been conceived by several individuals, but the statistician-psychometrician Karl G. Jöreskog, in my opinion, deserves to be recognized as the major developer of hypothesis-testing methods for analysis of nonexperimental data, particularly via LV models. Although Lawley (1940) and Anderson & Rubin (1956) provided complete statistical treatments of the factor analytic model, the first practical computer implementation of the statistical approach was obtained and applied to various important psychometric problems by Jöreskog (1967, 1969, 1971a). He also provided the first statistical development of Thurstone's second-order model, generalizing and applying it to various multivariate problems (Jöreskog 1970a,b, 1971b, 1973b). His LISREL computer implementation of the JKW model has become the standard of the field (Jöreskog & Sörbom 1978). The particular importance of Jöreskog's work lies in the balance achieved between statistical sophistication and concern for relevance to social science applications.

Generalizations and simplifications of the JKW model exist today (Bentler & Weeks 1979a). Although important psychometric and statistical issues in causal modeling remain to be solved, the field has progressed to the point where quite general causal structures can be dealt with on a routine basis without requiring investigators to study a whole host of seemingly unrelated techniques to deal with special situations. For example, general LV models can also deal with MV causal models. Consequently, it is not surprising that there exists a growing consensus about the relevance of causal modeling to such areas as economics (Aigner & Goldberger 1977), education (Anderson & Evans 1974, Cooley 1978), evaluation research (Bentler & Woodward 1979a), political science (Alker 1969), and sociology (Bielby & Hauser 1977). The reception by psychology, nonetheless, remains slow: a special issue on methodology of the Journal of Consulting and Clinical Psychology (August 1978) does not even mention the topic. This is unfortunate, since LV causal modeling finally provides a basis for quantifying and operationalizing well-known concepts of construct validity and nomological networks as spelled out by Cronbach & Meehl (1955) and Torgerson (1958); (Bentler 1978).

Importance of Latent Variable Causal Models

Causal models are applicable to experimental as well as non-experimental data. Although there appear to be no published applications to experimental research, causal modeling has great potential in this area (cf. Royce 1977). Consider, for example, an experiment designed to study the effects of levels of cognitive dissonance on attitude change. Dissonance is a hypothetical LV; it is not measured nor directly measurable. While the experimenter may assess the effectiveness of a manipulation and feel satisfied that any observed effect is due to dissonance if the manipulation checks show the desired effects, it is also possible that the manipulation induces additional states besides dissonance, such as anxiety. The researcher who has considered this side effect may attempt to assess anxiety and show that it is not differentially present in the experimental conditions. Anxiety, of course, is also an LV, so the experimenter will use some indicator of this construct. Analysis of covariance may be used to control for this extraneous variable if it is shown to affect attitude change. There is a problem: analysis of covariance does not make an appropriate statistical adjustment, since the LV is measured with error (Lord 1960). By creating an appropriate model of possible effects in the experiment, and using multiple indicators of LVs such as dissonance, anxiety, and even attitude change, it can be determined whether the observed effects occur as hypothesized or in other ways. Causal modeling in experimental contexts has been discussed by Alwin & Tessler (1974), Bagozzi (1977), Blalock (1971b), Costner (1971), Miller (1971), and D. Romer, H. S. Upshaw, and D. Koepke (unpublished); its role in clarifying univariate and multivariate analysis of variance has been spelled out by Rock et al. (1976, 1978); and its role in the analysis of covariance has been technically developed by Sörbom (1978).

The greatest contribution of causal modeling to psychology is liable to be in areas of quasi-experimental or nonexperimental research, where methods for testing theories are not well developed. Social programs can rarely adhere to pure experimental designs without some selection or attrition bias, which needs to be modeled to be understood (B. S. Barnow, G. G. Cain & A. S.

Goldberger, unpublished; Bentler & Woodward 1979a; Bloxom 1972; A. S. Goldberger, unpublished; Heckman 1979; Olsson 1978). Quasi-experimental designs can be analyzed meaningfully via structural models (Kenny 1975a, Linn & Werts 1977). More generally, correlational data can be analyzed to determine the correspondence of data to hypothesized processes and their invariance across populations (Larzelere & Mulaik 1977, McFatter 1979, Werts et al. 1976). Unfortunately, models based on MVs are more appropriate to problems of description and prediction than explanation and causal understanding: their parameters are unlikely to be invariant over various populations of interest. Thus, of particular importance, I believe, is the modeling of a process at the level of latent rather than measured variables, since the MVs only rarely correspond in a one-to-one fashion with the constructs of interest to the researcher, which will almost certainly be measured with error. As a consequence, conclusions about an MV model cannot be relied upon, since various theoretical effects will of necessity be estimated in a biased manner. They also will not replicate in other studies that are identical except for the level of precision or error in the variables. Thus, the main virtues of LV models are their ability to separate error from meaningful effects and the associated parametric invariance obtainable under various circumstances. An example that compares LV and MV models is given by Bentler & Huba (1979).

It is useful to distinguish between "true" LV models and some substitutes. In true LV models, there are more LVs than MVs, and the LVs, conceived as hypothetical constructs, are not directly assessible. Instead, it is their consequences or effects on MVs that are observed. In Figure 8.2a, for example, arrows go from LVs to MVs. In other models, the LVs are simply derived variables that could be measured directly in the study; hence, they are functionally equivalent to MVs. In such models, arrows go from MVs to the derived variables. Such substitute models, it seems to me, are not appropriate to causal modeling, since the uninteresting effects of random errors of measurement cannot be separated statistically from substantively important causal influences in these models. While true LV models permit the separation of error from meaningful effects, they are also subject to some ambiguity. Specifically, since LVs are statistical abstractions, precise inferences as to their substantive meaning may be equivocal (see a factor analytic text, or Burt 1976), and they are subject to certain problems of indeterminacy (McDonald & Mulaik 1979). The main source of ambiguity is that LVs cannot be measured directly. As in factor analysis, the MVs can be expressed as linear combinations of the LVs, but it is not possible to reverse the procedure to express the LVs as linear combinations of the MVs. These difficulties are not fatal to LV modeling, but they should be recognized.

Not all writers agree with my point of view. For example, Dempster (1971) considers linear combinations of observed variables as factors, and Schönemann & Steiger (1976) propose the use of such combinations instead of LVs in research. Similarly, Wold (1979) considers principal components and canonical variates as useful LVs. However, these derived variables do not have the requisite statistical and psychometric properties to replace LVs in causal modeling except in unusual circumstances. Of course, when it is difficult or impossible to utilize LVs, perhaps due to an inadequate research design, inadequate theoretical knowledge, or computational costs, derived variables can be used as a practical expediency (cf. A. E. Boardman, B. S. Hui & H. Wold, unpublished).

Types of Latent Variable Models

LV models can be distinguished in the major traditional contrast of factor analytic vs. simultaneous equation models. The latter can be further divided as recursive vs. nonrecursive models. Another distinction between models involves the contrast between models that ignore or account for the means of the MVs in the causal structure. A final distinction is concerned with the degree of abstractness of the LVs and the generality and simplicity of the mathematical structure within which they are embedded.

In factor analysis, LVs are related to each other symmetrically (see Figure 8.2a). There is no regression structure among LVs; instead they are simply correlated or independent (oblique vs. orthogonal factors). In LV models with simultaneous equations, LVs are also permitted to be regressed or predicted from other LVs in various ways (see Figure 8.3). Consequently, multiple regression with independent variables subject to error is contained in these models, and more complex models have several such relations. There are two well-known types of regression structures: recursive and nonrecursive. Somewhat paradoxically, nonrecursive structures are those that allow true "simultaneous" or reciprocal causation between LVs, while recursive structures do not. Recursive structures have been favored as easier to interpret causally (Strotz & Wold 1960). They generally are less difficult to estimate.

Another distinction between models involves the separability of first and second moment parameters. In separable models, the first moments (means) of the MVs are not structured in terms of the causal parameters. Consequently, they are effectively irrelevant to the causal modeling process, and the goal is the modeling of the second moments (covariances, correlations, or cross-products) of the MVs. These models are thus often called covariance structure models (Bock & Bargmann 1966, Browne 1974, Jöreskog 1978), and they apply in most instances. More general moment structure models also allow an interdependence of first and second moment parameters (Bentler 1973, 1976a, Jöreskog 1970a; Sörbom 1974, 1978). The MVs' means are decomposed into basic parameters that may also affect the covariance structure. These models are particularly appropriate to studies of multiple populations or groups of subjects, and to the analysis of experimental data. They have not been frequently applied.

Linear structural equation models with LVs have until recently been conceptualized as embodying only a first-order factor analytic measurement structure for the MVs. For example, the JKW model expresses manifest variables directly in terms of LVs. Thus, the LVs are removed by one level from the MVs, as in Figure 8.3. However, it is also easy to conceptualize measurement models that are more complex, in which there are several levels or orders of LVs. In such models, the influence of the higher order LV on the MVs is indirect. A general multi-level LV model that allows structured means was developed to deal with such situations (Bentler 1976a). Such a model blurs the distinction made above between factor and simultaneous equation LV models, because the various levels of LVs affect each other via regression structures. Weeks (1978, 1979) provided the first statistical development of a general model that allows multivariate regression structures on LVs of various types and levels, and he illustrated the approach by developing a longitudinal model considering general intelligence as a second-order LV. This LV has no direct effect on any MV; rather it affects primary factors that, in turn, affect MVs. Recent developments have been directed toward obtaining LV models that are both general and simple. These models allow causal influences across levels and types of LVs (primary or residual LVs at a given level), in addition to structured means. They generalize the JKW model (Bentler 1979, Bentler & Weeks 1979b, Lee & Bentler 1979a, Weeks 1978). Bentler & Weeks (1979a) discuss these results and clarify the interrelations among a variety of structured linear causal models.

A Guide to the Literature

There are no adequate introductions to LV modeling methods. The true importance of LV as compared to MV models is often not appreciated. The earlier literature (Blalock 1971a) contains discussions of specialized problems that provide little guidance for the general case; the more current literature typically assumes knowledge of matrix algebra; and important technical developments require extensive knowledge of multivariate analysis. Although many good books on multivariate analysis have appeared in the last decade, only one provides an introduction to LV causal modeling in the broad sense (Van de Geer 1971); it is now largely outmoded since it does not provide the required general models or statistics. Two methodology texts provide an introduction to the area (Hanushek & Jackson 1977, Namboodiri et al. 1975). Introductory overviews primarily of small MV models, are by Asher (1976), Duncan (1975), and MacDonald (1977); the books by Li (1975) and Heise (1975) have specialized viewpoints. Unfortunately, these sources do not place the field into a broad, modern framework. Although Kenny's (1979) text also does not utilize a modern and unified approach to the field, it has, among extant texts, the most comprehensive perspective for psychologists.

Virtually anything Jöreskog has written is relevant and worth reading, but it tends to be technical. See "Literature Cited" section of this chapter, and Joreskog (1974, 1977, 1978), Long (1976), and Werts et al. (1973). Since LVs are factors, introductions to factor analysis are useful (Bentler 1976b, Bohrnstedt 1979, Kim & Mueller 1978, Lawley & Maxwell 1971, Mulaik 1975). Two classic, though advanced, compendia of LV models and applications exist (Aigner & Goldberger 1977, Goldberger & Duncan 1973). Some specialized topics are found in Blalock et al. (1975). The psychology journal most attuned to these developments is Multivariate Behavioral Research.

Causal models with MVs are being increasingly utilized in psychological research. As mentioned above, the cross-lag panel correlation method (see Humphreys & Parsons 1979, Kenny 1975b) provided an early basis for such interest; in applied psychology, for example, this interest (Lawler 1968, Sheridan & Slocum 1977) has now extended into MV models generally (Billings & Wroten 1978, Feldman 1975, Sims & Szilagyi 1975, Wanous 1974, Young 1977). A sampling of recent psychological applications includes the study of creativity (Simonton 1977), drug use (Ginsberg & Greenley 1978, Kohn & Annis 1978, O'Donnell & Clayton 1979), educational and occupational aspirations (Rosen & Aneshensel 1978), epidemiology (Eaton 1975, 1977), love and attraction (Tesser & Paulhus 1976, Bentler & Huba 1979), psychopathology and reactions to drugs (Naditch 1974, 1975), schizophrenia (Mednick et al. 1978), self-esteem (Bachman & O'Malley 1977), sexual responsiveness and behavior (Bentler & Peeler 1979, Hornick 1978), and social indicators research (McKennell 1978). The role of such models in research is not without controversy, e.g. in the study of stimulus exposure effects (Birnbaum & Mellers 1979a,b, Moreland & Zajonc 1979). Sociological applications are referenced by Bielby & Hauser (1977).

The statistical approach transforms exploratory factor analysis into a confirmatory LV method to some extent. P. H. Ramsey (1978) studied intellectual performance in this way. More restricted confirmatory models have been proposed (Upshaw 1978) and studied in the area of attitudes (Bagozzi 1978) and personality (Bramble & Wiley 1974). The multitrait multimethod matrix represents a still more specialized structure having wide applicability (Alwin 1974, Kenny 1976, Schmitt et al. 1977, Werts & Linn 1970a, Werts et al. 1972). New models for such data have recently been proposed (Bentler & Lee 1979a). Possible similarities and differences in factor analytic causal structures across several sets of subjects were first studied in the domain of abilities (Bechtoldt 1974, Cohen et al. 1977, Evans & Poole 1975, Hyde et al. 1975, McGaw & Jöreskog 1971, Sörbom 1974). Complete LV models with both measurement and causal regression structures are only beginning to appear in psychology. Models of achievement and desegregation in naturalistic contexts have been developed (Maruyama 1977, Maruyama & Miller 1979, McGarvey 1978), and applications in quality of life (H. M. Allen, P. M. Bentler & B. A. Gutek, unpublished) and evaluation research have been described (Bentler & Woodward 1978, Magidson 1977, 1978). Traditional problems of social psychology are being studied with LV models (Bentler & Huba 1979, Bentler & Speckart 1979). In cognitive psychology, models of human memory processes are being evaluated (Woodward et al. 1978). A variety of technically sophisticated socioeconomic models appears in Aigner & Goldberger (1977) and Taubman (1977). Causal modeling is particularly relevant to longitudinal research, since the passage of time helps to eliminate possible competing causal explanations of phenomena. Thus, it is not surprising that numerous methodological and empirical investigations exist in this area (Bentler 1980; Jordan 1978; Jöreskog 1979; Jöreskog & Sörbom 1976, 1977; Kohn & Schooler 1978; Olsson & Bergman 1977; Rogosa 1979; Roskam 1976; Sörbom 1976; Weeks 1979; Werts & Hilton 1977; Werts et al. 1977, 1978; Wheaton et al. 1977).

Professional Issues

In my opinion, causal modeling methods deserve the recognition that analysis of variance receives in graduate psychology training. While these methods can be taught in an ad hoc fashion, it makes sense to generate a meaningful course sequence that includes an exposure to matrix notation, multivariate analysis, exploratory and confirmatory factor analysis, and structural equation MV and LV models. Computer exercises are essential. Such a teaching sequence is not a luxury, rather it is a necessity for adequate training of researchers who are faced with problems of drawing causal inferences, or testing theories, with nonexperimental data. Without having such training, psychologists will be faced with literature that is too technical to evaluate critically, and they will find econometricians, biometricians, and sociologists encroaching on familiar territory [e.g. disentangling various environmental and possibly genetic effects on important variables (Eaves & Eysenck 1975; Eaves et al. 1977; Goldberger 1977, 1978; Kadane et al. 1976; Loehlin 1978; Martin & Eaves 1977; Rao et al. 1976; Taubman 1977; Thomas 1977; A. B. Zonderman, W. Meredith, J. C. DeFries & S. G. Vandenberg, unpublished); simulation shows how such effects may be generated (Baltes et al. 1978)].

Many departments are not prepared to teach LV modeling, and they may not be willing to make the necessary human and financial investment. There are not many psychologists with appropriate multivariate training as well as necessary applied interests. Existing faculty can learn to use and teach the techniques; however, statistically oriented faculty tend to be suspicious of LVs, and factor analysts tend to be sympathetic to LVs but critical of the hypothesis-testing, statistical approach. In addition to costs associated with faculty allocation, causal modeling is an expensive undertaking. Publicly distributed programs are expensive to buy or lease. Single runs on complicated models can cost more than dozens of analysis of variance runs, and typical problems require many such runs. Universities and granting agencies, if they want to support this type of research, will have to be prepared to support computing budgets that are 10- to 50-fold greater than previously. Extremely large problems will probably remain out of reach until computer costs come down another order of magnitude. These cost considerations should not inhibit LV modeling; instead, they require us to face the facts that psychology is becoming more technical and that its computer budgets will have to approach those of engineering.

TECHNICAL DEVELOPMENTS

Mathematical Structures

The preceding review has stressed an existing unity and continuity in development of mathematical structures for causal modeling. In fact, developments have been complex and diverse, and the reader should be alerted to recent references. In the previous sections, LVs have been taken as random variables rather than fixed mathematical variables whose values require estimation. The literature on multiple regression with LVs considers both cases. Since the seminal review by Cochran (1970), LV regression has become an important topic for research (Chen 1979, Davies & Hutton 1975, Feldstein 1974, Goldberger 1972b, Levi 1977, Robinson 1974, Warren et al. 1974, Zellner 1970). Lawley & Maxwell (1973) and Rock et al. (1977) study the problem from factor analytic and JKW model perspectives. The distinction between random and mathematical LVs also underlies the relevant research tradition on structural and functional relationships, otherwise known as errors-in-variables or linear transformation models. Structural relations models are random LV models most closely associated with causal models as reviewed above. Recent references include Anderson (1976), Bhargava (1977), Cox (1976), Dolby (1976a,b), Florens et al. (1974), Gleser (1979), Gleser & Watson (1973), Moran (1971), Patefield (1977), Robertson (1974), Spiegelman (1979), and Sprent (1969). Work by Robinson (1977) and Theobald & Mallinson (1978) has the greatest relevance to the JKW development. A problem for the psychologist with much of this literature is its focus upon highly specialized cases; results can depend on such conditions as the nonnormality of LVs. One way a researcher can avoid the special estimation problems associated with specific situations is to make a minor change in experimental design, e.g. by using more indicators for an LV. Developments that are most directly applicable to LV causal models are econometric studies of simultaneous equation models with measurement error (Hausman 1977, Hsiao 1976, Geraci 1977), as well as LV time series models that cannot be reviewed here (J. Geweke & K. Singleton, unpublished). In general, the above approaches utilize a pre-JKW conception of LVs as MVs purified of error rather than generalized common and unique factors. There is no utilization of higher-level LVs.

Psychometric research continues to develop a variety of LV models (Bentler & Lee 1978a, 1979b; Browne 1979; Corballis 1973; Frederiksen 1974; Jöreskog & Goldberger 1975; Nishisato

1971; Please 1973; Wiley et al. 1973). The study of higher-level
LVs and more complex causal structures has typically been
associated with increasingly complex mathematical representations
(Jöreskog 1970a, 1977; Bentler 1976a); McDonald (1978) developed
another expression for the covariance structure of the Bentler
(1976a) model. It now appears that arbitrarily complex models
can be handled by very simple representations, based on the
idea of classifying all variables, including MVs and LVs, into
independent or dependent sets (Bentler 1979, Bentler & Weeks
1979a,b). Independent variables are not regressed on any other
variables in any equation. The remaining variables, including
all MVs in complete LV models, are dependent. A covariance
structure that does the work of the previous models is then
given by $\Sigma = GB^{-1}\Gamma\Phi\Gamma'B'^{-1}G'$, where G (which is known) selects
the MVs from the classified variables $B_0 = (I - B)$ contains regression weights for dependent variables regressed on each
other, Γ contains regression weights for dependent on independent variables, and Φ represents the covariance matrix of the
independent variables. In this conception, errors in variables
and errors in equations are treated equivalently, typically as
independent variables. An even simpler representation is to
redefine variables, taking $\Gamma = I$. B_0 rather than B is a basic
parameter matrix in these models. Under a different definition
of variables, these matrices appear in the Keesling-Wiley (Wiley
1973) and Jöreskog (1977) representations of the JKW model.
The advantage of the B_0 parameterization is that it translates
directly into path diagrams, while B generally does not. In my
view, the B_0 method is not only easier to teach, but it avoids
errors in signs of parameters that appear even in the second
revision of the LISREL manual (Jöreskog & Sörbom 1978).

The scope of models can be extended by judicious constraints imposed upon parameters. Jöreskog's (1969) innovation
of treating parameters as known or fixed, free to be estimated,
or constrained equal allows quite general models to be specialized
into restricted structures; Bentler & Weeks (1979b) and Weeks
(1978) also allow proportionality constraints, introduced by
Bentler & Weeks (1978). More general nonlinear constraints
are considered by Robinson (1977), S. Y. Lee (unpublished),
and Lee & Bentler (1979b). Constraints enable general LV
models to deal with significant MV problems not reviewed here
(Browne 1978). An important issue, not yet completely resolved,
is the relative virtue of alternative representations that allow
parameters to be either scale invariant or scale free, i.e. un-
changing or changing with specified rescalings of variables
(Bentler 1976a, 1977). Unique or MV variances are natural

candidates for scale freeness (Bentler 1976a, Krane & McDonald 1978); the remaining parameters are then invariant with respect to the scale of MVs. This issue is of some applied importance, because it appears in contexts of model specification and modification. It cannot simply be assumed that correlation rather than covariance matrices can be analyzed in a given situation (Swaminathan & Algina 1978). Without the use of scale invariant representations, equality constraints can generally be rationalized for covariance but not correlation structures. Model modification based on first derivative information (Sörbom 1975) is incorporated into various computer programs (Jöreskog & Sörbom 1978, Sörbom & Jöreskog 1976). These sources do not indicate that a derivative can be arbitrary in size, depending upon scaling considerations.

The models considered here require continuity of variables and linearity of relations. As a consequence, they cannot be used without controversy in the study of ordinal relations (Kim 1975; Leik 1975; Reynolds 1974; Smith 1974, 1978; Wilson 1974). While an adequate statistical theory is beginning to appear for models with dichotomous variables (Amemiya 1978; Christoffersson 1975, 1977; Heckman 1978; Mouchart 1977; Muthén 1977, 1978), models that combine arbitrary nonlinear measurement (McDonald 1967) and simultaneous equation systems (Amemiya 1977, Laffont 1977) have yet to appear. Such developments will no doubt incorporate features of latent trait theory (Samejima 1979).

By their very definition, LV models make use of hypothesized latent constructs. These constructs are considered random variables in the models considered here. Although such LVs have been central to psychometric theory for about 80 years, their utilization has not been without a history of conflict. The Hotelling-Thurstone disagreements about factors vs. components are well known; in recent years the "controversy" (Green 1976) has focused on LVs as factor scores. Schönemann, in particular, argues that LV models should not be utilized since LVs are not uniquely determined by the MVs, and since LVs can be constructed to have very peculiar properties indeed (Schönemann & Steiger 1978, Steiger 1979a). See McDonald & Mulaik (1979), Steiger (1979b), and Steiger & Schönemann (1978) for reviews of issues and relevant references, and Williams (1978) for a technical point of view. The literature is extensive and cannot be reviewed here, but two new points relevant to causal modeling need to be made. First, although the problem has been conceptualized as one of LV indeterminacy, it can be considered one of model indeterminacy equally as well. Bentler (1976a) showed how LV and MV models can be framed in a factor analytic context to yield identical covariance structures. The LVs and

MVs have different properties, but there is no empirical way of distinguishing between the models. Hence, the choice of representation is arbitrary, and the LV model may be preferred on the basis of LV's simpler properties. Second, although an infinite set of LVs can be constructed under a given LV model to be consistent with given MVs, the goodness-of-fit of the LV model to data (as indexed, for example, by a χ^2 statistic) will be identical under all possible choices of such LVs. Consequently, the process of evaluating the fit of a model to data and comparing the relative fit of competing models is not affected by LV indeterminacy. Thus, in my view, theory testing via LV models remains a viable research strategy in spite of factor indeterminacy. Nonetheless, indeterminacy can be associated with ambiguity of interpretation of LVs and statistical estimation problems (van Driel 1978), particularly in the context of inadequate research designs.

LV models cannot be statistically tested without an evaluation of the identification problem. Identifiability depends on the choice of mathematical representation as well as the particular specification in a given application, and it refers to the uniqueness of the parameters underlying the distribution of MVs. A variety of general theoretical studies of identification have been made in recent years (Deistler & Seifert 1978, Geraci 1976, Monfort 1978), but these studies are not very helpful to the applied researcher. While there exist known conditions that an observable process must satisfy in order to yield almost sure consistent estimability, it remains possible to find examples showing that identifiability does not necessarily imply the existence of a consistent estimator (Gabrielsen 1978) and that multiple solutions are possible for locally identified models (Fink & Mabee 1978, Jennrich 1978). In practice, then, it may be necessary to use empirical means to evaluate the situation: Jöreskog & Sörbom (1978) propose that a positive definite information matrix almost certainly implies identification, and McDonald & Krane (1977) state that parameters are unambiguously locally identified if the Hessian is nonsingular. Such a pragmatic stance implies that identification is, in practice, the handmaiden of estimation, a point of view also taken by Chechile (1977) from the Bayesian perspective. Although such an empirical stance may be adopted to evaluate complex models, it is theoretically inadequate. Identification is a problem of the population, independent of sampling considerations. Thus, data-based evaluations of identification may be incorrect, as shown in Monte Carlo work by McDonald & Krane (1979), who retract their earlier claims on unambiguous identifiability. Jennrich's (1978) multiple solutions in a factor analytic problem also serve to remind one that local identification is not a sufficient condition for global identification.

Statistical Problems

The econometric literature contains highly developed approaches to statistical estimation in MV simultaneous equation models. Although most estimators are chosen for their optimal large sample properties, in the sense of meeting the Cramér-Rao bound for minimum sampling variance, emphasis is also placed on finding estimators that have superior small sample properties while retaining desirable large sample efficiency. Consideration is given to finding estimators that retain useful statistical properties under violation of assumptions such as normality, and simplicity of computation is evaluated in comparing estimators. Recent references include Hendry (1976), J. B. McDonald (1977), Mehta & Swamy (1978), J. B. Ramsey (1978), Sargan (1978) and Zellner (1978). This literature has not filtered into psychology, although James & Singh (1978) provide an introduction to two-stage least-squares estimators. The econometric literature also provides a concern for testing model assumptions and restrictions (Kadane & Anderson 1977, Hausman 1978). In contrast, the statistical theory involved in LV simultaneous equation models has only been developed in rudimentary form. Only large sample theory has been developed to any extent, and the relevance of this theory to small samples has not been established. Although a certain amount of simulation work has been done on MV estimators to determine their empirical small sample properties (Intriligator 1978, pp. 416-20), virtually nothing is known about equivalent LV estimators.

Although the statistical theory associated with LV models based on multinormally distributed MVs already existed, Jöreskog 1967-1979) must be given credit for establishing that maximum likelihood (ML) estimation could be practically applied to LV models. While various researchers, including econometricians, were studying specialized statistical problems and searching for estimators that might be easy to implement, Jöreskog showed that complex models could be estimated by difficult ML methods based on a standard covariance structure approach. The most general alternative approach to estimation in LV and other models was developed by Browne (1974). Building upon the work of Jöreskog & Goldberger (1972) and Anderson (1973), who had developed generalized least-squares (GLS) estimators for the factor analytic model and for linear covariance structures,

Browne showed that a class of GLS estimators could be developed
that have many of the same asymptotic properties as ML estimators, i.e. consistency, normality, and efficiency. He also
developed the associated goodness-of-fit tests. Lee (1977)
showed that ML and GLS estimators are asymptotically equal.
Swain (1975), Geraci (1977), and Robinson (1977) introduced
additional estimators with optimal asymptotic properties. Some
of the Geraci and Robinson estimators can be easier to compute
than ML or GLS estimators; the Browne and Robinson estimators
do not necessarily require multivariate normality of the MVs to
yield their minimal sampling variances. Unfortunately, the empirical meaning of loosening the normality assumption is open to
question, since simple procedures for evaluating the less restrictive assumption (that fourth-order cumulants of the distribution
of the variables are zero) do not appear to be available. Although
certain GLS estimators are somewhat easier to compute than ML
estimators, there is some evidence that they may be more biased
than ML estimators (Jöreskog & Goldberger 1972, Browne 1974).
Virtually nothing is known about the relative robustness of these
estimators to violation of assumptions or about their relative
small sample properties.

On the assumption that milder statistical assumptions would
provide an estimation method with wider applicability, Wold
(1975, 1979; Noonan & Wold 1977) has been concentrating on
the development of a method that yields consistent but not
necessarily optimal estimators. A feature of his estimation
method involves the use of proxies for LVs. Wold feels that
his method as currently implemented yields estimates of parameters in true LV models, but in my opinion his promising method
will need elaboration to be able to move from MV to LV parameters. Such a development may be based upon the known equivalence of some MV and LV representations (Bentler 1976a) and
on computations more similar to the EM algorithm (Dempster et
al. 1977).

The evaluation and comparison of LV models based on the
above statistical developments are carried out within a traditional
statistical framework. It may be argued that a Bayesian framework is more likely to yield cumulative results than a likelihood
based approach (Chen 1979). A most delightful work espousing
such a point of view is the book by Leamer (1978), which attempts
to rationalize inference in nonexperimental MV contexts. Leamer
provides a rationale for such practices as comparing nonnested
models (see also Pesaran & Deaton 1978) and utilizing significance
levels that vary with sample size, which are questionable procedures within currently accepted statistical approaches to LV

models. In my opinion, statistical evaluation of models by goodness-of-fit tests should sometimes also be of secondary importance to determining the extent to which models account for moment data regardless of sample size (N). Research is needed to determine a more adequate indicator of goodness-of-fit that does not depend on N. A candidate is $\lambda^{1/n}$ based on the likelihood ratio λ, with $n = N$ or $n = N - 1$ depending on the covariance matrix; $\lambda^{1/n}$ can simplify to $[\det(S\hat{\Sigma}^{-1})]^{1/2}$. Bentler & Bonett (1979) propose an index that can be used with any fit function. For a general argument against statistical significance testing, see Carver (1978).

It has been proposed that standard concepts such as minimum variance that are generally used to support ML estimators provide an inadequate evaluation of the merit of an estimator (Rubin 1978). Nonetheless, the typical practitioner is faced with utilizing available methods such as ML to estimate parameters and evaluate models. Little is known about the robustness of these methods to violation of assumptions such as normality or large sample size. Methods for evaluating multinormality and detecting multivariate outliers exist (Barnett & Lewis 1978, Gnanadesikan 1977, Cox & Small 1978), but they seem not to be applied in practice. Although optimistic statements exist about the effects of nonnormality on correlations (Havlicek & Peterson 1977), the distribution of correlations has also been found to be quite sensitive to nonnormality (Kowalski 1972), thus implying difficulties in drawing inferences about structural parameters that generate correlations in nonnormal data. Fuller & Hemmerle (1966) concluded that ML estimates were quite robust to violation of normality assumptions in the factor model, but Olsson (1978) found that an erroneous need for more factors was typically indicated under conditions of skewing of variables, particularly if the skew was in inconsistent directions. Robust substitutes for moment data may be adopted, but their use cannot as yet be statistically rationalized in causal modeling. Although mean square error minimizing estimators (Maasoumi 1978, Winer 1978) have not been proposed for LV models, Darlington (1978) suggests that such estimators would not serve well in a causal modeling framework. (See Bentler & Woodward 1979b for a similar viewpoint, based on a different rationale.) S. Y. Lee (unpublished) investigated whether theorems on asymptotic statistics would hold in small samples. In simulation with a factor model, he found that the results were generally valid down to a sample size of about 80. Several people, including A. Boomsa and D. Bonnet, are investigating various statistical properties by Monte Carlo methods. Such efforts also need to be addressed to questions of power as well as to the distribution of various test statistics under typically accepted methods of model modification.

The recent literature on causal modeling has tended to ignore the older path analytic literature on decomposition of effects aimed at attributing dependent variable variance to antecedent variables. Alwin & Hauser (1975) provide an overview of this literature. MacDonald (1979) argues that a substantively meaningful decomposition of variance is possible in fully recursive path models. Such decompositions also deal with the difficult problem of decomposing variance among correlated predictors in a unique way, whether in regression (Green et al. 1978) or factor analytic contexts (Bentler 1968). Although contributions to this literature continue to be made (Carter & Nagar 1977), no generally statistically satisfactory answer to this problem has been forthcoming.

The statistical implementation of LV modeling hinges to a large extent on new developments in nonlinear programming. An early adoption of the Davidon-Fletcher-Powell method gave Jöreskog (1967) the first practicable implementation of ML estimation. The method is no longer used to produce standard error estimates since they are of questionable accuracy (Lee & Jennrich 1979), but it remains an effective optimization method. Lee & Jennrich (1979) provide the statistical and optimization justification for a Gauss-Newton iterative method that yields least squares as well as ML and GLS estimates (see also Dagenais 1978). Matrix calculus is useful to simplify such developments (Bentler & Lee 1975, McDonald 1976, McDonald & Swaminathan 1973, Swaminathan 1976); D. G. Nel (unpublished) provides the most comprehensive overview of the field, while Bentler & Lee (1978b) provide the most accessible unified presentation of basic results. New combinations of statistics and nonlinear optimization continue to be developed (Dempster et al. 1977, Jennrich & Ralston 1978, Lord 1975). This is both necessary and fortunate, since effective utilization of multivariate analysis with latent variables hinges upon simple, general, and inexpensive computer based implementations.

ACKNOWLEDGMENTS

Preparation of this chapter was facilitated by a Research Scientist Development Award (K02-DA00017) and a research grant from the US Public Health Service (DA01070). The literature search for this chapter was completed in December 1978.

The generous suggestions on organization and coverage of this chapter made by a dozen colleagues were very helpful to me, and the specific feedback on drafts of the manuscript provided by G. J. Huba, W. E. McGarvey, D. G. Weeks, J. A. Wingard, H. Wold, J. A. Woodward, and A. T. Yates improved the work substantially. K. S. Bentler, C. P. Chong, and B. R. Woodward provided invaluable bibliographic and editorial assistance, and J. A. Hetland provided outstanding support in manuscript production.

LITERATURE CITED

Aigner, D. J., Goldberger, A. S., eds. 1977. Latent Variables in Socioeconomic Models. Amsterdam: North-Holland. 383 pp.

Ajzen, I., Fishbein, M. 1977. Attitude-behavior relations: A theoretical analysis and review of empirical research. Psychol. Bull. 84:888-918.

Alker, H. R. 1969. Statistics and politics: The need for causal data analysis. In Politics and the Social Sciences, ed. S. M. Lipset, pp. 244-313. New York: Oxford. 328 pp.

Alwin, D. F. 1974. Approaches to the interpretation of relationships in the multitrait-multimethod matrix. In Sociological Methodology, 1973-74, ed. H. Costner, pp. 79-105. San Francisco: Jossey-Bass. 410 pp.

Alwin, D. F., Hauser, R. M. 1975. The decomposition of effects in path analysis. Am. Sociol. Rev. 40:37-47.

Alwin, D. F., Tessler, R. C. 1974. Causal models, unobserved variables, and experimental data. Am. J. Sociol. 80:58-86.

Amemiya, T. 1977. The maximum likelihood and the nonlinear three-stage least squares estimator in the general nonlinear simultaneous equation model. Econometrica 45:955-68.

Amemiya, T. 1978. The estimation of a simultaneous equation generalized probit model. Econometrica 46:1193-1205.

Anderson, J. G., Evans, F. B. 1974. Causal models in educational research: Recursive models. Am. Educ. Res. J. 11: 29-39.

Anderson, T. W. 1973. Asymptotically efficient estimation of covariance matrices with linear structure. Ann. Stat. 1: 135-41.

Anderson, T. W. 1976. Estimation of linear functional relationships: Approximate distributions and connections with simultaneous equations in econometrics. J. R. Stat. Soc. B 38:1-20. Discussion: 20-36.

Anderson, T. W., Das Gupta, S. Styan, G. P. H. 1972. A Bibliography of Multivariate Statistical Analysis. Edinburgh: Oliver & Boyd. 642 pp.

Anderson, T. W., Rubin, H. 1956. Statistical inference in factor analysis. Proc. 3rd Berkeley Symp. Math. Stat. Probl. 5:111-50.

Asher, H. B. 1976. Causal Modeling. Beverly Hills: Sage. 80 pp.

Bachman, J. G., O'Malley, P. M. 1977. Self-esteem in young men: A longitudinal analysis of the impact of educational and occupational attainment. J. Pers. Soc. Psychol. 35:365-80.

Bagozzi, R. P. 1977. Structural equation models in experimental research. J. Mark. Res. 14:209-26.

Bagozzi, R. P. 1978. The construct validity of the affective, behavioral, and cognitive components of attitude by analysis of covariance structures. Multivar. Behav. Res. 13:9-31.

Baltes, P. B., Nesselroade, J. R., Cornelius, S. W. 1978. Multivariate antecedents of structural change in development: A simulation of cumulative environmental patterns. Multivar. Behav. Res. 13:127-52.

Barnett, V., Lewis, T. 1978. Outliers in Statistical Data. New York: Wiley. 365 pp.

Bechtoldt, H. P. 1974. A confirmatory analysis of the factor stability hypothesis. Psychometrika 39:319-26.

Bentler, P. M. 1968. A new matrix for the assessment of factor contributions. Multivar. Behav. Res. 3:489-94.

Bentler, P. M. 1973. Assessment of developmental factor change at the individual and group level. In Life-span Developmental Psychology: Methodological Issues, ed. J. R. Nesselroade, H. W. Reese, pp. 145-74. New York: Academic. 364 pp.

Bentler, P. M. 1976a. Multistructure statistical model applied to factor analysis. Multivar. Behav. Res. 11:3-25.

Bentler, P. M. 1976b. Factor analysis. In Data Analysis Strategies and Designs for Substance Abuse Research, ed. P. M. Bentler, D. J. Lettieri, G. Austin, pp. 139-58. Washington: GPO. 226 pp.

Bentler, P. M. 1977. Factor simplicity index and transformations. Psychometrika 42:277-95.

Bentler, P. M. 1978. The interdependence of theory, methodology, and empirical data: Causal modeling as an approach to construct validation. In Longitudinal Research on Drug Use: Empirical Findings and Methodological Issues, ed. D. B. Kandel, pp. 267-302. New York: Wiley. 314 pp.

Bentler, P. M. 1979. Linear simultaneous equation systems with multiple levels and types of latent variables. In Systems Under Indirect Observation, ed. K. G. Jöreskog, H. Wold. In press.

Bentler, P. M. 1980. Structural equation models in longitudinal research. In Longitudinal Research in the United States, ed. S. A. Mednick, M. Harway. In press.

Bentler, P. M., Bonett, D. G. 1979. Significance tests and goodness of fit in the analysis of covariance structures. Los Angeles: Univ. California. 27 pp.

Bentler, P. M., Huba, G. J. 1979. Simple minitheories of love. J. Pers. Soc. Psychol. 37:124-30.

Bentler, P. M., Lee, S. Y. 1975. Some extensions of matrix calculus. Gen. Syst. 20:145-50.

Bentler, P. M., Lee, S. Y. 1978a. Statistical aspects of a three-mode factor analysis model. Psychometrika 43:343-52.

Bentler, P. M., Lee, S. Y. 1978b. Matrix derivatives with chain rule and rules for simple, Hadamard, and Kronecker products. J. Math. Psychol. 17:255-62.

Bentler, P. M., Lee, S. Y. 1979a. A statistical development of three-mode factor analysis. Br. J. Math. Stat. Psychol. 32: 87-104.

Bentler, P. M., Lee, S. Y. 1979b. Newton-Raphson approach to exploratory and confirmatory maximum likelihood factor analysis. J. Chin. Univ. Hong Kong. In press.

Bentler, P. M., Peeler, W. H. 1979. Models of female orgasm. Arch. Sex. Behav. 8:405-23.

Bentler, P. M., Speckart, G. 1979. An evaluation of models of attitude-behavior relations. Psychol. Rev. 86:452-64.

Bentler, P. M., Weeks, D. G. 1978. Restricted multidimensional scaling models. J. Math. Psychol. 17:138-51.

Bentler, P. M., Weeks, D. G. 1979a. Interrelations among models for the analysis of moment structures. Multivar. Behav. Res. 14:169-85.

Bentler, P. M., Weeks, D. G. 1979b. Linear simultaneous equations with latent variables. Los Angeles: Univ. Calif. 25 pp.

Bentler, P. M., Weeks, D. G. 1980. Multivariate analysis with latent variables. In Handbook of Statistics, Vol. 2, ed. P. R. Krishnaiah, L. Kanal. Amsterdam: North-Holland. In press.

Bentler, P. M., Woodward, J. A. 1978. A Head Start re-evaluation: Positive effects are not yet demonstrable. Eval. Q. 2:493-510.

Bentler, P. M., Woodward, J. A. 1979a. Nonexperimental evaluation research: Contributions of causal modeling. In Improving Evaluations, ed. L. Datta, R. Perloff, pp. 71-102. Beverly Hills: Sage.

Bentler, P. M., Woodward, J. A. 1979b. Regression on linear composites: Statistical theory and applications. Multivar.

Behav. Res. Monogr. 79-1. Ft. Worth: Texas Christian Univ. 58 pp.

Bhargava, A. K. 1977. Maximum likelihood estimation in a multivariate 'errors in variables' regression model with unknown error covariance matrix. Commun. Stat. Theory Methods A 6:587-601.

Bielby, W. T., Hauser, R. M. 1977. Structural equation models. Ann. Rev. Sociol. 3:137-61.

Billings, R. S., Wroten, S. P. 1978. Use of path analysis in industrial/organizational psychology: Criticisms and suggestions. J. Appl. Psychol. 63:677-88.

Birnbaum, M. H., Mellers, B. A. 1979a. Stimulus recognition may mediate exposure effects. J. Pers. Soc. Psychol. 37: 391-94.

Birnbaum, M. H., Mellers, B. A. 1979b. One-mediator model of exposure effects is still viable. J. Pers. Soc. Psychol. 37:1090-96.

Blalock, H. M. 1961. Correlation and causality: the multivariate case. Soc. Forces 39:246-51.

Blalock, H. M. 1963. Making causal inferences for unmeasured variables from correlations among indicators. Am. J. Sociol. 69:53-62.

Blalock, H. M. 1964. Causal Inferences in Nonexperimental Research. Chapel Hill: Univ. North Carolina. 200 pp.

Blalock, H. M., ed. 1971a. Causal Models in the Social Sciences. Chicago: Aldine-Atherton. 515 pp.

Blalock, H. M. 1971b. Causal models involving unmeasured variables in stimulus-response situations. See Blalock 1971a, pp. 335-47.

Blalock, H. M., Aganbegian, A., Borodin, F. M., Boudon, R., Capecchi, V., eds. 1975. Quantitative Sociology. New York: Academic. 643 pp.

Bloxom, B. 1972. Alternative approaches to factorial invariance. Psychometrika 37:425-40.

Bock, R. D., Bargmann, R. E. 1966. Analysis of covariance structures. Psychometrika 31:507-34.

Bohrnstedt, G. W. 1979. Measurement. In Handbook of Survey Research, ed. J. Wright, P. Rossi. New York: Academic. In press.

Boudon, R. 1965. A method of linear causal analysis: dependence analysis. Am. Sociol. Rev. 30:365-74.

Bramble, W. J., Wiley, D. E. 1974. Estimating content-acquiescence correlation by covariance structure analysis. Multivar. Behav. Res. 9:179-90.

Browne, M. W. 1974. Generalized least-squares estimators in the analysis of covariance structures. S. Afr. Stat. J. 8: 1-24.

Browne, M. W. 1978. The likelihood ratio test for the equality of correlation matrices. Br. J. Math. Stat. Psychol. 31:209-17.

Browne, M. W. 1979. The maximum likelihood solution in inter-battery factor analysis. Br. J. Math. Stat. Psychol. 32:75-86.

Burt, R. S. 1976. Interpretational confounding of unobserved variables in structural equation models. Sociol. Methods Res. 5:3-52.

Campbell, D. T., Stanley, J. C. 1963. Experimental and quasi-experimental designs for research on teaching. In Handbook of Research on Teaching, ed. N. L. Gage, 5:171-246. New York: Rand-McNally.

Carter, R. A. L., Nagar, A. L. 1977. Coefficients of correlation for simultaneous equation systems. J. Econometrics 6: 39-50.

Carver, R. P. 1978. The case against statistical significance testing. Harv. Educ. Rev. 48:378-99.

Chechile, R. 1977. Likelihood and posterior identification: Implications for mathematical psychology. Br. J. Math. Stat. Psychol. 30:177-84.

Chen, C. F. 1979. Bayesian inference for a normal dispersion matrix and its application to stochastic multiple regression analysis. J. R. Stat. Soc. B. In press.

Christoffersson, A. 1975. Factor analysis of dichotomized variables. Psychometrika 40:5-32.

Christoffersson, A. 1977. Two-step weighted least squares factor analysis of dichotomized variables. Psychometrika 42: 433-38.

Cochran, W. G. 1970. Some effects of errors of measurement on multiple correlation. J. Am. Stat. Assoc. 65:22-34.

Cohen, D., Schaie, K. W., Gribbin, K. 1977. The organization of spatial abilities in older men and women. J. Gerontol. 32: 578-85.

Cooley, W. W. 1978. Explanatory observational studies. Educ. Res. 7:9-15.

Corballis, M. C. 1973. A factor model for analysing change. Br. J. Math. Stat. Psychol. 26:90-97.

Corballis, M. C., Traub, R. E. 1970. Longitudinal factor analysis. Psychometrika 35:79-98.

Costner, H. L. 1971. Utilizing causal models to discover flaws in experiments. Sociometry 34:398-410.

Cox, D. R., Small, N. J. H. 1978. Testing multivariate normality. Biometrika 65:263-72.

Cox, N. R. 1976. The linear structural relation for several groups of data. Biometrika 63:231-37.

Cronbach, L. J., Meehl, P. E. 1955. Construct validity in psychological tests. Psychol. Bull. 52:281-302.

Dagenais, M. G. 1978. The computation of FIML estimates as iterative generalized least squares estimates in linear and nonlinear simultaneous equations models. Econometrica 46: 1351-62.

Darlington, R. B. 1978. Reduced-variance regression. Psychol. Bull. 85:1238-55.

Davies, R. B., Hutton, B. 1975. The effect of errors in the independent variables in linear regression. Biometrika 62: 383-91.

Deistler, M., Seifert, H. G. 1978. Identifiability and consistent estimability in econometric models. Econometrica 46:969-80.

Dempster, A. P. 1971. An overview of multivariate data analysis. J. Multivar. Anal. 1:316-46.

Dempster, A. P., Laird, N. M., Rubin, D. B. 1977. Maximum likelihood from incomplete data via the EM algorithm. J. R. Stat. Soc. B 39:1-38.

Dolby, G. R. 1976a. The ultrastructural relation: A synthesis of the functional and structural relations. Biometrika 63: 39-50.

Dolby, G. R. 1976b. A note on the linear structural relation when both residual variances are known. J. Am. Stat. Assoc. 71:352-53.

Duncan, O. D. 1966. Path analysis: Sociological examples. Am. J. Sociol. 72:1-16.

Duncan, O. D. 1969. Some linear models for two-wave, two-variable panel analysis. Psychol. Bull. 72:177-82.

Duncan, O. D. 1975. Introduction to Structural Equation Models. New York: Academic. 180 pp.

Eaton, W. W. 1975. Causal models for the study of prevalence. Soc. Forces 54:415-26.

Eaton, W. W. 1977. An addendum to causal models for the study of prevalence. Soc. Forces 56:703-6.

Eaves, L., Eysenck, H. 1975. The nature of extraversion: A genetical analysis. J. Pers. Soc. Psychol. 32:102-12.

Eaves, L. J., Martin, N. G., Eysenck, S. B. G. 1977. An application of the analysis of covariance structures to the psychogenetical study of impulsiveness. Br. J. Math. Stat. Psychol. 30:185-97.

Evans, G. T., Poole, M. E. 1975. Relationships between verbal and nonverbal abilities for migrant children and Australian children of low socio-economic status: Similarities and contrasts. Aust. J. Educ. 19:209-30.

Feldman, J. 1975. Considerations in the use of causal-correlational techniques in applied psychology. J. Appl. Psychol. 60:663-70.

Feldstein, M. 1974. Errors in variables: a consistent estimator with smaller MSE in finite samples. J. Am. Stat. Assoc. 69: 990-96.

Fink, E. L., Mabee, T. I. 1978. Linear equations and nonlinear estimation: A lesson from a nonrecursive example. Sociol. Methods Res. 7:107-20.

Fishbein, M., Ajzen, I. 1975. Belief, Attitude, Intention and Behavior: An Introduction to Theory and Research. Reading, Mass.: Addison-Wesley. 578 pp.

Florens, J. P., Mouchart, M., Richard, J. F. 1974. Bayesian inference in error-in-variables models. J. Multivar. Anal. 4:419-52.

Frederiksen, C. H. 1974. Models for the analysis of alternative sources of growth in correlated stochastic variables. Psychometrika 39:223-45.

Frisch, R. 1934. Statistical Confluence Analysis by Means of Complete Regression Systems. Oslo: Oslo Univ.

Fuller, E. L., Hemmerle, W. J. 1966. Robustness of the maximum likelihood estimation procedure in factor analysis. Psychometrika 31:255-66.

Gabrielsen, A. 1978. Consistency and identifiability. J. Econometrics 8:261-63.

Geraci, V. J. 1976. Identification of simultaneous equation models with measurement error. J. Econometrics 4:263-83.

Geraci, V. J. 1977. Estimation of simultaneous equation models with measurement error. Econometrica 45:1243-55.

Ginsberg, I. J., Greenley, J. R. 1978. Competing theories of marijuana use: A longitudinal study. J. Health Soc. Behav. 19:22-34.

Gleser, L. J. 1979. Estimation of a linear transformation: Large sample results. Ann. Stat. In press.

Gleser, L. J., Watson, G. S. 1973. Estimation of a linear transformation. Biometrika 60:525-34.

Gnanadesikan, R. 1977. Methods for Statistical Data Analysis of Multivariate Observations. New York: Wiley. 311 pp.

Goldberger, A. S. 1971. Econometrics and psychometrics: A survey of communalities. Psychometrika 36:83-107.

Goldberger, A. S. 1972a. Structural equation methods in the social sciences. Econometrica 40:979-1001.

Goldberger, A. S. 1972b. Maximum-likelihood estimation of regressions containing unobservable independent variables. Int. Econ. Rev. 13:1-15.

Goldberger, A. S. 1977. On Thomas's model for kinship correlations. Psychol. Bull. 84:1239-44.

Goldberger, A. S. 1978. Pitfalls in the resolution of IQ inheritance. In Genetic Epidemiology, ed. N. E. Morton, C. S. Chung, pp. 195-222. New York: Academic.

Goldberger, A. S., Duncan, O. D., eds. 1973. Structural Equation Models in the Social Sciences. New York: Academic. 358 pp.

Green, B. F. 1976. On the factor score controversy. Psychometrika 41:263-66.

Green, P. E., Carroll, J. D., DeSarbo, W. S. 1978. A new measure of predictor variable importance in multiple regression. J. Mark. Res. 15:356-60.

Guttman, L. 1977. What is not what in statistics. Statistician 26:81-107.

Hanushek, E. A., Jackson, J. E. 1977. Statistical Methods for Social Scientists. New York: Academic. 374 pp.

Hausman, J. A. 1977. Errors in variables in simultaneous equation models. J. Econometrics 5:389-401.

Hausman, J. A. 1978. Specification tests in econometrics. Econometrica 46:1251-71.

Havlicek, L. L., Peterson, N. L. 1977. Effects of the violation of assumptions upon significance levels of the Pearson r. Psychol. Bull. 84:373-77.

Healy, J. D. 1979. On Kristof's test for a linear relation between true scores of two measures. Psychometrika 44:235-38.

Heckman, J. J. 1978. Dummy endogenous variables in a simultaneous equation system. Econometrica 46:931-59.

Heckman, J. J. 1979. Sample selection bias as specification error. Econometrica 47:153-61.

Heise, D. R. 1975. Causal Analysis. New York: Wiley. 301 pp.

Hendry, D. F. 1976. The structure of simultaneous equations estimators. J. Econometrics 4:51-88.

Hornick, J. P. 1978. Premarital sexual attitudes and behavior. Sociol. Q. 19:534-44.

Hsiao, C. 1976. Identification and estimation of simultaneous equation models with measurement error. Int. Econ. Rev. 17:319-39.

Humphreys, L. G., Parsons, C. K. 1979. A simplex process model for describing differences between cross-lagged correlations. Psychol. Bull. 86:325-34.

Hyde, J. S., Geiringer, E. R., Yen, W. M. 1975. On the empirical relation between spatial ability and sex differences in other aspects of cognitive performance. Multivar. Behav. Res. 10:289-310.

Intriligator, M. D. 1978. Econometric Models, Techniques, and Applications. Englewood Cliffs: Prentice-Hall. 638 pp.

Isaac, P. D. 1970. Linear regression, structural relations, and measurement error. Psychol. Bull. 74:213-18.

James, L. R., Singh, B. K. 1978. An introduction to the logic, assumptions, and basic analytic procedures of two-stage least squares. Psychol. Bull. 85:1104-22.

Jennrich, R. I. 1978. Rotational equivalence of factor loading matrices with specified values. Psychometrika 43:421-26.

Jennrich, R. I., Ralston, M. L. 1978. Fitting nonlinear models to data. Ann. Rev. Biophys. Bioeng. 8:195-238.

Jordan, L. A. 1978. Linear structural relation, longitudinal data, and the crosslag idea. ASA Proc. Bus. Econ. Stat. 1:1-5.

Jöreskog, K. G. 1967. Some contributions to maximum likelihood factor analysis. Psychometrika 32:443-82.

Jöreskog, K. G. 1969. A general approach to confirmatory maximum likelihood factor analysis. Psychometrika 34:183-202.

Jöreskog, K. G. 1970a. A general method for analysis of covariance structures. Biometrika 57:239-51.

Jöreskog, K. G. 1970b. Estimation and testing of simplex models. Br. J. Math. Stat. Psychol. 23:121-45.

Jöreskog, K. G. 1971a. Simultaneous factor analysis in several populations. Psychometrika 36:409-26.

Jöreskog, K. G. 1971b. Statistical analysis of sets of congeneric tests. Psychometrika 36:109-33.

Jöreskog, K. G. 1973a. A general method for estimating a linear structural equation system. See Goldberger & Duncan 1973, pp. 85-112.

Jöreskog, K. G. 1973b. Analysis of covariance structures. In Multivariate Analysis-III, ed. P. R. Krishnaiah, pp. 263-85. New York: Academic. 410 pp.

Jöreskog, K. G. 1974. Analyzing psychological data by structural analysis of covariance matrices. In Contemporary Developments in Mathematical Psychology, ed. R. C. Atkinson, D. H. Krantz, R. D. Luce, P. Suppes, pp. 1-56. San Francisco: Freeman. 468 pp.

Jöreskog, K. G. 1977. Structural equation models in the social sciences: specification, estimation and testing. Applications of Statistics, ed. P. R. Krishnaiah, pp. 265-87. Amsterdam: North-Holland.

Jöreskog, K. G. 1978. Structural analysis of covariance and correlation matrices. Psychometrika 43:443-77.

Jöreskog, K. G. 1979. Statistical estimation of structural models in longitudinal-developmental investigation. In Longitudinal Methodology in the Study of Behavior and Human Development, ed. J. R. Nesselroade, P. B. Baltes. New York: Academic. In press.

Jöreskog, K. G., Goldberger, A. S. 1972. Factor analysis by generalized least squares. Psychometrika 37:243-60.

Jöreskog, K. G., Goldberger, A. S. 1975. Estimation of a model with multiple indicators and multiple causes of a single latent variable. J. Am. Stat. Assoc. 70:631-39.

Jöreskog, K. G., Sörbom, D. 1976. Statistical models and methods for test-retest situations. In Advances in Psychological and Educational Measurement, ed. D. N. M. deGrujter, L. J. Th. van der Kamp, pp. 135-57. New York: Wiley, 320 pp.

Jöreskog, K. G., Sörbom, D. 1977. Statistical models and methods for analysis of longitudinal data. See Aigner & Goldberger 1977, pp. 285-325.

Jöreskog, K. G., Sörbom, D. 1978. LISREL IV Users Guide. Chicago: Natl. Educ. Res. 93 pp.

Kadane, J. B., Anderson, T. W. 1977. A comment of the test of overidentifying restrictions. Econometrica 45:1027-31.

Kadane, J. B., McGuire, T. W., Sanday, P. R., Staelin, R. 1976. Models of environmental effects on the development of IQ. J. Educ. Stat. 1:181-231.

Kaiser, H. F. 1976. Review of D. N. Lawley, A. E. Maxwell, "Factor analysis as a statistical method" 2nd ed. Educ. Psychol. Meas. 36:586-88.

Keesling, W. 1972. Maximum likelihood approaches to causal flow analysis. PhD thesis. Univ. Chicago, Chicago, Ill. 114 pp.

Kenny, D. A. 1975a. A quasi-experimental approach to assessing treatment effects in the nonequivalent control group design. Psychol. Bull. 82:345-62.

Kenny, D. A. 1975b. Cross-lagged panel correlation: A test for spuriousness. Psychol. Bull. 82:887-903.

Kenny, D. A. 1976. An empirical application of confirmatory factor analysis to the multitrait multimethod matrix. J. Exp. Soc. Psychol. 12:247-52.

Kenny, D. A. 1979. Correlation and Causality. New York: Wiley, 288 pp.

Kim, J. O. 1975. Multivariate analysis of ordinal variables. Am. J. Sociol. 81:261-98, 84:448-56.

Kim, J. O., Mueller, C. W. 1978. Factor Analysis. Beverly Hills: Sage. 88 pp.

Kohn, M. L., Schooler, C. 1978. The reciprocal effects of the substantive complexity of work and intellectual flexibility: A longitudinal assessment. Am. J. Sociol. 84:24-52.

Kohn, P. M., Annis, H. M. 1978. Personality and social factors in adolescent marijuana use: A path-analytic study. J. Consult. Clin. Psychol. 46:366-67.

Kowalski, C. J. 1972. On the effects of nonnormality on the distribution of the sample product-moment correlation coefficient. Appl. Stat. 21:1-12.

Krane, W. R., McDonald, R. P. 1978. Scale invariance and the factor analysis of correlation matrices. Br. J. Math. Stat. Psychol. 31:218-28.

Kristof, W. 1973. Testing a linear relation between true scores of two measures. Psychometrika 38:101-11.

Laffont, J. J. 1977. A note on the asymptotic Cramer Rao bound in nonlinear simultaneous equation systems. Ann. Econ. Soc. Meas. 6:445-51.

Larzelere, R. E., Mulaik, S. A. 1977. Single-sample tests for many correlations. Psychol. Bull. 84:557-69.

Lawler, E. E. 1968. A correlational-causal analysis of the relationship between expectancy attitudes and job performance. J. Appl. Psychol. 52:462-68.

Lawley, D. N. 1940. The estimation of factor loadings by the method of maximum likelihood. Proc. R. Soc. Edinburgh 60: 64-82.

Lawley, D. N., Maxwell, A. E. 1971. Factor Analysis as a Statistical Method. London: Butterworth. 153 pp.

Lawley, D. N., Maxwell, A. E. 1973. Regression and factor analysis. Biometrika 60:331-38.

Leamer, E. E. 1978. Specification Searches: Ad Hoc Inference with Nonexperimental Data. New York: Wiley. 370 pp.

Lee, S. Y. 1977. Some algorithms for covariance structure analysis. PhD thesis. Univ. California, Los Angeles, Calif. 130 pp.

Lee, S. Y., Bentler, P. M. 1979a. Analysis of a general co-variance structure model. Hong Kong: Chinese Univ. 15 pp.

Lee, S. Y., Bentler, P. M. 1979b. Asymptotic theorems for constrained generalized least squares estimation in covariance structure models. Hong Kong: Chinese Univ. 15 pp.

Lee, S. Y., Jennrich, R. I. 1979. A study of algorithms for covariance structure analysis with specific comparisons using factor analysis. Psychometrika 44:99-113.

Leik, R. K. 1975. Causal models with nominal and ordinal data: Retrospective. In Sociological Methodology 1976, ed. D. R. Heise, pp. 271-75. San Francisco: Jossey-Bass. 299 pp.

Levi, M. D. 1977. Measurement errors and bounded OLS estimates. J. Econometrics 6:165-71.

Li, C. C. 1975. Path Analysis—A Primer. Pacific Grove, Calif.: Boxwood. 347 pp.

Linn, R. L., Werts, C. E. 1977. Analysis implications of the choice of a structural model in the nonequivalent control group design. Psychol. Bull. 64:229-34.

Loehlin, J. C. 1978. Heredity-environment analyses of Jencks's IQ correlations. Behav. Genet. 8:415-36.

Long, J. S. 1976. Estimation and hypothesis testing in linear models containing measurement error. Sociol. Methods Res. 5:157-206.

Lord, F. M. 1960. Large-sample covariance analysis when the control variable is fallible. J. Am. Stat. Assoc. 55:307-21.

Lord, F. M. 1973. Testing if two measuring procedures measure the same dimension. Psychol. Bull. 79:71-72.

Lord, F. M. 1975. Automated hypothesis tests and standard errors for nonstandard problems. Am. Stat. 29:56-59.

Maasoumi, E. 1978. A modified Stein-like estimator for the reduced form coefficients of simultaneous equations. Econometrica 46:695-703.

MacDonald, K. I. 1977. Path analysis. In The Analysis of Survey Data, Vol. 2, Model Fitting, ed. C. A. O'Muircheartaigh, C. Payne, pp. 81-104. New York: Wiley, 255 pp.

MacDonald, K. I. 1979. Interpretation of residual paths and decomposition of variance. Sociol. Methods Res. 7:289-304.

Magidson, J. 1977. Toward a causal model approach for adjusting for preexisting differences in the nonequivalent control group situation: A general alternative to ANCOVA. Eval. Q. 1:399-419.

Magidson, J. 1978. Reply to Bentler and Woodward: The .05 significance level is not all-powerful. Eval. Q. 2:511-20.

Martin, N. G., Eaves, L. J. 1977. The genetical analysis of covariance structure. Heredity 38:79-95.

Maruyama, G. 1977. A causal-model analysis of variables related to primary school achievement. PhD thesis. Univ. Southern California, Los Angeles, Calif.

Maruyama, G., Miller, N. 1979. Re-examination of normative influence processes in desegregated classrooms. Am. Educ. Res. J. In press.

Masters, W. H., Johnson, V. E. 1966. Human Sexual Response. Boston: Little, Brown, 366 pp.

McDonald, J. B. 1977. The K-class estimators as least variance difference estimators. Econometrica 45:759-63.

McDonald, R. P. 1967. Nonlinear factor analysis. Psychom. Monogr. 15. Chicago: Univ. Chicago Press.

McDonald, R. P. 1976. The McDonald-Swaminathan matrix calculus: Clarifications, extensions, and illustrations. Gen. Syst. 21:87-94.

McDonald, R. P. 1978. A simple comprehensive model for the analysis of covariance structures. Br. J. Math. Stat. Psychol. 31:59-72.

McDonald, R. P., Krane, W. R. 1977. A note on local identifiability and degrees of freedom in the asymptotic likelihood ratio test. Br. J. Math. Stat. Psychol. 30:198-203.

McDonald, R. P., Krane, W. R. 1979. A Monte Carlo study of local identifiability and degrees of freedom in the asymptotic likelihood ratio test. Br. J. Math. Stat. Psychol. In press.

McDonald, R. P., Mulaik, S. A. 1979. Determinacy of common factors: A nontechnical review. Psychol. Bull. 86:297-306.

McDonald, R. P., Swaminathan, H. 1973. A simple matrix calculus with applications to multivariate analysis. Gen. Syst. 18:37-54.

McFatter, R. M. 1979. The use of structural equation models in interpreting regression equations including suppressor and enhancer variables. Appl. Psychol. Meas. 3:123-35.

McGarvey, W. E. 1978. Longitudinal factors in school desegregation. PhD thesis. Univ. Southern California, Los Angeles, Calif. 160 pp.

McGaw, B., Jöreskog, K. G. 1971. Factorial invariance of ability measures in groups differing in intelligence and socio-economic status. Br. J. Math. Stat. Psychol. 24:154-68.

McKennell, A. C. 1978. Cognition and affect in perceptions of well-being. Soc. Indic. Res. 5:389-426.

Mednick, S. A., Schulsinger, F., Teasdale, T. W., Schulsinger, H., Venables, P. H., Rock, D. R. 1978. Schizophrenia in high-risk children: sex differences in predisposing factors. In Cognitive Defects in the Development of Mental Illness, ed. G. Sorban. New York: Brunner-Mazel.

Mehta, J. S., Swamy, P. A. V. B. 1978. The existence of moments of some simple Bayes estimators of coefficients in a simultaneous equation model. J. Econometrics 7:1-13.

Miller, A. D. 1971. Logic of causal analysis from experimental to nonexperimental designs. See Blalock 1971a, pp. 273-94.

Monfort, A. 1978. First-order identification in linear models. J. Econometrics 7:333-50.

Moran, P. A. P. 1971. Estimating structural and functional relationships. J. Multivar. Anal. 1:232-55.

Moreland, R. L., Zajonc, R. B. 1979. Exposure effects may not depend on stimulus recognition. J. Pers. Soc. Psychol. In press.

Mouchart, M. 1977. A regression model with an explanatory variable which is both binary and subject to errors. See Aigner & Goldberger 1977, pp. 49-66.

Mulaik, S. 1975. Confirmatory factor analysis. In Introductory Multivariate Analysis, ed. D. J. Amick, H. J. Walberg. Berkeley: McCutchan.

Muthén, B. 1977. Statistical methodology for structural equation models involving latent variables with dichotomous indicators. PhD thesis. Univ. Uppsala, Uppsala, Sweden.

Muthén, B. 1978. Contributions to factor analysis of dichotomous variables. Psychometrika 43:551-60.

Naditch, M. P. 1974. Acute adverse reactions to psychoactive drugs, drug usage, and psychopathology. J. Abnorm. Psychol. 83:394-403.

Naditch, M. P. 1975. Relation of motives for drug use and psychopathology in the development of acute adverse reactions to psychoactive drugs. J. Abnorm. Psychol. 84:374-85.

Namboodiri, N. K., Carter, L. F., Blalock, H. M. 1975. Applied Multivariate Analysis and Experimental Designs. New York: McGraw Hill. 688 pp.

Nishisato, S. 1971. Transform factor analysis: A sketchy presentation of a general approach. Jpn. Psychol. Res. 13:155-66.

Noonan, R., Wold, H. 1977. NIPALS path modeling with latent variables. Scand. J. Educ. Res. 21:33-61.

O'Donnell, J. A., Clayton, R. R. 1979. Determinants of early marihuana use. In Youth Drug Use, ed. G. M. Beschner, A. S. Friedman, pp. 63-110. Lexington, Mass.: Heath. 683 pp.

Olsson, U. 1978. Some data analytic problems in models with latent variables. PhD thesis. Univ. Uppsala, Uppsala, Sweden. 161 pp.

Olsson, U., Bergman, L. R. 1977. A longitudinal factor model for studying change in ability structure. Multivar. Behav. Res. 12:221-42.

Patefield, W. M. 1977. On the information matrix in the linear functional relationship problem. Appl. Stat. 26:69-70.

Pelz, D. C., Andrews, F. M. 1964. Detecting causal priorities in panel study data. Am. Sociol. Rev. 29:836-47.

Pesaran, M. H., Deaton, A. S. 1978. Testing non-nested nonlinear regression models. Econometrica 46:677-94.

Please, N. W. 1973. Comparison of factor loadings in different populations. Br. J. Math. Stat. Psychol. 26:61-89.

Ramsey, J. B. 1978. Nonlinear estimation and asymptotic approximations. Econometrica 46:901-29.

Ramsey, P. H. 1978. Factor analysis of the WAIS and twenty French-kit reference tests. Appl. Psychol. Meas. 2:505-17.

Rao, D. C., Morton, N. E., Elston, R. C., Yee, S. 1976. Causal analysis of academic performance. Behav. Genet. 7: 147-59.

Reynolds, H. T. 1974. Nonparametric partial correlation and causal analysis. Sociol. Methods Res. 2:376-92.

Robertson, C. A. 1974. Large-sample theory for linear structural relation. Biometrika 61:353-59.

Robinson, P. M. 1974. Identification, estimation and large-sample theory for regressions containing unobservable variables. Int. Econ. Rev. 15:680-92.

Robinson, P. M. 1977. The estimation of a multivariate linear relation. J. Multivar. Anal. 7:409-23.

Rock, D. A., Werts, C. E., Flaugher, R. L. 1978. The use of analysis of covariance structures for comparing the psychometric properties of multiple variables across populations. Multivar. Behav. Res. 13:403-18.

Rock, D. A., Werts, C. E., Linn, R. L. 1976. Structural equations as an aid in the interpretation of the non-orthogonal analysis of variance. Multivar. Behav. Res. 11:443-48.

Rock, D. A., Werts, C. E., Linn, R. L., Jöreskog, K. G. 1977. A maximum likelihood solution to the errors in variables and errors in equations model. Multivar. Behav. Res. 12:187-98.

Rogosa, D. 1979. Causal models in longitudinal research. See Jöreskog 1979.

Rosen, B. C., Aneshensel, C. S. 1978. Sex differences in the educational-occupational expectation process. Soc. Forces 57:164-86.

Roskam, E. E. 1976. Multivariate analysis of change and growth: Critical review and perspectives. See Jöreskog & Sörbom 1976, pp. 111-33.

Royce, J. R. 1977. Guest editorial: Have we lost sight of the original vision for SMEP and MBR? Multivar. Behav. Res. 12:135-41.

Rozelle, R. M., Campbell, D. T. 1969. More plausible rival hypotheses in the cross-lagged panel correlation technique. Psychol. Bull. 71:74-80.

Rubin, D. B. 1978. A note on Bayesian, likelihood, and sampling distribution inferences. J. Educ. Stat. 3:189-201.

Samejima, F. 1979. Latent trait theory and its applications. In Multivariate Analysis V, ed. P. R. Krishnaiah. Amsterdam: North-Holland.

Sargan, J. D. 1978. On the existence of the moments of 3SLS estimators. Econometrica 46:1329-50.

Schmitt, N., Coyle, B. W., Saari, B. B. 1977. A review and critique of analyses of multitrait-multimethod matrices. Multivar. Behav. Res. 12:447-78.

Schönemann, P. H., Steiger, J. H. 1976. Regression component analysis. Br. J. Math. Stat. Psychol. 29:175-89.

Schönemann, P. H., Steiger, J. H. 1978. On the validity of indeterminate factor scores. Bull. Psychon. Soc. 12:287-90.

Sheridan, J. E., Slocum, J. W. 1977. Causal inferences in motivation research: A reinterpretation of results from panel studies. J. Appl. Psychol. 62:510-13.

Simon, H. A. 1954. Spurious correlation: A causal interpretation. J. Am. Stat. Assoc. 49:467-79.

Simonton, D. K. 1977. Eminence, creativity, and geographic marginality: A recursive structural equation model. J. Pers. Soc. Psychol. 35:805-16.

Sims, H. P., Szilagyi, A. D. 1975. Leader structure and subordinate satisfaction for two hospital administrative levels: A path analysis approach. J. Appl. Psychol. 60:194-97.

Smith, R. B. 1974. Continuities in ordinal path analysis. Soc. Forces 53:200-29.

Smith, R. B. 1978. Nonparametric path analysis: Comments on Kim's "Multivariate Analysis of Ordinal Variables." Am. J. Sociol. 84: 437-48.

Sörbom, D. 1974. A general method for studying differences in factor means and factor structure between groups. Br. J. Math. Stat. Psychol. 27: 229-39.

Sörbom, D. 1975. Detection of correlated errors in longitudinal data. Br. J. Math. Stat. Psychol. 28: 138-51.

Sörbom, D. 1976. A statistical model for the measurement of change in true scores. See Jöreskog & Sörbom 1976, pp. 159-70.

Sörbom, D. 1978. An alternative to the methodology for analysis of covariance. Psychometrika 43: 381-96.

Sörbom, D., Jöreskog, K. G. 1976. COFAMM, Confirmatory Factor Analysis with Model Modification. Chicago: Natl. Educ. Res. 51 pp.

Spearman, C. 1904. The proof and measurement of association between two things. Am. J. Psychol. 15: 72-101.

Spiegelman, C. 1979. On estimating the slope of a straight line when both variables are subject to error. Ann. Stat. 7: 201-6.

Sprent, P. 1969. Models in Regression and Related Topics. London: Methuen. 173 pp.

Steiger, J. H. 1979a. The relationship between external variables and common factors. Psychometrika 44: 93-97.

Steiger, J. H. 1979b. Factor indeterminacy in the 1930's and the 1970's: Some interesting parallels. Psychometrika 44: 157-67.

Steiger, J. H., Schonemann, P. H. 1978. A history of factor indeterminacy. In Theory Construction and Data Analysis, ed. S. Shye, pp. 136-78. San Francisco: Jossey-Bass. 426 pp.

Strotz, R. H., Wold, H. O. A. 1960. Recursive vs. nonrecursive systems: An attempt at synthesis. Econometrica 28:417-27.

Swain, A. J. 1975. A class of factor analytic estimation procedures with common asymptotic sampling properties. Psychometrika 40:315-35.

Swaminathan, H. 1976. Matrix calculus for functions of partitioned matrices. Gen. Syst. 21:95-99.

Swaminathan, H., Algina, J. 1978. Scale-freeness in factor analysis. Psychometrika 43:581-83.

Taubman, P., ed. 1977. Kinometrics: Determinants of Socioeconomic Success Within and Between Families. Amsterdam: North-Holland. 324 pp.

Tesser, A., Paulhus, D. L. 1976. Toward a causal model of love. J. Pers. Soc. Psychol. 34:1095-1105.

Theobald, C. M., Mallinson, J. R. 1978. Comparative calibration, linear structural relationships and congeneric measurements. Biometrics 34:39-45.

Thomas, H. 1977. Kinship correlations and scientific nihilism: Reply to Goldberger. Psychol. Bull. 84:1245-48.

Torgerson, W. S. 1958. Theory and Methods of Scaling. New York: Wiley. 460 pp.

Tucker, L. R., Lewis, C. 1973. A reliability coefficient for maximum likelihood factor analysis. Psychometrika 38:1-10.

Tukey, J. W. 1954. Causation, regression, and path analysis. In Statistics and Mathematics in Biology, ed. O. K. Kempthorne, T. A. Bancroft, J. W. Gowen, J. L. Lush, pp. 35-66. Ames: Iowa State Univ. Press.

Tukey, J. W. 1977. Exploratory Data Analysis. Reading, Mass.: Addison-Wesley. 688 pp.

Turner, M. E., Stevens, C. D. 1959. The regression analysis of causal paths. Biometrics 15:236-58.

Upshaw, H. S. 1978. Confirmatory factor analysis for the measurement of attitude. In Einstellungsmessung, ed. F. Petermann. Göttingen: Verlag für Psychologie.

Van de Geer, J. P. 1971. Introduction to Multivariate Analysis for the Social Sciences. San Francisco: Freeman. 293 pp.

van Driel, O. P. 1978. On various causes of improper solutions in maximum likelihood factor analysis. Psychometrika 43:225-43.

Wanous, J. P. 1974. A causal-correlational analysis of the job satisfaction and performance relationship. J. Appl. Psychol. 59:139-44.

Warren, R. D., White, J. K., Fuller, W. A. 1974. An errors-in-variables analysis of managerial role performance. J. Am. Stat. Assoc. 69:886-93.

Weeks, D. G. 1978. Structural equation systems on latent variables within a second-order measurement model. PhD thesis. Univ. California, Los Angeles, Calif. 125 pp.

Weeks, D. G. 1979. A second-order longitudinal model of ability structure. Multivar. Behav. Res. In press.

Werts, C. E., Hilton, T. L. 1977. Intellectual status and intellectual growth, again. Am. Educ. Res. J. 14:137-46.

Werts, C. E., Jöreskog, K. G., Linn, R. L. 1972. A multitrait-multimethod model for studying growth. Educ. Psychol. Meas. 32:655-78.

Werts, C. E., Jöreskog, K. G., Linn, R. L. 1973. Identification and estimation in path analysis with unmeasured variables. Am. J. Sociol. 78:1469-84.

Werts, C. E., Linn, R. L. 1970a. Path analysis: Psychological examples. Psychol. Bull. 74:193-212.

Werts, C. E., Linn, R. L. 1970b. A general linear model for studying growth. Psychol. Bull. 73:17-22.

Werts, C. E., Linn, R. L., Jöreskog, K. G. 1977. A simplex model for analyzing academic growth. Educ. Psychol. Meas. 37:745-56.

Werts, C. E., Linn, R. L., Jöreskog, K. G. 1978. Reliability of college grades from longitudinal data. Educ. Psychol. Meas. 38:89-95.

Werts, C. E., Rock, D. A., Linn, R. L., Jöreskog, K. G. 1976. Comparison of correlations, variances, covariances, and regression weights with or without measurement error. Psychol. Bull. 83:1007-13.

Wheaton, B., Muthén, B., Alwin, D. F., Summers, G. F. 1977. Assessing reliability and stability in panel models. In Sociological Methodology, 1977, ed. D. R. Heise, pp. 84-136. San Francisco: Jossey-Bass. 320 pp.

Wiley, D. E. 1973. The identification problem for structural equation models with unmeasured variables. See Goldberger & Duncan 1973, pp. 69-83.

Wiley, D. E., Schmidt, W. H., Bramble, W. J. 1973. Studies of a class of covariance structure models. J. Am. Stat. Assoc. 68:317-23.

Williams, J. S. 1978. A definition for the common-factor analysis model and the elimination of problems of factor score indeterminacy. Psychometrika 43:293-306.

Wilson, T. P. 1974. On interpreting ordinal analogies to multiple regression and path analysis. Soc. Forces 53:120-23.

Winer, B. J. 1978. Statistics and data analysis: Trading bias for reduced mean squared error. Ann. Rev. Psychol. 29: 647-81.

Wold, H. 1956. Causal inference from observational data: A review of ends and means. J. R. Stat. Soc. A 119:28-61.

Wold, H. 1975. Path models with latent variables: The NIPALS approach. See Blalock et al. 1975, pp. 307-57.

Wold, H. 1979. Model construction and evaluation when theoretical knowledge is scarce: An example of the use of partial least squares. In Evaluation of Econometric Models, ed. J. Kmenta, J. Ramsey. New York: Seminar. In press.

Woodward, J. A., Geiselman, R. E., Beatty, J. 1978. Structural Equation Models of Human Memory. NSF Grant Proposal 79-10053.

Wright, S. 1934. The method of path coefficients. Ann. Math. Stat. 5:161-215.

Wright, S. 1960a. Path coefficients and path regressions: Alternative or complementary concepts? Biometrics 16:189-202.

Wright, S. 1960b. The treatment of reciprocal interaction, with or without lag, in path analysis. Biometrics 16:423-45.

Yates, A. T. 1979. Multivariate Exploratory Data Analysis: True Exploratory Factor Analysis. In press.

Young, J. W. 1977. The function of theory in a dilemma of path analysis. J. Appl. Psychol. 62:108-10.

Zellner, A. 1970. Estimation of regression relationships containing unobservable independent variables. Int. Econ. Rev. 11:441-54.

Zellner, A. 1978. Estimation of functions of population means and regression coefficients including structural coefficients. J. Econometrics 8:127-58.

9

Linear Structural Relationships

Willem E. Saris

The practice of using partial correlations for testing causal theories formulated as a set of linear structural equations was first introduced by Simon (1954) and Blalock (1961, 1964) in sociology and political science. Path analysis was shortly thereafter introduced by Duncan (1966) and Boudon (1968) with the same intention. We can say at this time that both approaches had to deal with the following problems:

1. it was not clear at first how to deal with reciprocal causation,
2. there was no solution dealing with measurement error,
3. estimation procedures were not very efficient,
4. clear criteria were lacking for determining when a model should be rejected.

In econometrics the problem of reciprocal causation has been studied extensively, and the problems of identification and estimation have been solved in principle. Various approaches are to be found in econometric textbooks such as Johnston (1972), Goldberger (1964), and Malinvaud (1970).

The econometric approaches have found solutions to these problems while concentrating on the problem of errors in the equations. The problem of measurement error, however, has

Reprinted from **Quality and Quantity**, Vol. 14 (1980), pp. 205-224, with the permission of Elsevier Scientific Publishing Company.

been more the concern of psychologists. In classical test theory (Nunnally, 1968; Lord and Novick, 1968) and in factor analysis (Harman, 1968; Lawley and Maxwell, 1971; Mulaik, 1973) linear models have been used to describe the relationships between observed and latent variables. However, no efficient procedures were developed for studying the relationships between the latent variables, as there was more interest in the measurement problem itself.

In any case, linear models were being used, for different purposes, in several disciplines. Because of this mutual interest, a conference was held in 1971 with representatives from all disciplines mentioned (see Goldberger and Duncan, 1973). During this conference a system of equations was discussed which could be used to represent the causal models of sociologists and political scientists, incorporating the idea of reciprocal causation as found in econometrics and taking into account the distinction between observed and latent variables made by psychologists.

The identification of this system was discussed by Wiley (1973), and the estimation and testing by Jöreskog (1973). The development of this general system and the program LISREL (Jöreskog and van Thillo, 1973) are the most important developments of the last ten years in the application of linear structural equation models in the social sciences. We discuss this system and its identification, estimation and testing. An example is given and some recent and future developments are indicated.

THE GENERAL SYSTEM

The endogenous variables η_1, η_2, ... η_m will be represented by a vector η.

The predetermined variables ξ_1, ξ_2, ... ξ_n will be represented by a vector ξ.

Assuming linear relationships between these variables, we can formulate the general nonrecursive model of equation (1):

$$B\eta = \Gamma\xi + \zeta \tag{1}$$

where B (m × m) and Γ (m × n) are matrices with parameters and ζ is a vector with disturbance terms.

As in econometric literature, it is assumed that $E(\xi\zeta') = 0$ and $E(\zeta) = 0$. In contrast with most econometric literature, it is not assumed that η and ξ are measured exactly.

The indicators of the endogenous variables y_1, y_2, ... y_p will be represented by the vector y, and the indicators of the predetermined variables x_1, x_2, ... x_q will be represented by the vector x.

Assuming again linear relationships between the indicators and the unmeasured constructs, we can formulate equations (2) and (3):

$$y = \mu + \Lambda_y \eta + \varepsilon \tag{2}$$

$$x = \nu + \Lambda_x \xi + \delta \tag{3}$$

where ε and δ are vectors with disturbance terms. Λ_y (p × m) and Λ_x (q × n) are parameter matrices.

As in the factor analysis literature, it is assumed that $E(\varepsilon) = 0$; $E(\delta) = 0$; $E(\eta\varepsilon') = 0$; $E(\xi\delta') = 0$; $E(\eta) = 0$; $E(\xi) = 0$; $E(\varepsilon\varepsilon') = \theta_\varepsilon^2$ and $E(\delta\delta') = \theta_\delta^2$.

Because of the combination of the different equations in one model, some extra assumptions are introduced: $E(\delta\varepsilon') = 0$; $E(\eta\delta') = 0$; $E(\xi\varepsilon') = 0$; $E(\zeta\delta') = 0$ and $E(\zeta\varepsilon') = 0$.

From equations (1), (2) and (3), it is clear that the general model is a combination of an econometric nonrecursive (or simultaneous equation) model and two factor-analytic models. This means that the econometric model has been extended to the situation in which the variables are not measured exactly, and the factor-analytic model has been extended to include cases in which one tries to explain covariances between factors on the basis of scores on other variables.

Special Cases

The most important advantage of the use of the general system is that a large variety of models can be specified as special cases of the general system. We cannot discuss this point in detail in this chapter (see Saris, 1978 and Jöreskog, 1973), but we can mention several familiar models which are special cases of this general system.

a. Models without unobservable variables for cross-sectional data. This class contains regression models, recursive models, nonrecursive models, and block-recursive models. For discussion of these models see such texts as Blalock (1964, 1971), Duncan (1975), or Heise (1975).
b. Models without unobservable variables for time-series data, including autoregressive process models, simultaneous growth models and econometric models. For discussion of these models, we refer to econometrics texts and Jöreskog (1978).

c. The measurement models of classical test theory, factor analysis, second-order factor analysis, multitrait-multimethod models and variance-covariance components models. For discussion of these models see Lord and Novick (1968), Lawley and Maxwell (1971), or Jöreskog (1969, 1974).
d. Multiple indicators-multiple causes models. This kind of model has been discussed by Hauser and Goldberger (1971) and Duncan, Haller and Portes (1968).
e. Models for panel data, including models either with unobservable variables or without unobservable variables. For discussion of these models see Heise (1970), Duncan (1969), Jöreskog and Sörbom (1977), or Jöreskog (1978).

As all these models are special cases of the general system the procedures for identification, estimation and testing of the general system also apply to each special case mentioned. This is of course a very attractive feature of the approach discussed here.

Identification

The parameters of the system can be found in the matrices: B, Γ, Λ_y, Λ_x, θ_ε, θ_δ, and $\Phi = E(\xi\xi')$ $\Psi = E(\zeta\zeta')$. Some of the elements of these matrices can be restricted to specific values or to values of other parameters but a certain number of the parameters (let us say s) have to be estimated from the data. When we bring all the parameters which have to be estimated from the data in a vector θ, then this vector has s elements.

If the vector z contains the measured variables y and x, it is possible to express the variance-covariance matrix of z in the parameters of the model, as in factor analysis and path analysis (Jöreskog and Van Thillo, 1973). If we call this variance-covariance matrix Σ then each element of Σ is a non-linear function of θ

$$\sigma_{ij} = f_{ij}(\theta) \tag{4}$$

Because z is a (p + q) vector, Σ contains $\frac{1}{2}(p + q)(p + q + 1)$ distinct elements, and there are $\frac{1}{2}(p + 1)(p + q + 1)$ equations (4). From this it follows that a necessary condition for identification is

$$s \leqslant \frac{1}{2}(p + q)(p + q + 1) \tag{5}$$

The sufficient conditions for identification of the model are quite complex. Some information can be found in Wiley (1973), Werts, Linn and Jöreskog (1973) and Geraci (1974, 1977). Wiley comments upon this problem.

> If an estimation program already completes a numerical estimate of the information matrix for the parameters, an evaluation of its rank for reasonable parameter values will give good assurance as to the identifiability of the parameters (pp. 82-3).

We will return to this remark when we discuss the program.

Estimation

For complex models such as the model discussed here, the use of the maximum likelihood estimation procedures has considerable advantages over other procedures like least squares. The advantages are that the distribution of the estimators for large samples in the case of maximum likelihood estimation is known to be multivariate normal, and that the estimators are best unbiased estimators. In the case of other procedures, one has to derive what the characteristics of the estimators are for the specific model (Silvey, 1970, ch. 4; Goldberger, 1964, pp. 115-25). This means that even if one derives the distribution of unbiased estimators according to another principle, they still could not be better than the maximum likelihood estimators in the case of large samples.

From this it follows that maximum likelihood estimators are very attractive for this model, because hypotheses concerning individual parameters can be tested, since the distribution of the estimators is normal and the variance is known. The variance-covariance matrix of the estimators is known to be equal to the inverse of the information matrix (I_θ), where I_θ can be calculated with

$$I_\theta = N \cdot E\left(\left(\frac{\delta \log L}{\delta \theta}\right)\left(\frac{\delta \log L}{\delta \theta}\right)'\right)$$

L is the likelihood function (see LISREL, below). N is the size of the sample (Silvey, 1970, pp. 35-43). Thus the variances of the estimators can be found on the diagonal of I_θ^{-1}.

Testing the Model

In the case of maximum likelihood estimation and large samples, it is possible to test not only hypotheses concerning individual parameters, but also hypotheses concerning the whole model. Let H_0 be the hypothesis that the parameters of the model are in a subset of the parameter space which is restricted by the specification that some parameters are fixed on specific values and others are hypothesized to be equal to other parameters. Let H_1 be the hypothesis that the parameters of the model are in a subset of the parameter space which is restricted only as much as is necessary for identification. When we represent the likelihood functions under the two hypotheses by L_0 and L_1, then the likelihood ratio (λ) can be defined in (6):

$$\lambda = \frac{\max L_0}{\max L_1} \qquad (6)$$

It is well known that in the case of a correct specification, $-2\log\lambda$ is asymptotically distributed as χ_r^2 where r is the number *or.* of restrictions not necessary for identification, of $\frac{1}{2}(p + q)$ (p + q + 1) - s (Silvey, 1970, ch. 7). One can also test a model against any model that is more general. In that case one has to estimate the models separately and calculate their χ^2 value with equation (6). The difference between the χ^2 values is again χ^2 distributed with the number of degrees of freedom equal to the difference in numbers of restrictions in the different models (Byron, 1972).

In this way one may not only test hypotheses concerning individual parameters, but also the whole theory against an alternative theory, an important advantage compared with ordinary path analysis.

LISREL

Given that the maximum likelihood estimators are best in the sense of minimum variance and unbiased for large samples, and that the characteristics of other estimators for this model are unknown, it is understandable that Jöreskog and van Thillo have developed a program which calculates maximum likelihood estimates when the appropriate assumptions are fulfilled. To define the likelihood function for this model, one has to introduce further assumptions concerning the distribution of the variables in the vector of the measured variables. The most common assumption is that the variables have a multivariate normal distribution.

If this assumption is made, the maximum likelihood estimates of the means of the observed variables are $\bar{y}$ and $\bar{x}$ which are independent of the other parameters; to find the maximum likelihood estimates of the other parameters, the likelihood function L for this model has to be maximized. This function will reach its maximum at the same values of the estimates as the function log L, which is usually preferred because of mathematical simplification. In the case of a multivariate normal distribution, log L for S is given by equation (7) (see Anderson, 1958, ch. 3):

$$\log L = -\frac{1}{2}(N - 1)[\log|\Sigma| + tr(S\Sigma^{-1})] - \frac{1}{2}(p + q)N\log(2\pi)$$

(7)

where S is the sample variance-covariance matrix of z.

In the LISREL program, however, (8) is minimized:

$$F = \log|\Sigma| + tr(S\Sigma^{-1}) - \log|S| - (p + q)$$

(8)

These two functions have the same first two terms except for a negative multiplication factor - $\frac{1}{2}(N - 1)$ in equation (7). The third term in equation (7) and the third and the fourth terms in equation (8) are constants and do not influence the point at which the extreme will be reached. Therefore, it follows that log L reaches its maximum when F reaches its minimum, and the values of the parameters are equal to the values of the maximum likelihood estimates. We cannot go into details of the program (see Jöreskog and van Thillo, 1973), but this is an overview of some basic features.

a. The researcher may specify the model he wants to use by fixing some parameters at chosen values (for example 0), and may specify that two or more parameters must have the same value. In the program, the vector θ will then be reduced to the free parameters which have to be estimated. When two parameters have to have equal values, only one parameter is kept in the vector θ.

b. The values of the parameters in θ are calculated with a numeric procedure developed by Fletcher and Powell (1963) using a program by Gruvaeus and Jöreskog (1970). The fact that the procedure is iterative means that the function is minimized step by step, calculating new values of the parameters in each step and checking whether or not the function has been sufficiently minimized by some criteria.

c. After the minimization has come to a normal end, the estimates of θ are known, and from equations like (4) the reproduced covariance matrix $\hat{\Sigma}$ can be calculated. Then the maximum of the logarithm of the likelihood function under H_0 can also be calculated (see Testing, above):

$$\max \log L_0 = -\frac{1}{2}(N - 1)[\log|\hat{\Sigma}| + \text{tr } (S\hat{\Sigma}^{-1})] - \frac{1}{2}(p + q)N\log(2\pi)$$
(9)

The maximum of the logarithm of the likelihood function under H_1 (see Testing, above) is also known (Lawley and Maxwell, 1971, p. 34):

$$\max \log L_1 = -\frac{1}{2}(N - 1)[\log|S| - (p + q)] - \frac{1}{2}(p + q)N\log(2\pi)$$
(10)

From (9) and (10), equation (11) follows:

$$-2\log\lambda = -(N - 1)\left[\log|\hat{\Sigma}| + \text{tr}(S\hat{\Sigma}^{-1}) - \log|S| - (p + q)\right]$$
(11)

When we compare this with (8), we see that $-2\log\lambda = -(N - 1)F_0$, where F_0 is the value of the function at the minimum under hypothesis H_0. Moreover, we know that $-2\log\lambda$ is distributed as χ_p^2. In this way the goodness-of-fit test is calculated from the minimum of the function.

d. As the variance-covariance matrix of the estimators is asymptotically equal to the inverse of the information matrix, LISREL estimates the standard errors and t-values by inverting the information matrix.

e. But if the information matrix is singular, which usually means that the model is not identified (Silvey, 1970, pp. 81-2) the program cannot compute the standard errors. In this case the program comments on the identifiability of this model.

f. Along with the values of the parameters, the goodness-of-fit test, the standard errors for the estimated parameters, and the t-values, the program LISREL also calculates the variance-covariance matrix of the unmeasured constructs, the residual matrix ($\hat{\Sigma}$ - S) and the standardized solution of the model (path coefficients).

ILLUSTRATION

   Zaal (1978) has studied the relationships between the variables: school results, educational background, work attitude, and intelligence. All these variables are not directly observable and not observable without measurement error. For each of them two observable indicators have been constructed. For intelligence and work attitude scores have been obtained for the kindergarten and the primary school levels while the scores for school results have been obtained at the primary school level and the scores for educational background at the kindergarten level.

   The following relationships have been hypothesized between the theoretical variables, recalling notational distinctions made above:

   (i)   The school results of a child in the primary school (η_2) are proportionally related to intelligence (η_3), to work attitude (η_1), and to the educational background of the child (ξ_2).

  (ii)   The work attitude of a child in the primary school (η_1) is proportionally related to its school results (η_2), to the work attitude it has learned in the kindergarten (ξ_1), and to its educational background (ξ_2).

 (iii)   The intellectual capacity (intelligence) in the primary school (η_3) is proportionally related to intellectual capacity in the kindergarten (ξ_3) and to educational background (ξ_2).

  (iv)   All other variables not included in the theory are assumed to have minor effects, to cancel each other out, and to be unrelated to the variables intelligence and work attitude at kindergarten level and unrelated to educational background.

   (v)   The predetermined variables may be related to each other.

   This verbal theory is presented in Figure 9.1. Following equation (1), statements (i), (ii) and (iii) can be formulated in the following matrix equation:

$$\begin{pmatrix} 1 & -\beta_{12} & 0 \\ -\beta_{21} & 1 & -\beta_{23} \\ 0 & 0 & 1 \end{pmatrix} \begin{pmatrix} \eta_1 \\ \eta_2 \\ \eta_3 \end{pmatrix} = \begin{pmatrix} \gamma_{11} & \gamma_{12} & 0 \\ 0 & \gamma_{22} & 0 \\ 0 & \gamma_{32} & \gamma_{33} \end{pmatrix} \begin{pmatrix} \xi_1 \\ \xi_2 \\ \xi_3 \end{pmatrix} + \begin{pmatrix} \zeta_1 \\ \zeta_2 \\ \zeta_3 \end{pmatrix}$$

$$(12)$$

FIGURE 9.1

A Theory of Success and Failure in School

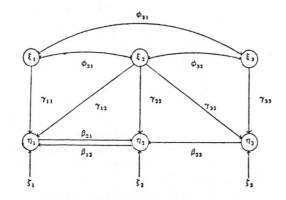

where η_1 = work attitude at the primary school
η_2 = school results at the primary school
η_3 = intellectual capacity at the primary school
ξ_1 = work attitude at the kindergarten
ξ_2 = educational background
ξ_3 = intellectual capacity at the kindergarten
β_{ij} = the effect of the jth endogenous variable on the ith endogenous variable
γ_{ij} = the effect of the jth predetermined variable on the ith endogenous variable
ϕ_{ij} = the correlation between the ith and jth predetermined variable
ζ_i = the disturbance term of the ith equation

Statement (iv) indicates that $E(\zeta) = 0$ and $E(\xi\zeta') = 0$ as usual, and that $E(\zeta\zeta') = \Psi$ is diagonal:

$$\psi = \begin{pmatrix} \psi_{11} & 0 & 0 \\ 0 & \psi_{22} & 0 \\ 0 & 0 & \psi_{33} \end{pmatrix} \tag{13}$$

Furthermore, according to (v):

$$\phi = \begin{pmatrix} \phi_{11} & \phi_{12} & \phi_{13} \\ \phi_{21} & \phi_{22} & \phi_{23} \\ \phi_{31} & \phi_{32} & \phi_{33} \end{pmatrix} \tag{14}$$

For each of the theoretical variables two indicators have been constructed. The variable "work attitude" has been measured by teachers' ratings which have been split into two categories. One category contains the rating scales with even numbers and the other contains the rating scales with odd numbers. Total scores for the two indicators have been obtained by summing over the scores on the rating scales in each category. These indicators will be denoted by y_1 and y_2 for the scores at the primary school and x_1 and x_2 at kindergarten level. Similarly, the variable "intellectual capacity" is measured by an intelligence test which has been split into two parts, the even and the odd-numbered items. The total score on each indicator is obtained by counting the number of correctly answered items. These two indicators at the primary school are denoted by y_5 and y_6 and the scores at kindergarten level are denoted by x_5 and x_6. The two indicators for "school results" are obtained from teachers' ratings of verbal and arithmetic achievement of the children. The indicators are denoted by y_3 and y_4. Finally, "educational background" is measured by education of the father and education of the mother. These indicators are denoted by x_3 and x_4.

Let us further assume the following measurement model

(vi) Each indicator is linearly related to the theoretical variable for which it is an indicator.

$E(\eta\delta) = 0 = E(\xi\epsilon)$ (vii) The measurement errors in the indicators are unrelated to the theoretical variables and to each other.

$E(\delta) = 0$ (viii) The measurement errors depend on a large number of independent variables which cancel each other out.

This measurement model is presented in Figures 9.2 and 9.3 and can be formulated in the following matrix equations:

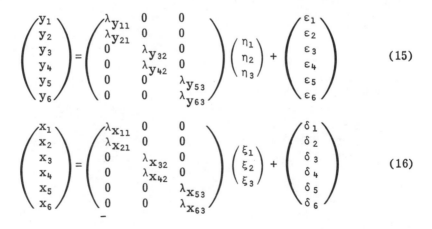

$$
\begin{pmatrix} y_1 \\ y_2 \\ y_3 \\ y_4 \\ y_5 \\ y_6 \end{pmatrix} = \begin{pmatrix} \lambda_{y11} & 0 & 0 \\ \lambda_{y21} & 0 & 0 \\ 0 & \lambda_{y32} & 0 \\ 0 & \lambda_{y42} & 0 \\ 0 & 0 & \lambda_{y53} \\ 0 & 0 & \lambda_{y63} \end{pmatrix} \begin{pmatrix} n_1 \\ n_2 \\ n_3 \end{pmatrix} + \begin{pmatrix} \varepsilon_1 \\ \varepsilon_2 \\ \varepsilon_3 \\ \varepsilon_4 \\ \varepsilon_5 \\ \varepsilon_6 \end{pmatrix} \tag{15}
$$

$$
\begin{pmatrix} x_1 \\ x_2 \\ x_3 \\ x_4 \\ x_5 \\ x_6 \end{pmatrix} = \begin{pmatrix} \lambda_{x11} & 0 & 0 \\ \lambda_{x21} & 0 & 0 \\ 0 & \lambda_{x32} & 0 \\ 0 & \lambda_{x42} & 0 \\ 0 & 0 & \lambda_{x53} \\ 0 & 0 & \lambda_{x63} \end{pmatrix} \begin{pmatrix} \xi_1 \\ \xi_2 \\ \xi_3 \end{pmatrix} + \begin{pmatrix} \delta_1 \\ \delta_2 \\ \delta_3 \\ \delta_4 \\ \delta_5 \\ \delta_6 \end{pmatrix} \tag{16}
$$

FIGURE 9.2

A Path Diagram of the Measurement Model for the Endogenous
Variables, Where η_i Represents the ith Theoretical Variable,
y_1 the ith Indicator and ε_i the Measurement Error in the ith
Indicator

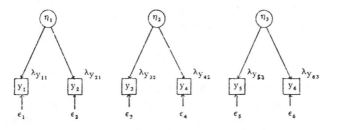

According to (vi) and (vii) $E(\varepsilon) = 0$, $E(\eta\varepsilon') = 0$, and $E(\varepsilon\varepsilon') =$
Θ_ε is diagonal. $E(\delta) = 0$, $E(\xi\delta') = 0$, and $E(\delta\delta') = \Theta_\delta$ is diagonal.

In order to establish the scale unit for the unobservable
variables, they are each arbitrarily expressed in the same scale
unit as their first observable variable. This can be done by
giving the   coefficient the value 1 <u>a priori</u>. As the tests for
intelligence and work attitude are parallel tests, it is expected
that the error variances at both points in time are the same.
Thus, the following restrictions on the parameter values can
be formulated:

FIGURE 9.3

A Path Diagram of the Measurement Model for the Predetermined
Variables, Where ξ_i Represents the ith Theoretical Variable,
x_1 the ith Indicator and δ_i the Measurement Error in the ith
Indicator

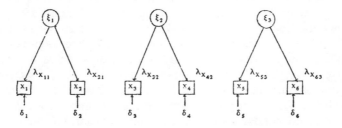

$$\lambda_{y_{11}} = \lambda_{y_{21}} = \lambda_{x_{11}} = \lambda_{x_{21}} = 1$$
$$\Theta_{\varepsilon_{11}} = \Theta_{\varepsilon_{22}} = \Theta_{\delta_{11}} = \Theta_{\delta_{22}}$$
$$\lambda_{y_{53}} = \lambda_{y_{63}} = \lambda_{x_{53}} = \lambda_{x_{63}} = 1$$
$$\Theta_{\varepsilon_{55}} = \Theta_{\varepsilon_{66}} = \Theta_{\delta_{55}} = \Theta_{\delta_{66}}$$

Such restrictions cannot be formulated for the other observed variables, as they are not parallel instruments but instead conform to the restricted factor analysis model (Jöreskog, 1969). Therefore we only state that

$$\lambda_{y_{32}} = 1 \text{ and } \lambda_{x_{32}} = 1.$$

Having formulated the theory as a special case of the general system we shall now discuss the analysis of the data of Zaal (1978), starting with this model. Zaal collected the data for 75 children at two points in time. The variance-covariance matrix for this data is presented in Table 9.1. Given this information, the program LISREL can be used to estimate the parameters of the model and to test the goodness-of-fit. For details of using LISREL, see Jöreskog and Sörbom (1978).

Table 9.2a gives the unstandardized coefficients with the standard errors and the χ^2 statistic for the goodness-of-fit of the model to Zaal's data. The goodness-of-fit test indicates that the model fits the data rather well so that the model is not rejected. Before we begin an interpretation of the estimated values of the parameters, we must be certain that the obtained values are unique. Otherwise the unidentified parameters cannot be interpreted. As above, however, the mere fact that the program could compute the standard errors of the parameters is sufficient indication that the model is identified and the results can be interpreted. Because of the small sample size, the standard errors for most parameters are relatively large. Only the parameters β_{23}, γ_{11}, γ_{32} and γ_{33} are larger than two times the standard error, while γ_{12} and γ_{22} are smaller than one standard error in absolute value. This means that the stability of the variables work attitude and intelligence are relatively large through time, while the educational climate has an effect mainly on intelligence at the second point in time, but not on the work attitude or the school results. In model 2b we have tested whether the parameters γ_{12} and γ_{22} could be fixed at zero without harming the fit. As indicated in Table 9.2b, this simplification indeed did not harm the fit of the model. The analysis is indecisive concerning the reciprocal causation between work attitude and school results. Neither parameter is very large in relation to its standard error. As they are not significant,

TABLE 9.1

The Sample Variance-Covariance Matrix for the Data of Zaal (1978)

	1	2	3	4	5	6
S_{YY}						
1	146.9					
2	126.2	150.4				
3	8.2	7.0	1.6			
4	6.7	6.1	1.3	1.7		
5	18.8	16.5	3.0	4.0	42.8	
6	24.4	19.1	3.4	3.8	32.7	40.6
S_{XY}						
1	69.6	58.6	4.9	4.2	28.2	38.2
2	53.1	47.1	3.5	3.1	25.3	31.0
3	0.6	-1.1	0.2	0.3	2.3	1.8
4	-0.0	-0.6	0.0	0.1	1.7	1.4
5	19.6	18.6	2.7	2.6	24.3	19.5
6	21.5	19.2	2.6	2.4	27.5	23.7
S_{XX}						
1	153.5					
2	122.5	129.1				
3	0.3	-1.6	1.5			
4	0.2	-0.1	0.5	0.7		
5	36.2	38.6	0.3	0.6	47.0	
6	44.0	41.4	0.4	0.0	37.4	50.6

one could fix them both on zero, but in fact then the model has a bad fit. An alternative would be to fix one of the two on zero, but the problem is that there is no theoretical or statistical evidence to determine such a choice. Only a new study with a larger sample could clarify this problem. In the meantime model 2b will serve as the final solution for this data.

We have shown that the program LISREL can estimate the parameters of a model with reciprocal causation and measurement error. Previously it would have been difficult to tackle a problem like this. LISREL does not only give the estimates, the standard errors and a goodness-of-fit test, but it also gives an indication of the identifiability of the model, an indication otherwise rather difficult to obtain. However, the results are

TABLE 9.2

Results of the Analysis of Zaal's Data

Parameters	(a) Original Model		(b) Corrected Model	
	Estimates	Standard Errors	Estimates	Standard Errors
λy_{42}	1.009	.140	1.006	.140
λx_{42}	.718	.237	.695	.232
β_{12}	3.045	2.123	2.388	1.985
β_{21}	.022	.022	.027	.019
β_{23}	.098	.031	.092	.025
γ_{11}	.367	.125	.388	.125
γ_{12}	-.957	1.588	.000[a]	—
γ_{22}	-.035	.182	.000[a]	—
γ_{32}	2.262	.813	2.225	.806
γ_{33}	.608	.095	.609	.095
ϕ_{11}	121.634	22.179	121.630	22.178
ϕ_{21}	-.032	1.448	-.073	1.472
ϕ_{22}	.828	.343	.854	.353
ϕ_{31}	41.023	10.155	41.036	10.156
ϕ_{32}	.445	.838	.443	.852
ϕ_{33}	37.897	7.249	37.906	7.250
ψ_{11}	78.979	17.566	82.367	18.717
ψ_{22}	.762	.228	0.742	.210
ψ_{33}	12.649	3.632	12.639	3.631
$\theta_{\varepsilon 11}$[b]	20.620	2.446	20.620	2.446
$\theta_{\varepsilon 33}$	.329	.154	.326	.155
$\theta_{\varepsilon 44}$	.424	.163	.426	.163
$\theta_{\delta 11}$	10.301	1.215	10.298	1.215
$\theta_{\delta 33}$	.698	.282	.672	.290
$\theta_{\delta 44}$	.296	.141	.310	.139

$\chi^2 = 50.784$ df = 53 $\chi^2 = 51.154$ df = 51

Pr. = .560   Pr. = .62

[a]Fixed value.
[b]The values of the constraint parameters are identical.

not always as simple as in this case. If a model is faulty, it is sometimes difficult to determine in which part of the model the problems arise. Recent studies have concentrated on such points. Furthermore, it is doubtful whether the assumptions made in LISREL are always realistic. More research is needed to study the effects of deviations from the assumptions.

RECENT DEVELOPMENTS

Since 1973 there have been several new developments, along with considerable simplification and improvement of the program LISREL (Jöreskog and Sörbom, 1977). The first is the extension of the program to include simultaneous analysis of several data sets. This development is important as one is usually interested in models which hold for several populations or for various points in time. In LISREL IV (Jöreskog and Sörbom, 1978) it is possible to test such hypotheses and less restrictive hypotheses in a very efficient way. This extension is straightforward, as the joint likelihood function for all sample variance-covariance matrices in the case of exclusive populations is simply the product of the individual likelihood functions. More detailed information concerning the extension can be found for a special case in Jöreskog (1971) and for the general system in Jöreskog and Sörbom (1978). Sörbom (1974) has shown how the means of the latent variables can also be estimated simultaneously, making comparisons between groups easier.

The second development concerns the correction of rejected models. The correction of a rejected model is essentially a theoretical problem. But when one has introduced all available theoretical knowledge into a model without reaching an acceptable fit to the data, one can no longer rely on theoretical knowledge. In such cases it seems useful to have a procedure which suggests corrections to the model. A procedure which minimizes the number of corrections is desirable as this would yield a new model resembling as much as possible the (existing) knowledge in the field. For this purpose Sörbom (1975) proposes using the first order partial derivatives of the function which is minimized in LISREL. When the model is correct, all derivatives must be zero. If the model is incorrect, large deviations from zero can occur for some restricted parameters. Sörbom suggests relaxing the restriction on the parameter having the highest derivative in absolute value. One can relax the model step by step until an acceptable fit has been obtained. In LISREL IV the derivatives can be obtained and this procedure applied. However,

Saris et al. (1979) show that this procedure is not always optimal as it does not take the interdependence of different estimates into account. A weighted procedure is suggested leading in many cases to simpler models. Although this procedure seems preferable one must realize that no guarantee can be given for the correctness of the derived models. Theoretical acceptability of the new models therefore remains a relevant criterion and only tests on new data can give an estimate of the model's validity.

A third development worth mentioning is the formulation of an estimation procedure for the scores of respondents on latent variables which preserves the structure existing among these variables. Such a procedure has been suggested by Anderson and Rubin (1956) and formulated by Saris et al. (1977). As of now LISREL is the only general program for the estimation of models with unmeasured variables. However, if one has estimates for the latent variables, one can just as well use econometric procedures like two-stage least squares. These have the advantage that they do not spread specification errors throughout the whole model, unlike full information procedures like LISREL. Such procedures may be advantageous in the early stages of research (Heise, 1974), provided that the estimation of the scores on the latent variables can be done without disturbing their variance-covariance matrix, a condition fulfilled by the new procedure.

Other relevant developments with respect to linear structural equation models concern measurement problems. The procedures discussed here assume at least interval measurement level, and this assumption cannot always be met in social science research. Efforts are being made to extend the LISREL model to lower-level variables. Although this problem has been solved in principle (Muthén, 1977), the solution can only be applied to models with a small number of observable variables because of its complexity. Further research is needed in this area.

On the other hand it has been found that measurement procedures from psycho physics can be applied in the social sciences, leading to ratio scales. An overview of this research is given by Hamblin (1973). Tests of these procedures in large-scale survey research have been reported by Rainwater (1973) and Saris et al. (1977). Although the results of these tests are, up to now, very promising, more research is needed in order to show how these methods can be efficiently used. In addition to measurement problems, attention must be directed toward other assumptions, for instance those of normality and linearity, and the consequences of deviations from them. Research on these points has only recently begun. The development of the general system and the program LISREL has been a major breakthrough in the field of linear structural equation models, but a lot of work is still left to be done.

REFERENCES

Anderson, T. W. (1958). An Introduction to Multivariate Statistical Analysis. New York: Wiley.

Anderson, T. W. and Rubin, H. (1956). 'Statistical inference in factor analysis', Proceedings of the 3rd Berkeley Symposium V:111-150.

Blalock, H. M. (1961). 'Correlation and causality: The multivariate case', Social Forces 39:246-251.

Blalock, H. M. (1964). Causal Inferences in Non-experimental Research. Chapel Hill: University of North Carolina Press.

Blalock, H. M. (1971). Causal Models in the Social Sciences. Chicago: Aldine-Atherton.

Boudon, R. (1968). 'A new look at correlation analysis', pp. 199-235 in Blalock, H. M. and Blalock, A. B. (eds.) Methodology in Social Research. New York: McGraw-Hill.

Byron, R. P. (1972). 'Testing for misspecification in econometric systems using full information maximum likelihood estimation', International Economic Review 13:1972.

Duncan, O. D. (1966). 'Path analysis: sociological examples', American Journal of Sociology 72:1-16.

Duncan, O. D., Haller, A. O. and Portes, A. (1968). 'Peer influences on aspirations; a reinterpretation', American Journal of Sociology 74:119-137.

Duncan, O. D. (1972). 'Unmeasured variables in linear models for panel analysis', in H. L. Costner (ed.) Sociological Methodology 1972. San Francisco: Jossey-Bass.

Duncan, O. D. (1975). Introduction to Structural Equation Models. New York: Academic Press.

Fletcher, R. and Powell, M. J. D. (1968). 'A rapidly convergent descent method for minimization', Computer Journal 2:
163-168.

Geraci, V. J. (1977). 'Identification of simultaneous equation
models with measurement error', in Aigner, D. J. and Goldberger, A. S. (eds.) Latent Variables in Socio-economic
Models. Amsterdam: North-Holland.

Geraci, V. J. and Goldberger, A. S. (1971). Simultaneity and
Measurement Error. University of Wisconsin, Social Systems
Research Institute. W. Papers EME 7125.

Goldberger, A. S. (1964). Econometric Theory. New York:
Wiley.

Goldberger, A. S. and Duncan, O. D. (1973). Structural
Equation Models in the Social Sciences. New York: Seminar
Press.

Hamblin, R. L. (1973). 'Social attitudes: Magnitude measurement and theory', in Blalock, H. M. Measurement in Social
Science. London: Macmillan.

Harman, H. H. (1968). Modern Factor Analysis. Chicago:
University of Chicago Press.

Hauser, R. M. and Goldberger, A. S. (1971). 'The treatment
of unobservable variables in path analysis', pp. 81-117 in
Costner, H. L. (ed.) Sociological Methodology 1971. San
Francisco: Jossey-Bass.

Heise, D. R. (1969). 'Separating reliability and stability in
test-retest correlations', American Sociological Review 34:
93-101.

Heise, D. R. (1974). 'Some issues in sociological measurement',
pp. 1-17 in Costner, H. L. (ed.) Sociological Methodology
1974. San Francisco: Jossey-Bass.

Johnston, J. J. (1972). Econometric Methods. 2nd edition.
New York: McGraw-Hill.

Jöreskog, K. G. (1969). 'A general approach to confirmatory
maximum likelihood factor analysis', Psychometrika 34:183-201.

Jöreskog, K. G. (1971). 'Simultaneous factor analysis in several populations', Psychometrika 36:409-426.

Jöreskog, K. G. (1973). 'A general method for estimating a linear structural equation system', in Goldberger, A. S. and Duncan, O. D. (eds.) Structural Equation Models in the Social Sciences. New York: Seminar Press.

Jöreskog, K. G. (1978). An Econometric Model for Multivariate Panel Data. Research Report 78-1 of the Dept. of Statistics, University of Uppsala.

Jöreskog, K. G. (1979). 'Statistical estimation of structural models in longitudinal developmental investigations', in Nesselroade, J. R. and Baltes, P. B. (eds.) Longitudinal Research in the Behavioral Sciences: Design and Analysis. (forthcoming).

Jöreskog, K. G. and Van Thillo, M. (1973). LISREL: a general computer program for estimating linear structural equation systems involving multiple indicators of measurement variables. Dept. of Statistics, University of Uppsala.

Jöreskog, K. G. and Sörbom, D. (1977). 'Statistical models and methods for analysis of longitudinal data', in Aigner, D. J. and Goldberger, A. S. Latent Variables in Socio-economic Models. Amsterdam: North-Holland.

Jöreskog, K. G. and Sörbom, D. (1978). LISREL IV: A General Computer Program for Estimation of Linear Structural Equation Systems by Maximum Likelihood Methods. Chicago: International Educational Services.

Lawley, D. N. and Maxwell, A. E. (1971). Factor Analysis as a Statistical Method. London: Butterworth.

Lord, F. M. and Novick, M. R. (1968). Statistical Theories of Mental Test Scores. London: Addison-Wesley.

Mulaik, S. A. (1972). The Foundations of Factor Analysis. New York: McGraw-Hill.

Nunnally, J. C. (1967). Psychometric Theory. London: McGraw-Hill.

Muthén, B. (1977). Statistical methodology for structural equation models involving latent variables with dichotomous indicators. Research Report 77-7, Uppsala, Dept. of Statistics, 1977.

Rainwater, L. (1971). What Money Buys: Inequality and the Social Meaning of Income. New York: Basic Books.

Saris, W. E., Bruinsma, C., Schoots, W. and Vermeulen, C. (1977). 'The use of magnitude estimation in large scale survey research', in Mens en Maatschappij: 369-395.

Saris, W. E., De Pijper, W. M. and Mulder, J. (1978). 'Optimal procedures for estimation of factor scores', Sociological Methods and Research 7: 85-105.

Saris, W. E., De Pijper, W. M. and Zegwaart, P. (1979). 'Detection of specification errors in linear structural equation models', in Schuessler, K. (ed.) Sociological Methodology 1979. San Francisco: Jossey-Bass.

Saris, W. E. (1978). The use of linear structural equation models in non-experimental research. Report of the department of methodology, Free University, Amsterdam.

Silvey, S. D. (1970). Statistical Inference. London: Penguin Books.

Sörbom, D. (1974). 'A general method for studying differences in factor means and factor structure between groups', British Journal of Mathematical and Statistical Psychology 27: 229-239.

Sörbom, D. (1975). 'Detection of correlated errors in longitudinal data', British Journal of Mathematical and Statistical Psychology 28: 138-151.

Simon, H. A. (1954). 'Spurious correlation: a causal interpretation', Journal of the American Statistical Association 49: 467-479.

Werts, C. F., Linn, R. L. and Jöreskog, K. G. (1973). 'Identification and estimation in path analysis with unmeasured variables', American Journal of Sociology 78: 1469-1484.

Wiley, D. E. (1973). 'The identification problem for structural equation models with unmeasured variables', in Goldberger, A. S. and Duncan, O. D. (eds.) Structural Equation Models in the Social Sciences. New York: Seminar Press.

Zaal, J. (1978). Sociaal emotioneel gedrag van kinderen van 5-7 jaar, beoordeeld door de leerkrachten. Unpublished doctoral dissertation.

10

Analysis of
Covariance Structures

Karl G. Jöreskog

INTRODUCTION

Analysis of covariance structures is the common term
for a number of techniques for analyzing multivariate data in
order to detect and assess latent (unobserved) sources of
variation and covariation in the observed measurements. The
techniques of covariance structure analysis are general and
flexible in that they can handle many types of covariance
structures useful especially in the behavioral and social
sciences. Although these techniques can be used for exploratory analysis they have been most successfully applied to confirmatory analysis where the type of covariance structure is
specified in advance. A covariance structure of a specified
kind may arise because of a specified substantive theory or
hypothesis, a given classificatory design for the measures,
known experimental conditions or because of results from
previous studies based on extensive data. Sometimes the
observed variables are ordered through time, as in longitudinal
studies, or according to linear or circular patterns, as in
Guttman's (1954) simplex and circumplex models or according
to a given causal scheme, as in path analysis.

This chapter reviews most of the important models and
techniques for analysis of covariance structures and illustrates

Reprinted from Scandinavian Journal of Statistics, Vol. 8
(1981), pp. 65-83. Reprinted by permission.

them with a few simple examples. The examples are introduced in the next section together with the main types of covariance structures. General covariance and correlation structures are defined in the third section. Special cases of general covariance structures are: the ACOVS-model (Jöreskog, 1970a, 1973a, 1974) and the LISREL-model (Jöreskog, 1973b, 1977; Jöreskog & Sörbom, 1978) and these are briefly described also. Approaches to the statistical problems of identification, estimation and testing are also considered in the third section. The analysis of the examples are continued in the fourth section. The last section discusses generalizations which permit simultaneous analysis of mean and covariance structures and simultaneous analysis of data from several populations.

This study draws on various material published previously by the author, in particular Jöreskog (1973a, 1974, 1978). Other important related material is contained in Bock & Bargmann (1966), Browne (1974, 1977), McDonald (1974, 1975, 1978) and Bentler & Weeks (1981).

Some Types of Covariance Structures

Variance and Covariance Components

Several authors (Bock, 1960; Bock & Bargmann, 1966; Wiley et al., 1973) have considered covariance structure analysis as an approach to study differences in test performance when the tests have been constructed by assigning items or subtests according to objective features of content or format to subclasses of a factorial or hierarchical classification.

Bock (1960) suggested that the scores of N subjects on a set of tests classified in a 2^n factorial design may be viewed as data from an $N \times 2^n$ experimental design, where the subjects represent a random mode of classification and the tests represent n fixed modes of classification. Bock pointed out that conventional mixed-model analysis of variance gives useful information about the psychometric properties of the tests. In particular, the presence of non-zero variance components for the random mode of classification and for the interaction of the random and fixed modes of classification provides information about the number of dimensions in which the tests are able to discriminate among subjects. The relative size of these components measures the power of the tests to discriminate among subjects along the respective dimensions.

The following example was given in an unpublished paper by Browne (1970).

Example 1: The Rod and Frame (RF) test is used as a measure of field dependence. A subject is seated in a darkened room on a chair which may be tilted to the left or to the right. In front of him is a luminous rod located in a luminous square frame. The chair, frame, and rod are tilted to prespecified positions. By operating push buttons connected to an electric motor the subject is to move the rod to the vertical position. The score on the trial is the angle of the rod from the vertical. This can assume positive and negative values. Each subject undergoes 12 trials. The last two columns of the design matrix A below give initial positions of the frame and chair for each trial. A value of +1 denotes that the position of the frame or chair was at +28° from the vertical, a value of -1 denotes that the angle was -28° and a value of 0 denotes that the initial position was vertical.

Table 10.1 shows a covariance matrix between trials of the RF-test obtained from a sample of 107 eighteen year old males.

One would like to estimate the variance components associated with the general bias, frame effect, chair effect and error.

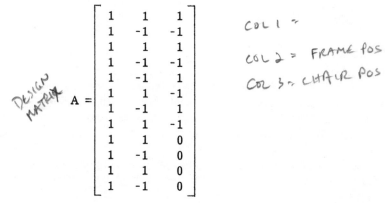

DESIGN MATRIX

$$A = \begin{bmatrix} 1 & 1 & 1 \\ 1 & -1 & -1 \\ 1 & 1 & 1 \\ 1 & -1 & -1 \\ 1 & -1 & 1 \\ 1 & 1 & -1 \\ 1 & -1 & 1 \\ 1 & 1 & -1 \\ 1 & 1 & 0 \\ 1 & -1 & 0 \\ 1 & 1 & 0 \\ 1 & -1 & 0 \end{bmatrix}$$

COL 1 =

COL 2 = FRAME POS

COL 3 = CHAIR POS

Let a, b and c be uncorrelated random components associated with general bias, frame effect and chair effect, respectively, and let e denote an error component uncorrelated with a, b and c and uncorrelated over trials. Let

$$u'_\nu = (a_\nu, b_\nu, c_\nu)$$

be the values of a, b and c for subject ν. Then the scores on the twelve trials for subject ν is

$$x_\nu = Au_\nu + e_\nu$$

$A = \xi$?

$u_\nu = \wedge_x$

TABLE 10.1

Inter Trial Covariance Matrix

	1	2	3	4	5	6	7	8	9	10	11	12
1	51.6											
2	-27.7	72.1										
3	38.9	-41.1	69.9									
4	-36.4	40.7	-39.1	75.8								
5	13.8	- 5.2	17.9	1.9	84.8							
6	-13.6	10.9	9.5	17.8	-37.4	91.1						
7	21.5	- 9.4	8.5	-13.1	59.7	-54.4	79.9					
8	-12.8	17.2	- 3.1	22.0	-43.3	52.7	-49.9	87.2				
9	11.0	- 8.9	19.2	-11.2	-12.6	21.9	-10.6	17.5	27.6			
10	- 4.5	10.2	- 7.6	12.7	20.4	-11.5	16.5	-14.8	- 8.8	19.9		
11	9.2	- 0.3	18.9	-13.6	- 3.9	19.0	- 8.3	13.1	17.7	- 2.8	27.3	
12	- 3.7	7.5	- 4.5	12.8	19.9	- 8.8	15.5	- 8.6	- 5.4	13.3	- 1.0	16.0

with covariance matrix

$$\Sigma = A\Phi A' + \sigma_e^2 I, \tag{1}$$

where

$$\Phi = \text{diag} \ (\sigma_a^2, \ \sigma_b^2, \ \sigma_c^2)$$

Equation (1) shows that all the 78 variances and covariances in Σ are <u>linear</u> functions of the four parameters σ_a^2, σ_b^2, σ_c^2 and σ_e^2. To see this explicitly, consider the covariance matrix generated by trials 1, 2, 5 and 9:

$$\Sigma = \begin{bmatrix} \sigma_a^2 + \sigma_b^2 + \sigma_c^2 + \sigma_e^2 & & & \\ \sigma_a^2 - \sigma_b^2 - \sigma_c^2 & \sigma_a^2 + \sigma_b^2 + \sigma_c^2 + \sigma_e^2 & & \\ \sigma_a^2 - \sigma_b^2 + \sigma_c^2 & \sigma_a^2 + \sigma_b^2 - \sigma_c^2 & \sigma_a^2 + \sigma_b^2 + \sigma_c^2 + \sigma_e^2 & \\ \sigma_a^2 + \sigma_b^2 & \sigma_a^2 - \sigma_b^2 & \sigma_a^2 - \sigma_b^2 & \sigma_a^2 + \sigma_b^2 + \sigma_e^2 \end{bmatrix}$$

This is an example of a <u>linear covariance structure</u>. If this structure holds, the four parameters can be solved in terms of the elements of Σ. For example, $\sigma_a^2 = \frac{1}{2}(\sigma_{41} + \sigma_{42})$, $\sigma_b^2 = \frac{1}{2}(\sigma_{41} + \sigma_{42} - \sigma_{21} - \sigma_{31})$, $\sigma_c^2 = \sigma_{31} - \sigma_{42}$, etc. There are many ways in which the four parameters can be solved in terms of the σ's. If the ten equations are consistent, however, all solutions are identical. In this case the parameters are <u>over-identified</u>.

Consider the covariance structure generated by the first four rows of A:

$$\Sigma = \begin{bmatrix} \sigma_a^2 + \sigma_b^2 + \sigma_c^2 + \sigma_e^2 & & & \\ \sigma_a^2 - \sigma_b^2 - \sigma_c^2 & \sigma_a^2 + \sigma_b^2 + \sigma_c^2 + \sigma_e^2 & & \\ \sigma_a^2 + \sigma_b^2 + \sigma_c^2 & \sigma_a^2 - \sigma_b^2 - \sigma_c^2 & \sigma_a^2 + \sigma_b^2 + \sigma_c^2 + \sigma_e^2 & \\ \sigma_a^2 - \sigma_b^2 - \sigma_c^2 & \sigma_a^2 + \sigma_b^2 + \sigma_c^2 & \sigma_a^2 - \sigma_b^2 - \sigma_c^2 & \sigma_a^2 + \sigma_b^2 + \sigma_c^2 + \sigma_e^2 \end{bmatrix}$$

In this case we can solve for $\sigma_a^2 = \frac{1}{2}(\sigma_{21} + \sigma_{31})$, say, but it is impossible to solve for σ_b^2 and σ_c^2 separately. Only the sum $\sigma_b^2 + \sigma_c^2$ is identified. This is an example of a <u>non-identified model</u> in which some parameters are underidentified and others are not. The reason for this is that the matrix A has rank 2 and not rank 3 as in the previous case. The example will be continued in the fourth section of this chapter.

In general, if A is of order $p \times r$ and of rank k, one may choose k independent linear functions, each one linearly dependent on the rows of A and estimate the covariance matrix of these functions. It is customary to choose linear combinations that are mutually uncorrelated but this is not necessary. Let L be the matrix of coefficients of the chosen linear functions and let K be any matrix such that $A = KL$. For example, K may be obtained from

$$K = AL'(LL')^{-1}.$$

The model may then be reparameterized to full rank by defining $u^* = Lu$. We then have $x = Au + e + KLu + e = Ku^* + e$. The covariance matrix of x is represented as

$$\Sigma = K\Phi^*K' + \Psi$$

where Φ^*, the covariance matrix of u^* is not necessarily diagonal and Ψ is the diagonal covariance matrix of e. The latter may be taken to be homogenous, if desired.

The above model assumes that all measurements are on the same scale. Wiley, Schmidt and Bramble (1973) suggested the study of a general class of components of covariance models which would allow different variables to be on different scales. The covariance matrix Σ will then be of the form

$$\Sigma = \Delta A\Phi A'\Delta + \Theta \text{ or } \Sigma = \Delta(A\Phi A' + \Psi)\Delta \qquad (2a\text{-}b)$$

The matrix $A(p \times k)$ is assumed to be known and gives the co-efficient of the linear functions connecting the manifest and latent variables, Δ is a $p \times p$ diagonal matrix of unknown scale factors, Φ is the $k \times k$ symmetric and positive definite covariance matrix of the latent variables and Ψ and Θ are $p \times p$ diagonal matrices of error variances.

Within this class of models eight different special cases are of interest. These are generated by the combination of the following set of conditions:

on Δ: $\begin{cases} \Delta = I \\ \Delta \neq I \end{cases}$

on Φ: $\begin{cases} \Phi \text{ is diagonal} \\ \Phi \text{ is not diagonal} \end{cases}$

on Ψ or Θ: $\begin{cases} \Psi \text{ or } \Theta = \sigma^2 I \\ \Psi \text{ or } \Theta \text{ general diagonal.} \end{cases}$

The classical formulation of the mixed model and its generalizations assume that $\Delta = I$. This is appropriate if the observed
variables are in the same metric as for example when the observed
variables represent physical measurements, time to criterion
measures, reaction times or items similarly scaled such as seman-
tic differential responses. However, if the observed variables
are measured in different metrics then the classical model would
not fit. In such cases the inclusion of Δ in the model as a general
diagonal matrix of scaling factors would provide a useful alternative specification. It should be pointed out that the elements
of Δ do not have to be related to the variances of the variables.

The classical components of variance model assume that Φ
is diagonal. However, there are usually no substantive reasons
for assuming this.

The two conditions on Ψ or Θ correspond to homogeneous
and heterogeneous error variances. If the variables are in the
same metric and if the measurement situation is sufficiently
similar from variable to variable then it would seem reasonable
to hypothesize that the variances of the errors of measurement
ought to be homogeneous, i.e., in (2a) we take $\Delta = I$ and
$\Theta = \sigma^2 I$.

If, on the other hand, the scale of measurement is the same
but the measurement situation from variable to variable is different enough to generate different kinds of error structures, then
the variances of the errors of measurement might differ systematically from variable to variable. For this situation it would seem
best to take $\Delta = I$ but leave Θ free in (2a). If the manifest
variables were in different metrics then clearly the error variances in the observed metric will most likely be heterogeneous.
One useful hypothesis to test in this context would be that the
standard deviations of the errors of measurement are proportional
to the rescaling factors. This would correspond to taking $\Psi = \sigma^2 I$ in (2b). When both Δ and Ψ are free, (2a) and (2b) are
equivalent.

Measurement Models

Most measurements employed in the behavioral and social
sciences contain sizeable errors of measurements and any adequate theory or model must take this fact into account. Of
particular importance is the study of congeneric measurements,
i.e., those measurements that are assumed to measure the same
thing.

Classical test theory (Lord & Novick, 1968) assumes that
a test score x is the sum of a true score τ and an error score e,
where e and τ are uncorrelated. A set of test scores $x_1, \ldots, x_p$

with true scores $\tau_1, \ldots, \tau_p$ is said to be congeneric if every pair of true scores τ_i and τ_j have unit correlation. Such a set of test scores can be represented as

$$x = \mu + \beta\tau + e,$$

where $x' = (x_1, \ldots, x_p)$, $\beta' = (\beta_1, \ldots, \beta_p)$ is a vector of regression coefficients, $e' = (e_1, \ldots, e_p)$ is the vector of error scores, μ is the mean vector of x and τ is a true score, for convenience scaled to zero mean and unit variance. The elements of x, e and τ are regarded as random variables for a population of examinees. Let $\theta_1, \ldots, \theta_p$ be the variances of $e_1, \ldots, e_p$, respectively, i.e., the error variances. The corresponding true score variances are $\beta_1^2, \ldots, \beta_p^2$. One important problem is that of estimating these quantities. The covariance matrix of x is

$$\Sigma = \beta\beta' + \Theta, \tag{3}$$

where $\Theta = \text{diag}(\theta_1, \ldots, \theta_p)$.

Parallel tests and tau-equivalent tests, in the sense of Lord & Novick (1968), are special cases of congeneric tests. Parallel tests have equal true score variances and equal error variances, i.e.,

$$\beta_1^2 = \ldots = \beta_p^2, \quad \theta_1 = \ldots = \theta_p.$$

Tau-equivalent tests have equal true score variances but possibly different error variances.

Parallel and tau-equivalent tests are homogenous in the sense that all covariances between pairs of test scores are equal. For parallel tests the variances are also equal. Scores on such tests are directly comparable, i.e., they represent measurements on the same scale. For tests composed of binary items this can hold only if the tests have the same number of items and are administered under the same time limits. Congeneric tests, on the other hand, need not satisfy such strong restrictions. They need not even be tests consisting of items but can be ratings, for example, or even measurements produced by different measuring instruments.

The previous model generalizes immediately to several sets of congeneric test scores. If there are q sets of such tests, with $m_1, m_2, \ldots, m_q$ tests respectively, we write $x' = (x_1', x_2', \ldots, x_q')$ where x_g', $g = 1, 2, \ldots, q$ is the vector of observed scores for the gth set. Associated with the vector

m = # of tests in each set

x_g there is a true score τ_g and vectors μ_g and β_g defined as before so that

$$x_q = \mu_g + \beta_g \tau_g + e_g.$$

As before we may, without loss of generality, assume that τ_g is scaled to zero mean and unit variance. If the different true scores $\tau_1, \tau_2, \ldots, \tau_q$ are all mutually uncorrelated, then each set of tests can be analyzed separately. However, in most cases these true scores correlate with each other and an overall analysis of the entire set of tests must be made. Let $p = m_1 + m_2 + \ldots m_q$ be the total number of tests. Then x is of order p. Let μ be the mean vector of x, and let e be the vector of error scores. Furthermore, let

$$\tau' = (\tau_1, \tau_2, \ldots, \tau_q)$$

and let B be the matrix of order p × q, paritioned as

$$B = \begin{bmatrix} \beta_1 & 0 & \cdots & 0 \\ 0 & \beta_2 & \cdots & 0 \\ \cdot & \cdot & \cdots & \cdot \\ \cdot & \cdot & \cdots & \cdot \\ 0 & 0 & & \beta_q \end{bmatrix}.$$

Then x is represented as

$$x = \mu + B\tau + e.$$

Let Γ be the correlation matrix of τ. Then the covariance matrix Σ of x is

$$\Sigma = B\Gamma B' + \Theta, \tag{4}$$

where Θ is a diagonal matrix of order p containing the error variances.

The correlation coefficient corrected for attenuation between two tests x and y is the correlation between their true scores. If, on the basis of a sample of examinees, the disattenuated correlation is near unity, the experimenter concludes that the two tests are measuring the same trait.

The following example is based on some data from Lord (1957).

Example 2: Two tests x_1 and x_2 are 15-item vocabulary tests administered under liberal time limits. Two other tests y_1 and

TABLE 10.2

Lord's Vocabulary Test Data Covariance Matrix (N = 649)

	x_1	x_2	y_1	y_2
x_1	86.3979			
x_2	57.7751	86.2632		
y_1	56.8651	59.3177	97.2850	
y_2	58.8986	59.6683	73.8201	97.8192

and y_2 are highly speeded 75-item vocabulary tests. The covariance matrix is given in Table 10.2. One would like to
estimate the disattenuated correlation between x and y and to
test whether this is one. One would also like to test whether
the two pairs of tests are parallel.

We set up the following measurement model

$$\begin{pmatrix} x_1 \\ x_2 \\ y_1 \\ y_2 \end{pmatrix} = \begin{pmatrix} \beta_1 & 0 \\ \beta_2 & 0 \\ 0 & \beta_3 \\ 0 & \beta_4 \end{pmatrix} \begin{pmatrix} \tau_x \\ \tau_y \end{pmatrix} + \begin{pmatrix} e_1 \\ e_2 \\ e_3 \\ e_4 \end{pmatrix}$$

with covariance matrix

$$\Sigma = \begin{pmatrix} \beta_1 & 0 \\ \beta_2 & 0 \\ 0 & \beta_3 \\ 0 & \beta_4 \end{pmatrix} \begin{pmatrix} 1 & \rho \\ \rho & 1 \end{pmatrix} \begin{pmatrix} \beta_1 & \beta_2 & 0 & 0 \\ 0 & 0 & \beta_3 & \beta_4 \end{pmatrix}$$

$$+ \begin{pmatrix} \theta_1 & 0 & 0 & 0 \\ 0 & \theta_2 & 0 & 0 \\ 0 & 0 & \theta_3 & 0 \\ 0 & 0 & 0 & \theta_4 \end{pmatrix}$$

$$= \begin{bmatrix} \beta_1^2 + \theta_1 \\ \beta_1\beta_2 & \beta_2^2 + \theta_2 \\ \beta_1\beta_3\rho & \beta_2\beta_3\rho & \beta_3^2 + \theta_3 \\ \beta_1\beta_4\rho & \beta_2\beta_4\rho & \beta_3\beta_4 & \beta_4^2 + \theta_4 \end{bmatrix} \qquad (5)$$

This is an example of a <u>non-linear covariance structure</u> in which the ten variances and covariances of the observed variables are non-linear functions of nine parameters. Each of these nine parameters are identified in terms of the σ's, except possibly for the sign in some of them as can easily be verified. For example, $\beta_1^2 = (\sigma_{21}\sigma_{31}/\sigma_{32})$, $\varrho^2 = (\sigma_{31}\sigma_{42}/\sigma_{21}\sigma_{43})$, $\theta_1 = \sigma_{11} - \beta_1^2$, etc.

In this model, x_1 and x_2 are congeneric measures of τ_x and y_1 and y_2 are congeneric measures of τ_y. The disattenuated correlation ϱ is the correlation between τ_x and τ_y. To analyze the data one can set up the four hypotheses:

H_1: $\beta_1 = \beta_2$, $\beta_3 = \beta_4$, $\theta_1 = \theta_2$, $\theta_3 = \theta_4$, $\varrho = 1$

H_2: $\beta_1 = \beta_2$, $\beta_3 = \beta_4$, $\theta_1 = \theta_2$, $\theta_3 = \theta_4$

H_3: $\sigma = 1$

H_4: β_1, β_2, β_3, β_4, θ_1, θ_2, θ_3, θ_4, and ϱ unconstrained

and estimate the model under each of these. Under hypotheses H_1, H_2 and H_3 the model involves <u>equality constraints</u> imposed on the parameters of the base model H_4. The analysis of the example will be continued in the fourth section of this chapter.

Path Analysis Models

Path analysis, due to Wright (1934), is a technique to assess the direct causal contribution of one variable to another in non-experimental investigations. The problem in general is that of estimating the parameters of a set of linear structural equations representing the cause and effect relationships hypothesized by the investigator. Traditionally the variables in the structural equation system were directly observed variables but recently several models have been studied which involve hypothetical constructs, i.e., latent variables, which, while not directly observed, have operational implications for relationships among observable variables (see, e.g., Werts & Linn, 1970; Hauser & Goldberger, 1971; Jöreskog & Goldberger, 1975). In some models, the observed variables appear only as effects (indicators) of the hypothetical constructs, while in others, the observed variables appear as causes or as both causes and effects of latent variables.

Suppose that two variables are used on two occasions, i.e., in a two-wave longitudinal design. Assume that the two variables measure the same latent variable η on two different occasions,

y_1 and y_2 measure η_1 on the first occasion and y_3 and y_4 measure η_2 on the second occasion.

The equations defining the measurement relations are

$$y_1 = \eta_1 + \varepsilon_1$$

$$y_2 = \lambda_1 \eta_1 + \varepsilon_2$$

$$y_3 = \eta_2 + \varepsilon_3$$

$$y_4 = \lambda_2 \eta_2 + \varepsilon_4$$

The main interest is in the stability of η over time. This can be studied by means of the structural equation

$$\eta_2 = \beta \eta_1 + \zeta,$$

the regression of η_2 on η_1. In particular, one is interested in whether β is close to one and ζ is small.

Let Ω be the covariance matrix of (η_1, η_2) and let Θ be the covariance matrix of $(\varepsilon_1, \varepsilon_2, \varepsilon_3, \varepsilon_4)$. If all the ε's are uncorrelated so that Θ is diagonal, the covariance matrix of (y_1, y_2, y_3, y_4) is

$$\Sigma = \begin{bmatrix} \omega_{11} + \theta_{11} & & & \\ \lambda_1 \omega_{11} & \lambda_1^2 \omega_{11} + \theta_{22} & & \\ \omega_{21} & \lambda_1 \omega_{21} & \omega_{22} + \theta_{33} & \\ \lambda_2 \omega_{21} & \lambda_1 \lambda_2 \omega_{21} & \lambda_2 \omega_{22} & \lambda_2^2 \omega_{22} + \theta_{44} \end{bmatrix}$$

The matrix Σ has ten variances and covariances which are functions of nine parameters. The model is in fact a reparameterization of that discussed in the previous subsection.

Often when the same variables are used repeatedly there is a tendency for the corresponding errors (the ε's) to correlate over time because of memory and other retest effects. Hence there is a need to generalize the above model to allow for correlations between ε_1 and ε_3 and also between ε_2 and ε_4. This means that there will be two non-zero covariances θ_{31} and θ_{42} in Θ. The covariance matrix of the observed variables will now be

$$\Sigma = \begin{bmatrix} \omega_{11} + \theta_{11} & & & \\ \lambda_1 \omega_{11} & \lambda_1^2 \omega_{11} + \theta_{22} & & \\ \omega_{21} + \theta_{31} & \lambda_1 \omega_{21} & \omega_{22} + \theta_{33} & \\ \lambda_2 \omega_{21} & \lambda_1 \lambda_2 \omega_{21} + \theta_{42} & \lambda_2 \omega_{22} & \lambda_2^2 \omega_{22} + \theta_{44} \end{bmatrix}$$

(6)

This Σ has its ten independent elements expressed in terms of eleven parameters. Hence it is clear that the model is not identified. In fact, none of the eleven parameters are identified without further conditions imposed. The loadings λ_1 and λ_2 may be multiplied by a constant and the ω's divided by the same constant. This does not change σ_{21}, σ_{32}, σ_{41} and σ_{43}. The change in the other σ's may be compensated by adjusting the Θ's additively. Hence to make the model identified one must fix one λ or one ω at a non-zero value or one Θ at some arbitrary value. However, the correlation between η_1 and η_2 is identified without any restrictions, since

$$\text{Corr } (\eta_1, \eta_2) = (\omega_{21}^2/\omega_{11}\omega_{22})^{\frac{1}{2}} = [(\sigma_{32}\sigma_{41})/(\sigma_{21}\sigma_{43})]^{\frac{1}{2}}.$$

This model may therefore be used to estimate this correlation coefficient and to test whether this is one. To make further use of the model it is necessary to make some assumption about the nature of the variables. For example, if it can be assumed that the two variables on each occasion are tauequivalent we can set both λ_1 and λ_2 equal to one. Then the model can be estimated and tested with one degree of freedom. If $\lambda_1 = \lambda_2$ the model is just identified.

While the above model is not identified as it stands it becomes so as soon as there is information about one or more background variables affecting η_1 or η_2 or both. To illustrate this an example of a longitudinal study analyzed in more detail by Wheaton et al. (1977) will be used.

Example 3: This study was concerned with the stability over time of attitudes such as alienation and the relation to background variables such as education and occupation. Data on attitude scales were collected from 932 persons in two rural regions in Illinois at three points in time: 1966, 1967 and 1971. The variables used for the present illustration are the Anomia subscale and the Powerlessness subscale, taken to be indicators of Alienation. This example uses data from 1967 and 1971 only. The background variables are the respondent's education (years of schooling completed) and Duncan's Socioeconomic Index (SEI). These are taken to be indicators of the respondent's socioeconomic status (SES). The sample covariance matrix of the six observed variables is given in Table 10.3

TABLE 10.3

Covariance Matrix for Variables in the Stability of
Alienation Example

	y_1	y_2	y_3	y_4	x_1	x_2
y_1	11.834					
y_2	6.947	9.364				
y_3	6.819	5.091	12.532			
y_4	4.783	5.028	7.495	9.986		
x_1	-3.839	-3.889	-3.841	-3.625	9.610	
x_2	-21.899	-18.831	-21.748	-18.775	35.522	450.288

Let

y_1 = Anomia 67

y_2 = Powerlessness 67

x_1 = Education

ξ = SES

y_3 = Anomia 71

y_4 = Powerlessness 71

x_2 = SEI

η_1 = Alienation 67

η_2 = Alienation 71

The model is then specified as

$$\begin{pmatrix} y_1 \\ y_2 \\ y_3 \\ y_4 \end{pmatrix} = \begin{pmatrix} 1 & 0 \\ \lambda_1 & 0 \\ 0 & 1 \\ 0 & \lambda_2 \end{pmatrix} \begin{pmatrix} \eta_1 \\ \eta_2 \end{pmatrix} + \begin{pmatrix} \varepsilon_1 \\ \varepsilon_2 \\ \varepsilon_3 \\ \varepsilon_4 \end{pmatrix},$$

$$\begin{pmatrix} x_1 \\ x_2 \end{pmatrix} = \begin{pmatrix} 1 \\ \lambda_3 \end{pmatrix} \xi + \begin{pmatrix} \delta_1 \\ \delta_2 \end{pmatrix}, \qquad x = \Lambda_x \xi + \delta$$

$$\begin{pmatrix} 1 & 0 \\ -\beta & 1 \end{pmatrix} \begin{pmatrix} \eta_1 \\ \eta_2 \end{pmatrix} = \begin{pmatrix} \gamma_1 \\ \gamma_2 \end{pmatrix} \xi + \begin{pmatrix} \zeta_1 \\ \zeta_2 \end{pmatrix}, \qquad B\eta = \Gamma\xi + \zeta \quad (7)$$

It is assumed that ζ_1 and ζ_2 are uncorrelated and that the scales for η_1, η_2 and ξ have been chosen to be the same as for y_1, y_3 and x_1, respectively.

Let ϕ = Var (ξ) and ψ_i = Var (ζ_i), $i = 1,2$ and let Ω be the covariance matrix of (η_1, η_2, ξ). It is obvious that there is a one-to-one correspondence between the six ω's in Ω and $(\phi, \beta,$

γ_1, γ_2, ψ_1, ψ_2). In terms of Ω the covariance matrix of (y_1, y_2, y_3, y_4, x_1, x_2) is

$$
\begin{bmatrix}
\omega_{11} + \Theta_{11} \\
\lambda_1\omega_{11} & \lambda_1^2\omega_{11} + \Theta_{22} \\
\omega_{21} + \Theta_{31} & \lambda_1\omega_{21} & \omega_{22} + \Theta_{33} \\
\lambda_2\omega_{21} & \lambda_1\lambda_2\omega_{21} + \Theta_{42} & \lambda_2\omega_{22} & \lambda_2^2\omega_{22} + \Theta_{44} \\
\omega_{31} & \lambda_1\omega_{31} & \omega_{32} & \lambda_2\omega_{32} & \omega_{33} + \Theta_{55} \\
\lambda_3\omega_{31} & \lambda_1\lambda_3\omega_{31} & \lambda_3\omega_{32} & \lambda_2\lambda_3\omega_{32} & \lambda_3\omega_{33} & \lambda_3^2\omega_{33} + \Theta_{66}
\end{bmatrix}
$$

$$(8)$$

The upper left 4×4 part is the same as (6). It is clear that the two last rows of Σ determines λ_1, λ_2, λ_3, ω_{31}, ω_{32}, ω_{33}, Θ_{55} and Θ_{66}. With λ_1 and λ_2 determined, the other parameters are determined by the upper left part. Altogether there are seventeen parameters to estimate. The example is continued in the fourth section of this chapter.

Simplex Models

Simplex models is a type of covariance structure which often occurs in longitudinal studies when the same variable is measured repeatedly on the same people over several occasions. The simplex model is equivalent to the covariance structure generated by a first-order non-stationary autoregressive process. Guttman (1954) used the term simplex also for variables which are not ordered through time but by other criteria. One of his examples concerns tests of verbal ability ordered according to increasing complexity. The typical feature of a simplex correlation structure is that the correlations decrease as one moves away from the main diagonal.

Jöreskog (1970b) formulated various simplex models in terms of the well-known Wiener and Markov stochastic processes. A distinction was made between a perfect simplex and a quasi-simplex. A perfect simplex is reasonable only if the measurement errors in the test scores are negligible. A quasi-simplex, on the other hand, allows for sizeable errors of measurement.

Consider p fallible variables y_1, y_2, $\ldots$, y_p. The unit of measurement in the true variables η_i may be chosen to be the same as in the observed variables y_i. The equations defining the model are then, taking all variables as deviations from their means,

$$y_i = \eta_i + \varepsilon_i, \quad i = 1, 2, \ldots, p,$$

$$\eta_i = \beta_i\eta_{i-1} + \zeta_i, \quad i = 2, 3, \ldots, p,$$

where the ε_i are uncorrelated among themselves and uncorrelated with all the η_i and where ζ_{i+1} is uncorrelated with η_i, $i = 1, 2$, ..., $p - 1$. The parameters of the model are $\omega_i = \text{Var}(\eta_i)$, $\Theta_i = \text{Var}(\varepsilon_i)$, $i = 1, 2, \ldots, p$ and $\beta_2, \beta_3, \ldots, \beta_p$. The residual variance $\text{Var}(\zeta_{i+1})$ is a function of ω_{i+1}, ω_i and β_{i+1}, namely $\text{Var}(\zeta_{i+1}) = \omega_{i+1} - \beta_{i+1}^2 \omega_i$, $i = 1, 2, \ldots, p - 1$. The covariance matrix of $y_1, y_2, \ldots, y_p$ is of the form, here illustrated with $p = 4$.

$$\Sigma = \begin{bmatrix} \omega_1 + \Theta_1 \\ \beta_2 \omega_1 & \omega_2 + \Theta_2 \\ \beta_2 \beta_3 \omega_1 & \beta_3 \omega_2 & \omega_3 + \Theta_3 \\ \beta_2 \beta_3 \beta_4 \omega_1 & \beta_3 \beta_4 \omega_2 & \beta_4 \omega_3 & \omega_4 + \Theta_4 \end{bmatrix}, \tag{9}$$

Consider first the perfect simplex, i.e., the case $\Theta_i = 0$, $i = 1, 2, \ldots, p$. Then

$$\Sigma^{-1} = \begin{bmatrix} \alpha_1 \\ \gamma_2 & \alpha_2 \\ 0 & \lambda_3 & \alpha_3 \\ 0 & 0 & \lambda_4 & \alpha_4 \end{bmatrix}$$

is tri-diagonal and it may be readily verified that there is a one-to-one correspondence between the parameters (ω_i, ω_2, ω_3, ω_4, β_2, β_3, β_4) of Σ and the parameters (α_1, α_2, α_3, α_4, γ_2, γ_3, γ_4) of Σ^{-1}. Although Σ is non-linear, Σ^{-1} is linear. Covariance structures in which Σ or Σ^{-1} has a linear structure were considered by Anderson (1969, 1970).

Now consider the quasi-simplex represented by (9). It is seen from (9) that although the product $\beta_2 \omega_1 = \sigma_{21}$ is identified, β_2 and ω_1 are not separately identified. The product $\beta_2 \omega_1$ is involved in the off-diagonal elements in the first column (and row) only. One can multiply β_2 by a constant and divide ω_1 by the same constant without changing the product. The change induced by ω_1 in σ_{11} can be absorbed in Θ_1 in such a way that σ_{11} remains unchanged. Hence $\Theta_1 = \text{Var}(\varepsilon_1)$ is not identified. For η_2 and η_3 we have

$$\omega_2 = \frac{\sigma_{32} \sigma_{21}}{\sigma_{31}},$$

$$\omega_3 = \frac{\sigma_{43} \sigma_{32}}{\sigma_{42}}$$

so that ω_2 and ω_3, and hence also θ_2 and θ_3, are identified. With ω_2 and ω_3 identified, β_3 and β_4 are identified by σ_{32} and σ_{43}. The middle coefficient β_3 is overidentified since

$$\beta_3\omega_2 = \frac{\sigma_{31}\sigma_{42}}{\sigma_{41}} = \sigma_{32}.$$

Since both ω_4 and θ_4 are involved in σ_{44} only, these are not identified but their sum σ_{44} is.

This analysis of the identification problem shows that for the "inner" variables y_2 and y_3, the parmeters ω_2, ω_3, θ_2, θ_3, and β_3 are identified, whereas there is an indeterminacy associated with each of the "outer" variables y_1 and y_4. To eliminate these indeterminacies one condition must be imposed on the parameters ω_1, θ_1 and β_2, and another on the parameters ω_4 and θ_4. Perhaps the most natural way of eliminating the indeterminacies is to set $\theta_1 = \theta_2$ and $\theta_4 = \theta_3$. To illustrate a simplex model, some data published by Humphreys (1968) and analyzed by Werts et al. (1978) will be used.

Example 4: The variables include eight semesters of gradepoint averages, high school rank and a composite score on the American College Testing tests for approximately 1,600 undergraduate students at the University of Illinois. The correlation matrix is given in Table 10.4. One would like to estimate the reliabilities of the gradepoint averages and to test whether the auto-regressive process is stationary.

The example is analyzed in the fourth section.

General Approaches to Analysis of
Covariance Structures

General Covariance and Correlation Structures

Any covariance structure may be defined by specifying that the population variances and covariances of the observed variables are certain functions of parameters θ_1, θ_2, ..., θ_t to be estimated from data: $\sigma_{ij} = \sigma_{ij}(\theta)$, or in matrix form $\Sigma = \Sigma(\theta)$. It is assumed that the functions $\sigma_{ij}(\theta)$ are continuous and have continuous first derivatives and that Σ is positive definite at every point θ of the admissible parameter space. The distribution of the observed variables is assumed to be multivariate with an unconstrained mean vector μ and covariance matrix $\Sigma(\theta)$ and is assumed to be sufficiently well described by

TABLE 10.4

Correlations among Grade Point Averages, High School Rank and An Aptitude Test

	y_0	y_0'	y_1	y_2	y_3	y_4	y_5	y_6	y_7	y_8
y_0	1.000									
y_0'	.393	1.000								
y_1	.387	.375	1.000							
y_2	.341	.298	.556	1.000						
y_3	.278	.237	.456	.490	1.000					
y_4	.270	.255	.439	.445	.562	1.000				
y_5	.240	.238	.415	.418	.496	.512	1.000			
y_6	.256	.252	.399	.383	.456	.469	.551	1.000		
y_7	.240	.219	.387	.364	.445	.442	.500	.544	1.000	
y_8	.222	.173	.342	.339	.354	.416	.453	.482	.541	1.000

Note: y_0 is high school rank, y_0' ACT composite score, and y_1 through y_8 are eight semesters' grade-point averages.

217

the moments of first and second order, so that additional information about Θ contained in moments of higher order may be ignored. In particular this will hold if the distribution is multivariate normal. The condition that the mean vector μ is unconstrained will be relaxed in the last section where also mean structures will be considered.

A correlation structure is defined by specifying that the population correlations ρ_{ij} of the observed variables are functions $\rho_{ij} = \rho_{ij}(\Theta)$ of Θ. Such a correlation structure is treated as a covariance structure by specifying that

$$\Sigma = D_\sigma P(\Theta) D_\sigma \tag{10}$$

where D_σ is a diagonal matrix of population standard deviations $\sigma_1, \sigma_2, \ldots, \sigma_p$ of the observed variables, which are regarded as free parameters, and $P(\Theta)$ is the correlation matrix. The covariance structure (10) has parameters $\sigma_1, \sigma_2, \ldots, \sigma_p, \Theta_1, \Theta_2, \ldots, \Theta_t$. The standard deviations $\sigma_1, \sigma_2, \ldots, \sigma_p$ as well as Θ must be estimated from data and the estimate of σ_i does not necessarily equal the corresponding standard deviation in the sample.

Identification

Before an attempt is made to estimate the parameters Θ, the identification problem must be resolved. The identification problem is essentially whether or not Θ is uniquely determined by Σ. Every Θ in the admissible parameter space generates a Σ but two or more Θ's may possibly generate the same Σ. The whole model is said to be identified if for any two vectors Θ_1 and Θ_2 in a region of the parameter space, locally or globally, $\Theta_1 \neq \Theta_2$ implies that $\Sigma(\Theta_1) \neq \Sigma(\Theta_2)$, i.e., if Σ is generated by one and only one Θ. This means that all parameters are identified. However, even if the whole model is not identified some parameters can still be identified. Consider the set of all parameter vectors Θ generating the same Σ. If a parameter Θ_i has the same value in all such vectors, this parameter is identified. For parameters which are identified the methods to be described will yield consistent estimators. If a model is not completely identified, appropriate restrictions may be imposed on Θ to make it so, and the choice of restrictions may affect the interpretation of the results of an estimated model.

Identifiability depends on the choice of model. To examine the identification problem for a particular model consider the equations

$$\sigma_{ij} = \sigma_{ij}(\Theta), i \leqslant j. \tag{11}$$

There are $(\frac{1}{2})p(p + 1)$ equations in t unknown parameters Θ. Hence a necessary condition for identification of all parameters is that

$$t \leqslant (\frac{1}{2})p(p + 1).$$

If a parameter Θ can be determined from Σ by solving the equations (11) or a subset of them, this parameter is identified; otherwise it is not. Often some parameters can be determined from Σ in several ways, i.e., by using different sets of equations. This gives rise to overidentifying conditions on Σ which must hold if the model is true. Since the equations (11) are often non-linear, the solution of the equations is often complicated and tedious and explicit solutions for all Θ's seldom exist. In the previous section we discussed the identification problem in terms of some specific examples.

There are various ways in which the computer program may be used to check the identification status of the model. If the GLS or ML methods are used for estimation (see the next subsection) the information matrix may be obtained and checked for positive definiteness. If the model is identified then the information matrix is almost certainly positive definite. If the information matrix is singular, the model is not identified and the rank of the information matrix may indicate which parameters are not identified. Another procedure which may also be used when other methods of estimation are used is the following. Choose a set of reasonable values for the parameters and compute Σ. Then run the program with this Σ as input matrix and estimate Θ. If this results in the same estimated values as those used to generate Σ, then it is most likely that the model is identified. Otherwise, those parameters which gave a different value are probably not identified.

Estimation

The population is characterized by the mean vector μ, which is unconstrained, and the covariance matrix Σ which is a function of Θ. In practice Θ is unknown and must be estimated from a sample of N independent observations on the random vector x of order p. Let $S = (s_{ij})$ be the usual sample covariance matrix of order $p \times p$, based on $n = N - 1$ degrees of freedom. The information provided by S may also be represented by a correlation matrix $R = (r_{ij})$ and a set of standard deviations $s_1, s_2, \ldots, s_p$, where $s_i = (s_{ii})^{\frac{1}{2}}$ and $r_{ij} = s_{ij}/s_i s_j$. In many

applications both the origin and the unit in the scales of measurement are arbitrary or irrelevant and then only the correlation
matrix may be of interest. In such cases one takes S to be the
correlation matrix R in what follows.

Since the mean vector is unconstrained, and higher moments
are ignored, the estimation problem may be regarded as a problem
of how to fit a matrix Σ of the form $\Sigma(\Theta)$ to the observed covariance matrix S. Although a number of different methods of
an ad hoc nature have been used in specific cases there appears
to be only three methods which can be used in general. These
are the unweighted least squares (ULS) method, which minimizes

$$U = (\tfrac{1}{2}) \, tr \, (S - \Sigma)^2, \tag{12}$$

the generalized least squares (GLS) method, which minimizes

$$G = (\tfrac{1}{2}) \, tr \, (I - S^{-1}\Sigma)^2, \tag{13}$$

and the maximum likelihood (ML) method, which minimizes

$$M = tr \, (\Sigma^{-1}S) - \log|\Sigma^{-1}S| - p. \tag{14}$$

Each function is to be minimized with respect to Θ.

All three functions U, G and M may be minimized by basically
the same algorithm. The notation $F = F(S, \Sigma)$ will be used for
any one of the three functions. The GLS and ML methods are
scale-free in the sense that $F(S, \Sigma) = F(DSD, D\Sigma D)$ for any
diagonal matrix of positive scale-factors; ULS does not have
this property. With ULS, an analysis of S and of DSD yield
results which may not be properly related. When x has a multivariate normal distribution both GLS and ML yield estimates
that are efficient in large samples. Both GLS and ML require
a positive definite covariance matrix S or correlation matrix R;
ULS will work even on a matrix which is non-Gramian.

Under the assumption that x has a multinormal distribution
or that S has a Wishart distribution, M in (6) is a transform of
the log-likelihood function for the sample, hence its association
to the maximum likelihood method. Jöreskog and Goldberger
[1972] derived the expression for G from Aitken's [1934-35]
principle of generalized least squares using estimated asymptotic
variances and covariances of the elements of S under multinormality of x. Browne [1974] justified GLS under the slightly more
general assumption that the elements of S have an asymptotic
normal distribution. Since the variances and covariances in S
are generally correlated and have unequal variances, it would

seem that ULS uses the wrong metric in measuring deviations between S and Σ. Nevertheless, ULS produces consistent estimators under more general assumptions than those which have been used to justify ML and GLS.

All three functions U, G and M are members of a general class of weighted least squares (WLS) functions (see Browne, 1974 and Lee, 1979). Let A be a symmetric positive definite weight matrix. Then WLS minimizes the function

$$F = (\tfrac{1}{2}) \, \text{tr} \, \{[(S - \Sigma(\Theta)]A\}^2$$

with respect to Θ.

In ULS, $A = I$. Browne (1974) showed, under general regularity conditions, that if A is a random matrix converging in probability to Σ^{-1} then the estimator $\hat{\Theta}$ which minimizes F is consistent and asymptotically efficient. A convenient choice of A is $A = S^{-1}$ yielding the GLS function G in (13). Maximum likelihood estimators will be obtained if $A = \Sigma_0^{-1}$ where $\Sigma_0 = \Sigma(\Theta_0)$, Θ_0 being the true Θ. Since Θ_0 is unknown this is not operational. However, ML estimates will be obtained if F is minimized iteratively and $A = \Sigma^{-1}(\Theta)$ is updated in each iteration. The estimator of Θ which minimizes F in this way is identical to that which minimizes M in (14). Browne (1974) showed that the GLS and ML estimators are asymptotically equivalent. Computationally there is a slight advantage with GLS compared with ML since the weight matrix is constant through the iterations. When the covariance structure is linear, GLS has a great advantage for then the function G is exactly quadratic in Θ and can therefore be minimized in one step.

The derivatives of F are

$$\partial F/\partial \Theta_i = \text{tr} \, [A(\Sigma - S)A \partial \Sigma/\partial \Theta_i], \tag{15}$$

where $A = I$ in ULS, $A = S^{-1}$ in GLS and $A = \Sigma^{-1}$ in ML. Assuming that S converges in probability to Σ and ignoring terms of order $\Sigma - S$, the second derivatives are approximately

$$\partial^2 F/\partial \Theta_i \partial \Theta_j = \text{tr} \, [A \partial \Sigma/\partial \Theta_i A \partial \Sigma/\partial \Theta_j]. \tag{16}$$

Note that (15) and (16) require only the first derivatives of the covariance structure functions $\Sigma(\Theta)$. In ML and GLS, (16) yields the elements of the information matrix which is positive definite at every point Θ of the admissible parameter space, if Θ is identified.

The function $F(\theta)$ may be minimized numerically by Fisher's scoring method (see e.g. Rao, 1973) or the method of Davidon-Fletcher-Powell (see Fletcher & Powell, 1963; see also Gruvaeus & Jöreskog, 1970). Lee & Jennrich (1979) and Lee (1979) suggested that F can be minimized by the Gauss-Newton algorithm but this is equivalent to Fisher's scoring algorithm when the second-order derivatives are approximated by (16).

The minimization starts at an arbitrary starting point $\theta^{(1)}$ and generates successively new points $\theta^{(2)}$, $\theta^{(3)}$, ..., such that $F(\theta^{(s+1)}) < F(\theta^{(s)})$ until convergence is obtained.

Let $g^{(s)}$ be the gradient vector $\partial F/\partial\theta$ at $\theta = \theta^{(s)}$ and let $E^{(s)}$ be the matrix whose elements are given by (16) evaluated at $\theta = \theta^{(s)}$. Then Fisher's scoring method computes a correction vector $\delta^{(s)}$ by solving the symmetric equation system

$$E^{(s)}\delta^{(s)} = g^{(s)} \tag{17}$$

and then computes the new point as

$$\theta^{(s+1)} = \theta^{(s)} - \delta^{(s)}. \tag{18}$$

This requires the computation of $E^{(s)}$ and the solution of (17) in each iteration which is often quite time consuming. An alternative is to use the method of Davidon-Fletcher-Powell, which evaluates only the inverse of $E^{(1)}$ and in subsequent iterations E^{-1} is improved, using information built up about the function, so that ultimately E^{-1} converges to an approximation of the inverse of $\partial^2 F/\partial\theta\partial\theta'$ at the minimum.

In GLS and ML, $(2/N)$ times the inverse of the information matrix E, given by (8) and evaluated at the minimum of F, provides an estimate of the asymptotic covariance matrix of the estimators $\hat{\theta}$ of θ. The square root of the diagonal elements of $(2/N)E^{-1}$ are large-sample estimates of the standard errors of the $\hat{\theta}$'s.

Unfortunately no statistical theory is available for computing standard errors for ULS estimators. Such standard errors may be obtained by jackknifing but this requires extensive computation.

Assessment of Fit

When the number of independent parameters in θ is less than the total number of variances and covariances in Σ, i.e., when $t < (\frac{1}{2})p(p + 1)$, the model imposes conditions on Σ which must hold if the model is true. In GLS and ML, the validity of these conditions, i.e., the validity of the model, may be tested by a likelihood ratio test. The logarithm of the likelihood ratio is simply $(N/2)$ times the minumum value of the function F. Under the model, this is distributed, in large samples, as a χ^2 distribution with degrees of freedom equal to

$$d = (\tfrac{1}{2})p(p + 1) - t. \tag{19}$$

Tests of Structural Hypotheses

Once the validity of the model has been reasonably well established, various structural hypotheses within the model may be tested. One can test hypotheses of the forms

(i) that certain Θ's are fixed equal to assigned values and/or
(ii) that certain Θ's are equal in groups.

Each of these two types of hypotheses leads to a covariance structure $\Sigma(\nu)$ where ν is a subset of $u < t$ elements of Θ. Let F_ν be the minimum of F under the structural hypothesis and let F_Θ be the minimum of F under the general model. Then $(N/2)(F_\nu - F_\Theta)$ is approximately distributed as χ^2 with $t - u$ degrees of freedom.

The Use of χ^2 in Exploratory Studies

The values of χ^2 should be interpreted very cautiously because of the sensitivity of χ^2 to various model assumptions such as linearity, additivity, multinormality, etc., but also for other reasons. In most empirical work many of the models considered may not be very realistic. If a sufficiently large sample were obtained, the test statistic would, no doubt, indicate that any such model is statistically untenable. The model should rather be that $\Sigma(\Theta)$ represents a reasonable approximation to the population covariance matrix. From this point of view the statistical problem may not be one of testing a given hypothesis (which a priori may be considered false) but rather one of fitting various models with different numbers of parameters and to decide when to stop fitting. In other words, the problem is to extract as much information as possible out of a sample of given size without going so far that the result is affected to a large extent by "noise." It is reasonable and likely that more information can be extracted from a large sample than from a small one. In such a problem it is the difference between χ^2 values that matters rather than the χ^2 values themselves. In an exploratory study, if a value of χ^2 is obtained which is large compared to the number of degrees of freedom, the fit may be examined by an inspection of the residuals, i.e., the discrepancies between observed and reproduced values. Often the results of an analysis, an inspection of residuals or other considerations will suggest ways to relax the model somewhat by introducing more parameters. The new model usually yields a smaller χ^2. If the drop in χ^2 is large compared to the difference in degrees of freedom, this is an indication that the change made in the model represents a real improvement. If, on the other hand, the drop in χ^2 is close to the difference in number of degrees of freedom, this is an indication that the improvement in fit is obtained by "capitalizing on chance" and the added parameters may not have any real significance or meaning.

Often it is not possible, or even desirable, to specify the model completely since there may be other models which are equally plausible. In such a situation it is necessary to have a technique of analysis which will give information about which of a number of alternative models is (are) the most reasonable. Also, if there is sufficient evidence to reject a given model due to poor fit to the data, the technique should be designed to suggest which part of the model is causing the poor fit. The examples of the fourth section illustrate the assessment of fit of a model and strategies for model modification.

The ACOVS Model

The approach to analysis of covariance structures described in the previous sections is completely general in the sense that any covariance structure can be handled. This approach has not been used much in practice because it requires the specification, by means of programmed subroutines, of the functions $\sigma_{ij}(\Theta)$ and $\partial\sigma_{ij}/\partial\Theta$ for each application. Another approach, which has been found to be extremely useful in practice, is to assume a definite form for Σ but one which is still so general and flexible that it can handle most problems arising in practice. Two such approaches have been developed: the ACOVS-model and the LISREL-model.

The ACOVS-model (Jöreskog, 1970a, 1973a, 1974) assumes that Σ has the following form

$$\Sigma = B(\Lambda\Phi\Lambda' + \Psi^2)B' + \Theta^2, \tag{20}$$

where $B(p \times q)$, $\Lambda(q \times r)$, the symmetric matrix $\Phi(r \times r)$ and the diagonal matrices $\Psi(q \times q)$ and $\Theta(p \times p)$ are parameter matrices. The covariance structure (20) will arise if the vector of observed variables $x(p \times 1)$ has the form

$$x = \mu + B\Lambda\xi + B\zeta + e \tag{21}$$

where μ is the mean vector of x and ξ, ζ, and e are uncorrelated random vectors of latent (unobserved) variables with zero means and covariance matrices Φ, Ψ^2 and Θ^2, respectively.

When applying the model (20), the number of variables p is given by the data, and q and r are given by the particular application that the investigator has in mind. In any such application any parameter in B, Λ, Φ, Ψ, or Θ may be known a priori and one or more subsets of the remaining parameters may have identical but unknown values. Thus, parameters are of three kinds: (a) fixed parameters that have been assigned given values, (b) constrained parameters that are unknown, but equal to one or more other parameters, and (c) free parameters that are unknown and not constrained to be equal to any other parameter. The advantage of such an approach is the great generality and flexibility obtained by the various specifications that may be imposed. Thus the general model contains a wide range of specific models. The examples considered in the second section are all of the form of (20) or can be reparameterized to be of this form. Model (1) uses $B = I$, $\Theta = 0$, $\Lambda = A$ and equality constraints on the diagonal elements of Ψ. Models (2a-b) use $B = \Delta$ in addition. Models (3) and (4) use $\Lambda = I$ and $\Psi = 0$. Model (6) is of the form (20) with $B = I$, $\Psi = 0$ and

$$\Lambda = \begin{pmatrix} 1 & 0 \\ \lambda_1 & 0 \\ 0 & 1 \\ 0 & \lambda_2 \end{pmatrix} \quad \Phi = \begin{pmatrix} \omega_{11} & \\ \omega_{21} & \omega_{22} \end{pmatrix}$$

$$\Theta = \begin{pmatrix} \Theta_{11} & & & \\ 0 & \Theta_{22} & & \\ \Theta_{31} & 0 & \Theta_{33} & \\ 0 & \Theta_{42} & 0 & \Theta_{44} \end{pmatrix}$$

The path analysis model (8) and the simplex model (9) require reparameterization before they can be written in the form of (20), see Jöreskog (1970b). One disadvantage with the reparameterization is that one will not get estimates of the original parameters. Although estimates of the original parameters can easily be computed afterwards it is more difficult to obtain standard errors of the original parameters. For this reason the LISREL model has been developed to accommodate the original parameters of path models and simplex models, but as will be seen, the ACOVS-model is a submodel of the LISREL-model, so that many other models may also be fitted.

The LISREL Model

The LISREL model (Jöreskog, 1973b, 1977; Jöreskog & Sörbom, 1978) considers random vectors $\eta' = (\eta_1, \eta_2, \ldots, \eta_m)$ and $\xi' = (\xi_1, \xi_2, \ldots, \xi_n)$ of latent dependent and independent variables, respectively and the following system of linear structural relations

$$B\eta = \Gamma\xi + \zeta \tag{22}$$

STRUCTURAL MODEL

where $B(m \times m$ and $\Gamma (m \times n)$ are coefficient matrices and $E(\zeta) = 0$. It is furthermore assumed that ζ is uncorrelated with ξ and that B is nonsingular.

The vectors η and ξ are not observed but instead vectors $y' = (y_1, y_2, \ldots, y_p)$ and $x' = (x_1, x_2, \ldots, x_q)$ are observed such that

MEASUREMENT MODEL

$$y = \Lambda_y\eta + \varepsilon, \tag{23}$$

$$x = \Lambda_x\xi + \delta \tag{24}$$

where ε and δ are vectors of errors of measurement in y and x, respectively. We take y and x to be measured as deviations from their means. The matrices $\Lambda_y(p \times m)$ and $\Lambda_x(q \times n)$ are regression matrices of y on η and of x on ξ, respectively. It is convenient to refer to y and x as the observed variables and η and ξ as the latent variables. The errors of measurement are assumed to be uncorrelated with η, ξ and ζ but may be correlated among themselves.

Let $\Phi(n \times n)$ and $\Psi(m \times m)$ be the covariance matrices of ξ and ζ, respectively, and let Θ_ε and Θ_δ be the covariance matrices of ε and δ, respectively. Then it follows, from the above assumptions, that the covariance matrix $\Sigma[(p + q) \times (p + q)]$ of $z = (y', x')'$ is

$$\Sigma = \begin{pmatrix} Q & \Lambda_y B^{-1}\Gamma\Phi\Lambda_x' \\ \Lambda_x\Phi\Gamma'B'^{-1}\Lambda_y' & \Lambda_x\Phi\Lambda_x' + \Theta_\delta \end{pmatrix} \tag{25}$$

where

$$Q = \Lambda_y(B^{-1}\Gamma\Phi\Gamma'B'^{-1} + B^{-1}\Psi B'^{-1})\Lambda_y' + \Theta_\varepsilon$$

The elements of Σ are functions of the elements of Λ_y, Λ_x, B, Γ, Φ, Ψ, Θ_ε and Θ_δ. In applications some of these elements are fixed and equal to assigned values. In particular, this is so for elements of Λ_y, Λ_x, B and Γ, but we shall allow for fixed values in the other matrices also. For the remaining nonfixed elements of the eight parameter matrices one or more subsets may have identical but unknown values. Thus, the elements in Λ_y, Λ_x, B, Γ, Φ, Ψ, Θ_ϵ and Θ_δ are of three kinds:

(i) <u>fixed parameters</u> that have been assigned given values
(ii) <u>constrained parameters</u> that are unknown but equal to one or more other parameters and
(iii) <u>free parameters</u> that are unknown and not constrained to be equal to any other parameter.

Equations (22), (23) and (24), with the accompanying assumptions, define the general LISREL model. Equations (23) and (24) constitute the measurement model and equation (22) constitutes the structural equation model.

The computer program for LISREL (see Jöreskog & Sörbom, 1978) is structured in such a way that one can handle all kinds of reasonable submodels in a simple manner. For example, by specifying that there are no y- and η-variables, the model becomes

$$x = \Lambda_x \xi + \delta, \tag{26}$$

which is the classical factor analysis model. By specifying that there are no x- and no ξ-variables, the model reduces to

$$\left. \begin{array}{l} y = \Lambda_y \eta + \epsilon \\ B\eta = \zeta \end{array} \right\}. \tag{27}$$

With this model we also have an ordinary factor analysis model, but in this case we can handle relations among the factors by specifying a structure of the B-matrix. For example, by specifying a B which generates an autoregressive structure one can handle simplex models of the form (9) and more general ones, see e.g. Jöreskog & Sörbom (1977) and Jöreskog (1979).

By specifying that there are no x-variables and that B is an identity matrix we get

$$\left. \begin{array}{l} \eta = \Gamma\xi + \zeta \\ y = \Lambda_y \eta + \epsilon \end{array} \right\} \tag{28}$$

or equivalently, cf. (21)

$$y = \Lambda_y(\Gamma\xi + \zeta) + \epsilon, \tag{29}$$

which is a second-order factor analysis model equivalent to the ACOVS model (Jöreskog, 1974). In a similar manner we can get a model for interdependent systems by specifying that the Λ_y and Λ_x matrices are identity matrices and that δ and ε are zero, i.e., the model

$$By = \Gamma x + \zeta, \tag{30}$$

often used in econometrics.

Analysis of the Examples

Example 1: Repeated Trials of the Rod and Frame Test

Estimation of the variance components according to model (1) gives

$$\hat{\sigma}_a^2 = 4.08 \quad \hat{\sigma}_b^2 = 13.83 \quad \hat{\sigma}_c^2 = 28.64 \quad \hat{\sigma}_e^2 = 21.60$$

However, examination of the fit of the model to the data reveals that the fit is very poor: $\chi^2 = 319.4$ with 74 degrees of freedom. We shall therefore seek an alternative model which accounts for the data better. This is obtained by structuring the error component e.

There are six distinct experimental conditions among the twelve trials, each one repeated twice. Let τ_i, $i = 1, 1, \ldots, 6$ be random components associated with these experimental conditions. Then

$$x_{i\alpha} = \tau_i + e_{i\alpha}$$

where $\alpha = 1, 2$ indexes the two replications. This simply means that one should allow the error variances to be different for different experimental conditions but still equal within replications of the same condition. An analysis according to this model gives $\chi^2 = 179.6$ with 69 degrees of freedom. The reduction in χ^2 clearly indicates that the error variances depend on the experimental condition.

The ML-estimates of the variance components are now, with standard errors below the estimates,

$$\hat{\sigma}_a^2 = 4.18 \quad \hat{\sigma}_b^2 = 11.26 \quad \hat{\sigma}_c^2 = 27.10$$
$$(0.75) \qquad\quad (1.72) \qquad\quad (4.24)$$

$$\hat{\sigma}^2_{e_1} = 22.17 \qquad \hat{\sigma}^2_{e_2} = 34.61 \qquad \hat{\sigma}^2_{e_3} = 29.07 \qquad \hat{\sigma}^2_{e_4} = 37.32$$
$$\qquad (2.71) \qquad\qquad (3.89) \qquad\qquad (3.34) \qquad\qquad (4.30)$$

$$\hat{\sigma}^2_{e_1} = 11.74 \qquad \hat{\sigma}^2_{e_6} = 5.09$$
$$\qquad (1.41) \qquad\qquad (0.67)$$

The results indicate that most of the variance in the trials is associated with the chair effect. The variance due to the frame effect is less than half of this and the variance due to general bias is still smaller. The error variances are generally quite large except for the two experimental conditions in which the chair is already vertical.

Example 2: The Disattenuated Correlation between Two Vocabulary Tests

The measurement model set up in the second section is specified as a LISREL model with no y- and no η-variables as in (26). All four models H_1, H_2, H_3 and H_4 can be estimated in one run.

The results are shown in Table 10.5. Each hypothesis is tested against the general alternative that Σ is unconstrained. To consider various hypotheses that can be tested, the four χ^2 values of Table 10.5 are recorded in a 2 × 2 table as in Table 10.6. Test of H_1 against H_2 gives $\chi^2 = 35.40$ with 1 degree of freedom. An alternative test is H_3 against H_4, which gives $\chi^2 = 35.51$ with 1 degree of freedom. Thus, regardless of whether we treat the two pairs of tests as parallel or congeneric, the hypothesis $\varrho = 1$ is rejected. There is strong evidence that the unspeeded and speeded tests do not measure the same trait. The hypothesis of parallelness of the two pairs of tests can also

TABLE 10.5

Summary of Analyses

Hypothesis	No. par	χ^2	d.f.	P
H_1	4	37.33	6	0.00
H_2	5	1.93	5	0.86
H_3	8	36.21	2	0.00
H_4	9	0.70	1	0.40

TABLE 10.6

Tests of Hypotheses

	Parallel	Congeneric	
$\varrho = 1$	$\chi_6^2 = 37.33$	$\chi_2^2 = 36.21$	$\chi_4^2 = 1.12$
$\varrho \neq 1$	$\chi_5^2 = 1.93$	$\chi_2^2 = 0.70$	$\chi_4^2 = 1.23$
	$\chi_1^2 = 35.40$	$\chi_1^2 = 35.51$	

be tested by means of Table 10.6. This gives $\chi^2 = 1.12$ or $\chi^2 = 1.23$ with 4 degrees of freedom, depending on whether we assume $\varrho = 1$ or $\varrho \neq 1$. Thus we cannot reject the hypothesis that the two pairs of tests are parallel. It appears that H_2 is the most reasonable of the four hypotheses. The maximum likelihood estimate of ϱ under H_2 is $\hat{\varrho} = 0.899$ with a standard error of 0.019. An approximate 95% confidence interval for ϱ is $0.86 < \varrho < 0.94$. The substantive matter is discussed further in Jöreskog (1974).

Example 3: The Stability of Alienation

The model (7) is an example of the general LISREL form (22), (23) and (24). We distinguish between the two models

(A) $\Theta_{31} = \Theta_{42} = 0$
(B) Θ_{31} and Θ_{42} free

The maximum likelihood estimates of the parameters with standard errors in parenthesis are given in Table 10.7. The stability of alienation over time is reflected in the parameter β. The influence of SES on Alienation at the two occasions is significant in model A. The coefficient for 1967, γ_1, is -0.614 with a standard error of 0.056 and for 1971, γ_2, it is -0.174 with a standard error equal to 0.054. The negative sign of the λ-coefficients indicates that for high socioeconomic status the alienation is low and vice versa. However, the overall fit of the model A is not acceptable: χ^2 with six degrees of freedom equals 71.5. Model B is intuitively more plausible. As can be seen from Table 10.7 the inclusion of Θ_{31} and Θ_{42} results in a model with an acceptable overall fit. A test of the hypothesis that both Θ_{31} and Θ_{42} are zero yields $\chi^2 = 66.8$ with 2 degrees of freedom so that this hypothesis must be rejected.

TABLE 10.7

Maximum Likelihood Estimates for Models A and B
(The standard errors of the estimates are given within parenthesis)

Parameter	Model A	Model B
λ_1	0.889 (.041)	0.979 (.062)
λ_2	0.849 (.040)	0.922 (.060)
λ_3	5.329 (.430)	5.221 (.422)
β	0.705 (.054)	0.607 (.051)
γ_1	-0.614 (.056)	-0.575 (.056)
γ_2	-0.174 (.054)	-0.227 (.052)
ψ_1	5.307 (.473)	4.846 (.468)
ψ_2	3.742 (.388)	4.089 (.405)
ϕ	6.666 (.641)	6.803 (.650)
Θ_{11}	4.015 (.343)	4.735 (.454)
Θ_{22}	3.192 (.271)	2.566 (.404)
Θ_{33}	3.701 (.373)	4.403 (.516)
Θ_{44}	3.625 (.292)	3.074 (.435)
Θ_{31}	—	1.624 (.314)
Θ_{42}	—	0.339 (.261)
Θ_{55}	2.944 (.500)	2.807 (.508)
Θ_{66}	260.982 (18.242)	264.809 (18.154)
χ^2	71.470	4.730
d.f.	6	4

Example 4: A Simplex Model for Academic Performance

The quasi-simplex model underlying the covariance structure (9) is a LISREL model of the form (27) with no x- and no ξ-variables and with $\Lambda_y = I$ and

$$B = \begin{pmatrix} 1 & 0 & 0 & 0 \\ -\beta_2 & 1 & 0 & 0 \\ 0 & -\beta_3 & 1 & 0 \\ 0 & 0 & -\beta_4 & 1 \end{pmatrix}$$

Using first only the measures of grade-point averages in Table 10.4, estimation of a quasi-Markov simplex gives the following correlations between the true academic achievements $\eta_2, \eta_3, \ldots, \eta_7$. These correlations are

	η_2	η_3	η_4	η_5	η_6	η_7
η_2	1.000					
η_3	0.838	1.000				
η_4	0.812	0.969	1.000			
η_5	0.724	0.865	0.892	1.000		
η_6	0.677	0.809	0.834	0.935	1.000	
η_7	0.619	0.740	0.763	0.855	0.914	1.000

Here every correlation ρ_{ij} with $|i - j| > 1$ is the product of the correlations just below the diagonal. For example, $\rho(\eta_5, \eta_2) = 0.838 \times 0.969 \times 0.892 = 0.724$. These correlations form a perfect Markov simplex. The goodness of fit test of the model gives $\chi^2 = 23.91$ with 15 degrees of freedom. This represents reasonably good fit considering the large sample size. The reliabilities of the semester grades y_2, y_3, ..., y_7 can also be obtained directly from the solution in which the η's are standardized. The reliabilities are

y_2	y_3	y_4	y_5	y_6	y_7
0.569	0.575	0.562	0.584	0.581	0.608

A test of the hypothesis that these are equal gives $\chi^2 = 2.17$ with 5 degrees of freedom, so that this hypothesis is not rejected by the data despite the large sample size.

In this example the correlations $\rho(\eta_1, \eta_j)$, $j \neq 1$ and $\rho(\eta_i, \eta_8)$, $i \neq 8$ and the reliabilities of y_1 and y_8 are not identified. However, in view of the above test of equality of reliabilities it seems reasonable to assume that all reliabilities or equivalently all error variances in the standardized solution are equal for y_1 through y_8. This assumption makes it possible to estimate the intercorrelations among all the η's.

Assuming that x_0 and x_0' are indicators of pre-college academic achievement η_0 which is assumed to influence the true academic achievement in the first semester η_1, one can estimate again the quasi-Markov simplex and show how this use of x_0 and x_0' helps identify the parameters of the model. The only parameter which is now not identified is θ_8, the error variance in y_8. This gives a $\chi^2 = 36.92$ with 28 degrees of freedom. If we assume that the reliabilities of all the semester grades are equal, all parameters are identified and the goodness of fit becomes 45.22 with 34 degrees of freedom. The difference 8.30 with 6 degrees of freedom provides another test of equality of the reliabilities. Finally a test of the hypothesis that the whole process is stationary, i.e., that

$$\beta_2 = \beta_3 = \ldots = \beta_8$$

$$\Theta_2 = \Theta_3 = \ldots = \Theta_8$$

gives $\chi^2 = 12.82$ with 11 degrees of freedom so that this hypothesis cannot be rejected. There is good evidence that the whole Markov process is stable over time.

Generalizations

In this section we consider two different generalizations of the models and methods discussed in the previous sections. One generalization allows for mean structures as well as covariance structures; the other concerns the analysis of data from several independent samples.

Mean Structures As Well

Suppose that not only the covariance matrix Σ but also the mean vector μ is a function of the parameter vector Θ. The easiest way to generalize the development 3.1-3.5 to this situation is to consider the augmented vector $x^{*'} = (x', 1)$ where the last variable is a fixed variable which is constant equal to one for all observations, and use matrices of moments about zero instead of covariance matrices. Let $M[(p + 1) \times (p + 1)]$ be the sample moment matrix of x^*, i.e.,

$$M = \begin{pmatrix} M_{xx} & \bar{x} \\ \bar{x}' & 1 \end{pmatrix} = \begin{pmatrix} S + \bar{x}\bar{x}' & \bar{x} \\ \bar{x}' & 1 \end{pmatrix}$$

where, as before, $\bar{x}$ is the sample mean vector of x and S the sample covariance matrix of x and

$$M_{xx} = (1/N) \sum_{\alpha=1}^{N} X_\alpha X_\alpha',$$

x_α being the αth observation of x.

The corresponding population moment matrix is

$$\Omega = \begin{pmatrix} \Sigma + \mu\mu' & \mu \\ \mu' & 1 \end{pmatrix}$$

Estimates of θ may be obtained by fitting Ω to M using $F(\Omega, M)$ where F is any one of the three fitting functions (12), (13) and (14). Jöreskog and Sörbom (1980) showed that (14) leads to maximum likelihood estimates in the usual sense.

The LISREL-Model with Structured Means

The LISREL model has been defined by (22), (23) and (24) in which all random variables were assumed to have zero means. This assumption will now be relaxed and it will be shown that the LISREL IV computer program can be used also to estimate the same three equations even if they include constant intercept terms. This is possible if we introduce a fixed variable whose observations are all equal to one and analyze the sample moment matrix instead of the sample covariance matrix.

The LISREL model is now defined by the following three equations corresponding to (22), (23) and (24), respectively:

$$B\eta = \alpha + \Gamma\xi + \zeta, \tag{31}$$

$$y = \nu_y + \Lambda_y\eta + \epsilon, \tag{32}$$

$$x = \nu_x + \Lambda_x\xi + \delta, \tag{33}$$

where α, ν_y and ν_x are vectors of constant intercept terms. As before, we assume that ζ is uncorrelated with ξ, ϵ is uncorrelated with η and that δ is uncorrelated with ξ. However, it will not be necessary to assume that ϵ is uncorrelated with δ as we did previously. We also assume, as before, that $E(\zeta) = 0$, $E(\epsilon) = 0$ and $E(\delta) = 0$, but it is not assumed that $E(\xi)$ and $E(\eta)$ are zero. The mean of ξ, $E(\xi)$ will be a parameter denoted by κ. The mean of η, $E(\eta)$ is obtained from (31) as

$$E(\eta) = B^{-1}(\alpha + \Gamma\kappa) \tag{34}$$

By taking expectation of (32) and (33) we find the mean vectors of the observed variables to be

$$E(y) = \nu_y + \Lambda_y B^{-1}(\alpha + \Gamma\kappa) \tag{35}$$

$$E(x) = \nu_x + \Lambda_x\kappa. \tag{36}$$

In general, in a single population, all the mean parameters ν_y, ν_x, α and κ will not be identified without further conditions imposed. However, in simultaneous analysis of data from several groups (see next subsection), simple conditions (see e.g.

Jöreskog & Sörbom [1980]) can be imposed to make all the mean parameters identified. We shall not be concerned with identification here but merely confine ourselves to show how the model (31)-(33) can be written in the form of (22)-(24) which is the form used by the program.

The LISREL specification of (31)-(33) is as follows. We treat y and x as y-variables and η and ξ as η-variables in LISREL sense. In addition, we use a single fixed x-variable equal to 1. This variable 1 is also used as the first η-variable. We can then write the model in the form of (22) and (23) as

$$\begin{pmatrix} 1 & 0' & 0' \\ 0 & B & -\Gamma \\ 0 & 0 & I \end{pmatrix} \begin{pmatrix} 1 \\ \eta \\ \xi \end{pmatrix} = \begin{pmatrix} 1 \\ \alpha \\ \kappa \end{pmatrix} 1 + \begin{pmatrix} 0 \\ \zeta \\ \xi - E(\xi) \end{pmatrix} \tag{37}$$

$$\begin{pmatrix} y \\ x \end{pmatrix} = \begin{pmatrix} \nu_y & \Lambda_y & 0 \\ \nu_x & 0 & \Lambda_x \end{pmatrix} \begin{pmatrix} 1 \\ \eta \\ \xi \end{pmatrix} + \begin{pmatrix} \epsilon \\ \delta \end{pmatrix} \tag{38}$$

Note that the mean parameter vectors α and κ appear in the Γ-matrix in LISREL and ν_y and ν_x appear in the first column of Λ_y in LISREL.

The matrix Σ in (25) should now be interpreted as the population moment matrix of the vector $(y', x', 1) = (z', 1)$, say, where z corresponds to y-variables in LISREL and 1 corresponds to an x-variable. The LISREL estimates are obtained by minimizing the function $F(\theta)$, where S is now the sample moment matrix.

Simultaneous Analysis in Several Groups

The LISREL model (31), (32) and (33) can be used to analyze data from several groups simultaneously according to LISREL models for each group with some or all parameters constrained to be equal over groups. Examples of such simultaneous analysis have been given by Jöreskog (1971), McGaw & Jöreskog (1971), Sörbom (1974, 1975, 1976, 1981) and Jöreskog & Sörbom (1980).

Consider a set of G populations. These may be different nations, states or regions, culturally or socioeconomically different groups, groups of individuals selected on the basis of some known or unknown selection variable, groups receiving different treatments, etc. In fact, they may be any set of mutually exclusive groups of individuals which are clearly defined. It is assumed that a number of variables have been measured in a random sample of individuals from each population.

It is assumed that a LISREL model of the form (31), (32) and (33) holds in each group. The model for group g is defined by the parameter matrices $\Lambda_y^{(g)}$, $\Lambda_x^{(g)}$, $B^{(g)}$, $\Gamma^{(g)}$, $\Phi^{(g)}$, $\Psi^{(g)}$ $\Theta_\varepsilon^{(g)}$, $\Theta_\delta^{(g)}$, where the superscript (g) refers to the gth group, $g = 1, 2, \ldots, G$. Each of these matrices may contain fixed, free and constrained parameters as before. If there are no constraints between groups, each group can be analyzed separately. However, if there are constraints between groups, the data from all groups must be analyzed simultaneously to get efficient estimates of the parameters. For example, if the measurement properties of the observed variables are the same in all groups one would postulate that

$$\Lambda_y^{(1)} = \Lambda_y^{(2)} = \ldots = \Lambda_y^{(G)}$$

$$\Lambda_x^{(1)} = \Lambda_x^{(2)} = \ldots = \Lambda_x^{(G)}$$

and perhaps also that

$$\Theta_\varepsilon^{(1)} = \Theta_\varepsilon^{(2)} = \ldots = \Theta_\varepsilon^{(G)}$$

$$\Theta_\delta^{(1)} = \Theta_\delta^{(2)} = \ldots = \Theta_\delta^{(G)}$$

The possible differences between groups would then be represented by differences in the distributions of the latent variables, i.e., by $\Theta^{(g)}$ and $\Psi^{(g)}$. By postulating

$$B^{(1)} = B^{(2)} = \ldots = B^{(G)}$$

$$\Gamma^{(1)} = \Gamma^{(2)} = \ldots = \Gamma^{(G)}$$

one can test the hypotheses that also the structural relations are invariant over groups.

In general, any degree of invariance can be tested, from the one extreme where all parameters are assumed to be invariant over groups to the other extreme when there are no constraints between groups.

To estimate the models simultaneously we minimize the fitting function

$$F = \sum_{g=1}^{G} (N_g/N)F(S_g, \Sigma_g)$$

where F, as before, is any one of the functions (12), (13) and (14) and where N_g is the number of observations in group g and $N = N_1 + N_2 + \ldots + N_G$. When G = 1, this reduces to the same fitting function as before. When the observed variables have a multinormal distribution, F is minus $(N/2)$ times the logarithm of the likelihood function except for an additive constant.

The χ^2 goodness-of-fit measure is defined as before. This is a test of the hypotheses that the LISREL model holds in each group including all constraints, against the general alternative that all $\Sigma^{(g)}$ are unconstrained positive definite, g = 1, 2, ..., G. The degrees of freedom are

$$d = (\tfrac{1}{2})G(p + q)(p + q + 1) - t,$$

where t is the total number of independent parameters estimated in all groups.

While this approach to simultaneous analysis in several groups has been formulated in terms of LISREL models with unconstrained mean vectors it can be generalized in the same way to models with constraints on the means. In fact, the LISREL IV computer program (Jöreskog & Sörbom, 1978) can handle models combining the features of these last two subsections.

ACKNOWLEDGEMENT

The research reported in this chapter has been supported by the Bank of Sweden Tercentenary Foundation under project Methodology of Evaluation Research (project director: Karl G. Jöreskog).

REFERENCES

Aitken, A. C. (1934-35). On least squares and the linear combination of observations. Proceedings of the Royal Society of Edinburgh 55, 42-48.

Anderson, T. W. (1969). Statistical inference for covariance matrices with linear structure. In Multivariate analysis II, (ed. P. R. Krishnaiah). pp. 55-66. Academic Press, New York.

Anderson, T. W. (1970). Estimation of covariance matrices which are linear combinations or whose inverses are linear combinations of given matrices. In Essays in probability and statistics, pp. 1-24 (ed. R. C. Bose et al.). University of North Carolina Press, Chapel Hill, North Carolina.

Bentler, P. M. & Weeks, D. G. Multivariate analysis with latent variables. In Handbook of statistics, vol. 2 (ed. P. R. Krishnaiah & L. Kanal). North Holland Publishing Co., Amsterdam (in press).

Bock, R. D. (1960). Components of variance analysis as a structural and discriminal analysis for psychological tests. British Journal of Statistical Psychology 13, 151-163.

Bock, R. D. & Bargmann, R. E. (1966). Analysis of covariance structures. Psychometrika 31, 507-534.

Browne, M. W. (1970). Analysis of covariance structures. Paper presented at the annual conference of the South African Statistical Association.

Browne, M. W. (1964). Generalized least squares estimators in the analysis of covariance structures. South African Statistical Journal 8, 1-24. (Reprinted in D. J. Aigner & A. S. Goldberger (Eds.) (1977). Latent variables in socio-economic models, pp. 302-326. North Holland Publishing Co., Amsterdam.

Browne, M. W. (1977). The analysis of patterned correlation matrices by generalized least squares. British Journal of Mathematical and Statistical Psychology 30, 113-124.

Fletcher, R. & Powell, M. J. D. (1963). A rapidly convergent descent method for minimization. The Computer Journal 6, 163-168.

Gruveaus, G. & Jöreskog, K. G. (1970). A computer program for minimizing a function of several variables. Research Bulletin 70-14. Educational Testing Service, Princeton, N.J.

Guttman, L. (1954). A new approach to factor analysis: the radex. In Mathematical thinking in the social sciences (ed. P. F. Lazarsfeld), pp. 258-348. Columbia University Press, New York.

Hauser, R. M. & Goldberger, A. S. (1971). The treatment of unobservable variables in path analysis. In Sociological methodology 1971 (ed. H. L. Costner), pp. 81-117. Jossey-Bass, London.

Humphreys, L. G. (1968). The fleeting nature of college academic success. Journal of Educational Psychology 59, 375-380.

Jöreskog, K. G. (1970a). A general method for analysis of covariance structures. Biometrika 57, 239-251.

Jöreskog, K. G. (1970b). Estimation and testing of simplex models. British Journal of Mathematical and Statistical Psychology 23, 121-145.

Jöreskog, K. G. (1971). Simultaneous factor analysis in several populations. Psychometrika 36, 409-426.

Jöreskog, K. G. (1973a). Analysis of covariance structures. In Multivariate analysis. III (ed. P. R. Krishnaiah), pp. 263-285. Academic Press, New York.

Jöreskog, K. G. (1973b). A general method for estimating a linear structural equation system. In Structural equation models in the social sciences (ed. A. S. Goldberger & O. D. Duncan), pp. 85-112. Seminar Press, New York.

Jöreskog, K. G. (1974). Analyzing psychological data by structural analysis of covariance matrices. In Contemporary developments in mathematical psychology. II (ed. R. C. Atkinson, D. H. Krantz, R. D. Luce & P. Suppes), pp. 1-56. W. H. Freeman & Co., San Francisco.

Jöreskog, K. G. (1977). Structural equation models in the social sciences: Specification, estimation and testing. In Applications of statistics (ed. P. R. Krishnaiah), pp. 265-286. North Holland Publishing Co., Amsterdam.

Jöreskog, K. G. (1978). Structural analysis of covariance and correlation matrices. Psychometrika 43, 443-477.

Jöreskog, K. G. (1979). Statistical estimation of structural models in longitudinal-developmental investigations. In Longitudinal research in the study of behavior and development (ed. J. R. Nesselroade & P. B. Baltes), pp. 303-352. Academic Press, New York.

Jöreskog, K. G. & Goldberger, A. S. (1972). Factor analysis by generalized least squares. Psychometrika 37, 243-260.

Jöreskog, K. G. & Goldberger, A. S. (1975). Estimation of a model with multiple indicators and multiple causes of a single latent variable. Journal of the American Statistical Association 10, 631-639.

Jöreskog, K. G. & Sörbom, D. (1977). Statistical models and methods for analysis of longitudinal data. In Latent variables in socio-economic models (ed. D. J. Aigner & A. S. Goldberger), pp. 235-285. North Holland Publishing Co., Amsterdam.

Jöreskog, K. G. & Sörbom, D. (1978). LISREL IV—A general computer program for estimation of a linear structural equation system by maximum likelihood methods. International Educational Services. Chicago.

Jöreskog, K. G. & Sörbom, D. (1980). Simultaneous analysis of longitudinal data from several cohorts. Research Report 80-5. Department of Statistics, University of Uppsala.

Lee, S. Y. (1979). The Gauss-Newton algorithm for the weighted least squares factor analysis. The Statistician 27, 103-114.

Lee, S. Y. & Jennrich, R. I. (1979). A study of algorithms for covariance structure analysis with specific comparisons using factor analysis. Psychometrika 44, 99-113.

Lord, F. M. (1957). A significance test for the hypothesis that two variables measure the same trait except for errors of measurement. Psychometrika 22, 207-220.

Lord, F. M. & Novick, M. E. (1968). Statistical theories of mental test scores. Addison-Wesley Publishing Co. Reading.

McDonald, R. P. (1974). Testing pattern hypotheses for covariance matrices. Psychometrika 39, 189-201.

McDonald, R. P. (1975). Testing pattern hypotheses for correlation matrices. Psychometrika 40, 253-255.

McDonald, R. P. (1978). A simple comprehensive model for the analysis of covariance structures. British Journal of Mathematical and Statistical Psychology 31, 59-72.

McGaw, B. & Jöreskog, K. G. (1971). Factorial invariance of ability measures in groups differing in intelligence and socioeconomic status. British Journal of Mathematical and Statistical Psychology 24, 154-168.

Rao, C. R. (1973). Linear statistical inference and its applications. 2nd ed. Wiley, New York.

Sörbom, D. (1974). A general method for studying differences in factor means and factor structures between groups. British Journal of Mathematical and Statistical Psychology 27, 229-239.

Sörbom, D. (1975). Detection of correlated errors in longitudinal data. British Journal of Mathematical and Statistical Psychology 28, 138-151.

Sörbom, D. (1976). A statistical model for the measurement of change in true scores. In Advances in psychological and educational measurement (ed. D. N. M. de Gruijter and L. J. Th. van der Kamp), pp. 159-169. Wiley, New York.

Sörbom, D. (1978). An alternative to the methodology for analysis of covariance. Psychometrika 43, 381-396.

Sörbom, D. (1981). Structural equation models with structured means. To appear in Systems under indirect observation: Causality, structure, prediction (ed. K. G. Jöreskog and H. Wold). North-Holland Publishing Co., Amsterdam.

Werts, C. E. & Linn, R. L. (1970). Path analysis: Psychological examples. Psychological Bulletin 67, 193-212.

Werts, C. E., Linn, R. L. & Jöreskog, K. G. (1978). Reliability of college grades from longitudinal data. Educational and Psychological Measurement 38, 89-95.

Wheaton, B., Muthén, B., Alwin, D. & Summers, G. (1977). Assessing reliability and stability in panel models. In

Sociological methodology 1977 (ed. D. R. Heise), pp. 84-136, Jossey-Bass, San Francisco.

Wiley, D. E., Schmidt, W. H. & Bramble, W. J. (1973). Studies of a class of covariance structure models. Journal of the American Statistical Association 68, 317-323.

Wright, S. (1934). The method of path coefficients. The Annals of Mathematical Statistics 5, 161-215.

11

Discussion of
Karl G. Jöreskog's Chapter

Erling B. Andersen, Harri Kiiveri,
Petter Laake, D. R. Cox,
and Tore Schweder

ERLING B. ANDERSEN

My comments on Professor Jöreskog's excellent and clear presentation will consist of three parts. Firstly I shall briefly mention the parallel theory for discrete data, which is now known under the name "latent structure analysis." Secondly I shall discuss a number of problems connected with model selection and model checking. Thirdly and finally I shall present two examples, which I hope can give rise to a fruitful discussion.

1. Several of Professor Jöreskog's models are based on the interplay between latent variables and observed or manifest variables. The statistical problems are characterized by the fact that we have a relatively simple structure in the latent space and a somewhat more complex structure in the observed variables.

Without pretending to be a complete list, we have at least four important type of models, which are essential both for the continuous models, as for example LISREL, and their discrete counterparts.

(i) Simple latent structure models; for example one-factor models.

(ii) Models for comparison of several populations. (Section "Simultaneous Analysis in Several Groups" of Chapter 10.)

(iii) Models for correlated latent variables. (Examples 2

Reprinted from Scandinavian Journal of Statistics, Vol. 8 (1981), pp. 83-89. Reprinted by permission.

report

(iii) Models for correlated latent variables. (Examples 2 and 3 of Chapter 10.)

(iv) Longitudinal models. (Example 4 of Chapter 10.)

All these models are also treated in latent structure analysis for discrete data. Andersen (1980) gives a survey of the field.

2. Two quotations from Professor Jöreskog's chapter seem to suggest that the model is primarily checked by checking that the empirical covariance matrix has the structure prescribed by the model. In Chapter 10 it is claimed that: "if the covariance structure is accepted, the model is valid." And it is said that: "The result of an analysis, an inspection of the residuals or other considerations suggest ways to relax the model." (Here an analysis means an analysis of the covariance matrix.) Thus the main emphasis seems to be on the covariance matrix. This among other things raises the question of whether the modelling will take place in the covariance matrix or in the original data. Not only where the modelling starts but also where it ends.

From the examples given in Chapter 10 one gets the impression that remodelling to give a better fit is performed as reformulations of the covariance matrix. Many applied statisticians would undoubtedly feel it more relevant to go back to the original data for example by means of an analysis of the residuals in order to discover possible model reformulations.

The ambiguity in the formulation of a model is illustrated by the heavy attention one has to pay to identification problems. Personally I do not like the idea of computerized identification as described and feel that the identification of the model should be an integral and quite natural part of the process of formulating the model.

That the models are closely connected to the assumption of a normal distribution is very obvious. To claim that only the first two moments are considered is in reality the same thing. In some cases this assumption does not seem appropriate. One case in point is example 3, where one could question whether the variable "Education" is normally distributed.

A relevant question could be: Is the model choice based on the type of data at hand? Has it thus been considered to transform the data before the covariance matrix is computed?

3. Finally the two examples: The <u>first</u> example is constructed data, where two of the models fit the same data. (The data are constructed in such a way that we have a complete fit.) In Table 11.1 is shown the covariance matrix between the amount of iron, copper, silver and gold found in the tombs of 37 of Nebuchadnezzar's generals.

TABLE 11.1

Covariances between the Metal Found in 37 Ancient Tombs

Variable		Iron	Copper	Silver	Gold
1	Iron	104.4	81.9	46.1	62.9
2	Copper	81.9	109.8	47.6	65.0
3	Silver	46.1	47.6	80.0	74.7
4	Gold	62.9	65.0	74.7	127.2

These data are fit by the model with two correlated latent variables ($\beta_1 = 8.9$, $\beta_2 = 9.2$, $\beta_3 = 7.4$, $\beta_4 = 10.1$, $\theta_1 = \theta_2 = \theta_3 = \theta_4 = 25.2$, $\rho = 0.7$). The theory is here that the activities of war and peace take place at different times and require different—although correlated—talents. Since iron and copper are usually found in weapons and silver and gold in coins or jewelry, variables 1 and 2 relate to war and variables 3 and 4 to peace. The data thus bears evidence that this theory is correct.

But the data are also fitted by the longitudinal model ($\beta_2 = 1.03$, $\beta_3 = 0.56$, $\beta_4 = 1.36$, $\theta_1 = \theta_2 = \theta_3 = \theta_4 = 25.2$, $\omega_1 = 79.2$, $\omega_2 = 84.6$, $\omega_3 = 54.8$, $\omega_4 = 102.0$). Here one can theorize that a general goes through four phases: First a lower rank officer, then a high ranking officer, then a business man and finally an influential person at court. The four types of metal seem to associate naturally to these four phases. Since the model fits the data, also this theory is confirmed by the data. If the two theories are alternative and the metal found is the evidence available, how should we then choose or judge between the two theories.

The point I would like to make is that different models may fit the same set of data and statisticians would generally tend to claim that they have verified or confirmed a theory if the derived statistical model fit the data collected to check the theory.

The second example is given in Table 11.2. The table shows the empirical covariances between examination average after high school and examination average after one and two years of study in economics at the University of Copenhagen. The figures are based on a random sample of 27 students.

For these data I have tried to fit the longitudinal simplex model. Since we have 6 variances/covariances and 6 parameters

TABLE 11.2

Covariances between Examination Averages for 27 Economics Students

Average	Average		
	High School	1st Year	2nd Year
High school	0.792	0.600	0.137
1st year	0.600	1.265	0.600
2nd year	0.137	0.600	1.030

in the model (assuming $\theta_1 = \theta_2 = \theta_3$) we can fit directly. We get, when the model is

$$\begin{bmatrix} \omega_1 + \theta & & \\ \beta_2 \omega_1 & \omega_2 + \theta & \\ \beta_2 \beta_3 \omega_1 & \beta_3 \omega_2 & \omega_3 + \theta \end{bmatrix}$$

the following estimates

$$\hat{\beta}_3 = 0.137/0.600 = 0.228$$

$$\hat{\omega}_2 = 0.600/0.228 = 2.631$$

$$\hat{\theta} = 1.265 - 2.631 = -1.367!!$$

Hence the estimates do not belong to the range of the parameter's pace.

The maximum likelihood estimates are undoubtedly given by (although I did not check)

$$\hat{\theta} = 0$$

$$\hat{\omega}_1 = 0.792$$

$$\hat{\beta}_2 = 0.600/0.792 = 0.758$$

$$\hat{\omega}_2 = 1.265$$

$$\hat{\omega}_3 = 1.030$$

$\hat{\beta}_3 = 0.600/1.265 = 0.474.$

Since these are real data I do not know what to do and how to estimate, and, therefore, how to check my model. The example also raises the question of what to do when the likelihood function is maximized (or equivalently some convenient distance measure is minimized) at the boundary of the permissible range of the parameters.

Reference

Andersen, E. B. (1980). Latent Structure Analysis. A Survey, Research Report. Department of Statistics University of Copenhagen.

HARRI KIIVERI

Most of my comments will be concerned with covariance structures which arise from models with unobserved variables.

1. The Lisrel model without means can be written as

$$\begin{bmatrix} I & 0 & -\Lambda_y & 0 \\ & I & 0 & -\Lambda_x \\ & B & -\Gamma \\ & & & I \end{bmatrix} \begin{bmatrix} Y \\ X \\ \eta \\ \xi \end{bmatrix} = \begin{bmatrix} \varepsilon \\ \delta \\ \zeta \\ \xi \end{bmatrix}$$

i.e. $U^T Z = e$ $\underset{\sim}{U}' \underset{\sim}{z} = \underset{\sim}{e}$

and $\text{Cov (e)} = \begin{bmatrix} \Theta_\varepsilon & & & 0 \\ & \Theta_\delta & & \\ & & \Psi & \\ 0 & & & \Phi \end{bmatrix}$

= A (say).

Assuming that B and A are nonsingular and denoting the covariance matrix of Z by Σ we get:

$\Sigma^{-1} = U A^{-1} U^T.$

If all the variables were observed and had a joint Gaussian distribution, it is not difficult to see that the structure on Σ^{-1} gives rise to the following factorization of the joint density.

$$f(z) = g_0(\xi)g_1(x, \xi)g_2(\eta,\xi)g_3(y, \eta)$$

which can be put into the form

$$f(z) = f_0(\xi)f_1(x|\xi)f_2(\eta|\xi)f_3(y|\eta)$$

where the terms on the right hand side are conditional densities. This will be abbreviated as

$$f = (\xi)(x|\xi)(\eta|\xi)(y|\eta).$$

Depending on the structure in A and U further factorization is possible. For example, the model for study of stability of alienation with correlated "errors" has a structure which gives

$$f = (\xi)(x_1|\xi)(x_2|\xi)(\eta_1|\xi)(\eta_2|\eta_1\xi)(y_1y_3|\eta_1\eta_2)$$
$$x(y_2y_4|\eta_1\eta_2).$$

Such factorizations can be described by means of graphs and rules given for reading off the resulting conditional independences (Kiiveri & Speed, 1980).

There is also a connection with other conditional independence models for contingency tables and Gaussian data (Speed, 1978; Wermuth, 1976; Dempster, 1972). Viewing the LISREL model in this light gives a means of extending the model to other types of data (i.e. discrete or mixtures of discrete and continuous). For example, the discrete version of a simple test score model for tests $X_1, \ldots, X_p$ (having levels $r_1, r_2, \ldots, r_p$ respectively) assumed to be measuring an underlying trait τ might be specified as

$$p(X_1 = x_1, \ldots, X_p = x_p, \tau = t) = p_0(t) \prod_{i=1}^{p} p_i(X_i = x_i|t)$$

where τ could be discrete or continuous. This type of model for discrete data has been discussed by Goodman (1973, 1974).

2. The added complication of unobserved variables could be handled using incomplete data theory (Sundberg, 1974; Dempster et al., 1977). The complete data would be

$$\begin{bmatrix} S_{YY} & & & \\ S_{XY} & S_{XX} & & \\ S_{\eta Y} & S_{\eta X} & S_{\eta\eta} & \\ S_{\xi Y} & S_{\xi X} & S_{\xi\eta} & S_{\xi\xi} \end{bmatrix} \text{ with } \begin{bmatrix} \bar{Y} \\ \bar{X} \\ \bar{\eta} \\ \bar{\xi} \end{bmatrix}$$

if the means were included. Only S_{YY}, S_{XY}, $S_{XX}\bar{X}$, $\bar{Y}$ are observed. Function evaluations and derivative calculations can be done in a nice way using incomplete data theory and expressions for 2nd derivatives can also be given. The E-M algorithm (Dempster et al., 1977) although slow can be used to find a local maximum of the likelihood function of the observed data even if all the parameters are not identified. However it would be quicker to use this to produce initial values for some other algorithm.

3. From the examples given by Professor Jöreskog it appears that the one to oneness of the mapping $\theta \to \Sigma(\theta)$ must be checked on a case by case basis or in a crude way numerically. It would be nice to have a more general procedure and perhaps the matrix calculus of (McDonald & Swaminathan, 1973) along with incomplete data theory can be exploited in this direction. A numerical determination of the rank of $\partial\Sigma/\partial\theta$ can determine a set of parameters which need to be fixed to obtain (local) identifiability (McDonald & Krane, 1979.)

4. Finally I would like to touch on some theoretical points. There do not seem to be any existence of uniqueness results for the maximum likelihood estimates, and one wonders whether the special sort of incompleteness and structure could be used to produce some results in a curved exponential family context.

It seems as if testing a model should be possible without arbitrarily fixing unidentified parameters, since any parameter in the equivalence class of the maximum point gives the same value of the likelihood. Also I am not clear on the effect of the particular fixed values chosen on the relative magnitudes of the parameters or the standard errors.

In closing I would like to thank Professor Jöreskog for his survey and particularly for his examples on the application of the theory to data. I like the trick used to estimate means in LISREL and wonder what other "devious" means may be employed to fit seemingly "non-LISREL" models.

References

Darroch, J. N., Lauritzen, S. L. & Speed, T. P. (1980). Log-linear models for contingency tables and Markov fields over graphs. Annals of Statistics 8 (to appear).

Dempster, A. P. (1972). Covariance selection. Biometrics 28, 157-175.

Dempster, A. P., Laird, N. M. & Rubin, P. B. (1977). Maximum likelihood from incomplete data via the EM algorithm. J. Roy. Stat. Soc. Ser. B. 39, 1-21.

Goodman, L. A. (1973). The analysis of multidimensional contingency tables when some variables are posterior to others: a modified path analysis approach. Biometrika 60, 179-192.

Goodman, L. A. (1974). The analysis of systems of qualitative variables when some of the variables are unobservable. Part I—a modified latent structure approach. American Journal of Sociology 79, 1179-1259.

Kiiveri, H. T. & Speed, T. P. (1980). The structural analysis of multivariate data: a review. Submitted to Sociological methodology.

McDonald, R. P. & Krane, W. R. (1979). A Monte-Carlo study of local identifiability and degrees of freedom in the asymptotic likelihood ratio test. British Journal of Mathematical and Statistical Psychology 32, 121-132.

McDonald, R. P. & Swaminathan, H. (1973). A simple matrix calculus with applications to multivariate analysis. General Systems 18, 37-54.

Speed, T. P. (1978). Relations between models for spatial data, contingency tables and Markov fields on graphs. Proceedings of the conference on spatial patterns and processes (ed. R. L. Tweedie). Supplement to Advances in Applied Probability.

Sundberg, R. (1974). Maximum likelihood theory for incomplete data from an exponential family. Scand. J. Statist. 1, 49-58.

Wermuth, N. (1976). Analogies between multiplicative models in contingency tables and covariance selection. Biometrics 32, 95-108.

Wermuth, N. (1980). Linear recursive equations, covariance selection and path analysis. J. Amer. Statist. Assoc. (to appear).

PETTER LAAKE

1. First of all I want to thank Professor Jöreskog for his inspiring study on application and estimation in covariance structures. It certainly contained a lot of interesting and useful ideas. I have got my interest and experience in covariance structure analysis from cooperation with social scientists, and my discussion is probably influenced by that.

2. There can be no doubt that measurements in the social sciences may contain measurement errors. The statistical model ought to take this into consideration, especially, of course, when the parameters of interest are related to the true variables. Models which relate every measurement with its true, latent, variable and an error term, and relate the latent variables in a structural form are extremely useful. This is for instance the case in the quasi-simplex models. Then, the interpretation of the results is easy since the units of measurement in the observed variable are the same as in the latent variable, and there is no question of validity. This is not necessarily the case in the more advanced covariance structures.

To be more specific I will comment on one of Professor Jöreskog's examples, namely Example 3 which is an application of the LISREL-model. The model specifies the relationship between latent variables (socioeconomic status [SES] and alienation) and relates the latent variables and the measurements in a factor analysis model. The parameters of main interest are those of the relation between the latent variables, namely the parameters describing the influence of SES on alienation in 1967 and 1971 and the parameter describing the influence of alienation in 1967 on alienation 1971. Since all the relations are given explicitly the analysis seems to be confirmatory or in statistical terms that we want to estimate parameters of a given structural relationship. Then, the social scientists will be interested in the estimated size of the parameters regardless of their statistical significance. This is so because the parameters are included in the model according to our interest in them and because the question of statistical significance may be a question of a sufficiently large sample size. To interpret the estimated size of the parameter we usually relate it to the scale of the variables and their distributions and not to the standard error of the estimate. But latent variables are in general not related to any scale. So how do we interpret some of the main results of Example 3, for instance that the coefficients for SES on alienation 1967 and 1971 are -0.575 and -0.227, respectively? One possibility is to make the scale of the latent variable equal to the one of a measurement by fixing one element of the factor loading matrix equal to 1. This solves the problem of identification, but the problem of interpretation is not fully solved since the latent variable still influences other measurements. Another possibility, which could be studied in more detail, is to use the factor model for estimation of factor scores (see for instance Lawley & Maxwell [1971, p. 106]). Then, the interpretation of the coefficients can be related to the empirical distribution of the factor scores. Further, it is possible to relate the coefficients directly to the measurements through the factor model.

Professor Jöreskog's opinion on the application of covariance structure analysis seems to be that they most successfully have been applied to confirmatory analysis, but that they also can be used for exploratory analysis. I certainly agree with him that for instance the LISREL-model is useful when a well-established theory of the problem exists. This is so because we in some way must have demonstrated the validity of the factor model, for instance by earlier use of the latent variables in similar context. If the validity fails, the analysis of the structural relationship among the latent variables may be meaningless. On the other hand it is indicated that LISREL may be used for exploratory analysis. This use of it may be doubtful. Let us say that the relationship among the latent variables is given, but for a given vector of measurements we are free to choose the factor model. It is well known that all orthogonally rotated factor loading matrices are statistically identical with regard to the model. The interpretation of the factors and the coefficients of the main relationship will, however, depend upon the chosen factor loading matrix. This can make it difficult for the social scientist to get an understanding of the structure of the measurements. Further, the factor model itself will not demonstrate the validity of the latent variables resulting from the exploratory analysis. To give a latent variable a name is, of course, not necessarily enough to make it useful.

Social scientists often deal with complex social theories and try to build complex statistical models. I believe that it not necessarily will give us an uninteresting or unrealistic analysis of a problem even if we disregard structural relationships which are known to exist. This may be especially relevant in an exploratory analysis, when the aim is to get an understanding of the structure of a given set of measurements. Interesting information that is easy to interpret may, for instance, be gained if multivariate regression analysis, which disregards measurement errors, is considered. If we go back to Example 3, it will be interesting to know if conclusions similar to those of

Jöreskog are gained if a multivariate regression model is studied instead of the LISREL-model?

3. Example: Predicting the labour force in Norway (Helgeland, 1980). Next I will give an example of the use of a covariance structured model related to the simplex model which is being studied at the Central Bureau of Statistics of Norway. To be specific let η_t be the unobservable labour force at time t. In the Bureau η_t has to be estimated by a sample survey. Let therefore

$$y_t = \eta_t + \varepsilon_t,$$

where ε_t is the sampling error. Var ε_t can be found using the design of the sample survey. We now assume that $\{\eta_t\}$ is an ARMA-process

$$\eta_t = \sum_{i=1}^{p} \beta_i \eta_{t-i} + \sum_{j=1}^{q} \alpha_j a_{t-j},$$

with $\alpha_0 = 1$ and a_t's independent. The observations are the estimated labour force $y_1, y_2, \ldots, y_T$. Since the units of measurement in the latent variable η are the same as in the observed variable y, the interpretation is straightforward. The aim of the study is not only estimate the α's and β's, but to predict

$$E\{\eta_T | y_1, y_2, \ldots, y_T\}.$$

This is actually a LISREL-specified model, and numerical methods similar to those described by Jöreskog can be used for estimation. The asymptotic theory does not carry over, however, since no replications are made at time t. It is easy to imagine that models like this may be relevant in Central Bureaus of Statistics.

4. I have now a few more technical comments on Jöreskog's chapter. An important problem for the applicability of the LISREL specification is the efficiency of the numerical technique for maximum likelihood estimation of the parameters. It is well known that the efficiency depends upon the start value for iteration, but we may never know if we actually have reached a global maximum of the likelihood function. It is easy to imagine situations where this may be difficult, especially when the sample sizes are small and multiple solutions to the likelihood equations exist. In situations like these I think it may be a

good idea to try different start values for iteration and hope for every iteration to give the same solution.

Another problem is the one of restricted parameter estimation. Some parameters of the covariance structure are known to be greater or equal to zero. A numerical technique can in general produce an iterate which lies outside the constraint space. When for instance a variance estimate tends to zero the approach is usually to fix the variance at a given level during the following iterations. This may be a questionable procedure since it may cause the procedure to converge to a point that is not even a local maximum (Bard, 1974, p. 151). Thus, the problem is whether it is possible to consider numerical techniques that take the constraint into consideration and still iterate efficiently.

5. There exist alternatives for parameter estimations in the models. The most commonly used seem to be maximum likelihood estimation (ML) and generalized least squares estimation (GLS). I certainly agree with Professor Jöreskog in that GLS is computationally easier and thus probably more attractive from a practical point of view. They both produce asymptotically efficient estimates. Both Jöreskog & Goldberger (1972) and later Browne (1974) report numerical results that could indicate that there is a bias in the GLS-estimates compared to the ML-estimates. This could indicate that GLS-estimates really are biased in small samples or that they may have larger variances. To make the computationally more attractive GLS-method even more attractive to the user I think their small samples properties ought to be studied in some more detail.

6. The choice of a model is always a difficult problem. In a confirmatory analysis the model fitting is to me of minor interest and the only relevant testing seems to be the one of testing for correlation in residuals to detect any possible shortcoming of the model. In an exploratory analysis the model fitting is important and controversial. Professor Jöreskog uses the likelihood ratio test in his model choice. The properties of the test may be questionable. The test is likely, for example, even in large samples, to be sensitive to departures from normality. This is in my opinion not necessarily an important objection since the test statistic is mainly used to study the drop in fit when alternative models are introduced. I think, however, there may be some objections to it from a practical point of view.

Firstly, a straightforward use of the likelihood ratio test gives preference to models with many parameters. A model with many parameters will give a nice fit, which may be of importance to the statistician, but not necessarily to the social scientist who then will get a more complex structure to explain. Thus, I think, some kind of penalty function to keep the number of parameters at a low level could be introduced in addition to the maximum likelihood function. This is done by, for instance, Hannan & Quinn (1979) and Schwarz (1978), mainly for application in time series analysis. The procedures can, however, be used for any model choice when maximum likelihood estimation is used.

Secondly, the likelihood ratio test is not diagnostic which may be an important drawback to the use of it. When we reject a model we usually want to know where it is misspecified. In Example 3 we will, for instance, probably observe that the elements of $(S - \hat{\Sigma})$ at the position of the covariances of y_1, y_3 and y_2, y_4 are large. Thus, without predistinguishing between the models A and B, the analysis of the residuals will guide us to respecify to model B. To conclude, I think other test statistics which may be based on the residuals could contain more information on why models are rejected.

7. Finally, I will once more thank Professor Jöreskog for his fine overview. In my daily work I find it extremely useful that Professor Jöreskog has cleared up concepts which were in actual use in sociology and psychology and expressed them in a language natural to the statisticians. Together with his work on estimation methods, Professor Jöreskog has certainly made the covariance structured models valuable for statistical analysis in the social sciences.

References

Bard, Y. (1974). Nonlinear parameter estimation. Academic Press, New York.

Browne, M. W. (1974). Generalized least squares estimators in the analysis of covariance structures. South African Statistical Journal 8, 1-24.

Jöreskog, K. G. & Goldberger, A. S. (1972). Factor analysis by generalized least squares. Psychometrika 37, 243-260.

Hannan, E. J. & Quinn, B. G. (1979). The determination of the order of an autoregression. J. Roy. Stat. Soc. B 41, 190-195.

Helgeland, J. (1980). Unpublished manuscript. Central Bureau of Statistics of Norway.

Lawley, D. N. & Maxwell, A. E. (1971). Factor analysis as a statistical method. Butterworth & Co., London.

Schwarz, G. (1978). Estimating the dimension of a model. Ann. Statist. 6, 461-464.

D. R. COX

I congratulate Professor Jöreskog both on this impressively lucid and comprehensive study and also on the important body of work which it summarizes.

Three questions follow. Have Bartlett correction factors been evaluated for any of the likelihood ratio tests involved? Such corrections may appreciably improve the large-scale chi-squared approximations. Can marginal likelihoods be found for any of the problems with structure in the means as well as in the covariances? Has Professor Jöreskog suggestions on how to examine whether information on dependency is adequately summarized in a covariance matrix? The suggestion of Cox & Small (1978) for finding two derived directions corresponding to maximum curvature seems reasonable although its implementation is cumbersome computationally.

If the full data for Example 1 became available the possibility would arise of an analysis based on standard factorial contrasts and their interaction with subjects.

Reference

Cox, D. R. & Small, N. J. H. (1978). Testing multivariate normality. Biometrika 65, 263-272.

TORE SCHWEDER

The following comment is not meant to reduce the value of the very clear and informative exposition Jöreskog has given of the analysis of covariance structure. My intention is actually not to focus on the analysis of covariance structure as such, but rather to fix the attention for a while on the difficult concept of "causality" which you have touched upon repeatedly.

The causal relationships which are modelled in path analysis may be called static as opposed to dynamic causal relationships. The causal variables in your models have for each individual a

FIGURE 11.1

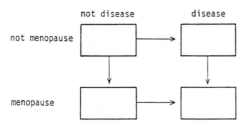

fixed value. The variation is over individuals, not also over time. And your concern is in some sense to estimate the correlation between the cause variable and the effect variable. Typical such static causal variables are sex and intelligence, the latter being a latent variable.

This static concept of causality is quite limited. In ordinary language, causality is often used in a more dynamic sense. And also in a sense more open for human action. The type of concept I have in mind is exemplified in the statement "smoking does cause lung cancer," rephrased "the probability of getting lung cancer increases when smoking is increased." It is hard to change your own sex or intelligence, but your level of smoking should be a matter of choice.

Such a dynamic concept of causality must be defined in the framework of stochastic processes. This has been done by Aalen et al. (1979) in their study of the relationship between menopause and the skin disease pustulosis planetaris. They modelled causality in terms of the transition intensities in the two-dimensional stochastic process with state space indicated in Figure 11.1.

Menopause being a cause of the disease is defined as the transition intensity of getting the disease being dependent on having had menopause or not.

As opposed to correlation, the concept of transition intensity—or drift in a continuous state space—is defined in a time context, and is a non-symmetric concept. When the situation may be modelled as a multivariate stochastic process—or rather one separate multivariate stochastic process for each individual of the population, it may be possible to formulate various hypotheses of causation in terms of the component transition intensities. In Schweder (1970) the concept of local dependence was introduced for this purpose. Statistical methodology has been developed by Aalen et al. (1979) for the two-dimensional situation mentioned above, and may be further developed along those lines.

As mentioned above, my intention with this comment has been to focus attention on a conceptual framework and a statistical methodology for "causal analysis" which in many situations, I believe, is more appropriate than path analysis or other methods based on correlation.

References

Aalen, O. et al. (1980). Interaction between life history events. Nonparametric analysis for prospective and retrospective data in the presence of censoring. Scand. J. Statist. 7, 161-171.

Schweder, T. (1970). Composable Markov processes. Journal of Applied Probability 7(2), 400-10.

12

Reply

Karl G. Jöreskog

I want to thank all the discussants for their constructive
comments on my chapter. The discussants have raised several
issues which are important in the practical applications of covariance structure analysis and I will therefore state my position
on these issues here.

Model specification and purpose of analysis. The model
should be based on a substantive theory or hypothesis, on a
given design for known experimental conditions or on known
results from previous studies. The model is supposed to explain
or account for the jointly dependent variables and their inter-
relationships. The purpose of the analysis is to test the validity
and adequacy of the model and to estimate its parameters. If
the model is misspecified the analysis should indicate this and
suggest how the model should be modified.

Identification. Before the model is estimated by means of
data the identification of the model must be examined. Although
the computer program LISREL IV can estimate models which are
not identified, I recommend to deal with non-identified models
by adding appropriate conditions so as to make all the parameters
identified. In interpreting the results of an analysis it is necessary to have a complete understanding of the identification status
of each parameter of the model. Since no general and practically
useful necessary and sufficient conditions for identification are
available for the general LISREL model, I suggest that the

Reprinted from Scandinavian Journal of Statistics, Vol. 8
(1981), pp. 89-92. Reprinted by permission.

identification problem be studied on a case by case basis by examining the equations (11). General necessary and sufficient conditions for identification have been given for some special classes of LISREL models, namely for factor analysis models by Howe (1955), Dunn (1973), Jennrich (1978), see also Jöreskog & Sörbom (1979, pp. 40-43), for simultaneous structural equation models, see e.g. Goldberger (1964, pp. 306-318) and for structural equation models with measurement errors by Geraci (1976). These conditions may be applied to the measurement model part (23) and (24) and to the structural equation model part (22) separately. It is a good idea to check these conditions.

For many users of LISREL IV the identification problem may be too difficult to resolve. There may be a tendency to run the model even though the identification of it is unclear. Therefore it is a good thing that the computer program checks the positive definiteness of the information matrix. One should be aware, however, that this check is not one hundred percent reliable although experience indicates that it is nearly so. The check depends on the estimated point in the parameter space at which the information matrix is evaluated and on the numerical accuracy by which the information matrix is inverted.

One and the same model, whether it is completely identified or not, may have two or more equivalent parametrizations even within the class of LISREL models. Which parametrization to use is a matter of preference and is related to the interpretation of the result. One example of this is Professor Andersen's first example and it is instructive to consider this in some detail. As demonstrated elsewhere (Jöreskog, 1974, p. 52) the two models (5) and (9) are equivalent when the number of observed variables is four. To see this let me write (5) with λ's instead of β's and ψ's instead of θ's:

$$\Sigma = \begin{bmatrix} \lambda_1^2 + \psi_1 & & & \\ \lambda_1\lambda_2 & \lambda_2^2 + \psi_2 & & \\ \lambda_1\lambda_3\varrho & \lambda_2\lambda_3\varrho & \lambda_3^2 + \psi_3 & \\ \lambda_1\lambda_4\varrho & \lambda_2\lambda_4\varrho & \lambda_3\lambda_4 & \lambda_4^2 + \psi_4 \end{bmatrix}$$

This model is identified and has nine independent parameters, i.e., the model has one degree of freedom. The parameter values $\lambda_1 = 8.9$, $\lambda_2 = 9.2$, $\lambda_3 = 7.4$, $\lambda_4 = 10.4$, $\psi_1 = \psi_2 = \psi_3 = \psi_4 = 25.2$, $\varrho = 0.7$ given by Professor Andersen are therefore unique.

As explained in Chapter 10, model (9) is not identified. The three parameters β_2, ω_1 and θ_1 are only determined by two equations:

$$\sigma_{11} = \omega_1 + \theta_1$$

$$\sigma_{21} = \beta_2 \omega_1$$

and the two parameters ω_4 and θ are only determined by the single equation:

$$\sigma_{44} = \omega_4 + \theta_4.$$

All the other parameters are uniquely identified. The parameter values given by Professor Andersen are to some extent arbitrary. For example the following set of parameter values $\beta_2 = 1.00$, $\beta_3 = 0.56$, $\beta_4 = 1.36$, $\theta_1 = 22.5$, $\theta_2 = \theta_3 = 25.2$, $\theta_4 = 10.0$, $\omega_1 = 81.9$, $\omega_2 = 84.6$, $\omega_3 = 54.8$, $\omega_4 = 117.2$ will also reproduce identically the same Σ.

That the two models are equivalent may be seen for example by fixing β_2 at one and absorbing θ_4 into ω_4. This gives the covariance structure

$$\Sigma = \begin{bmatrix} \omega_1 + \theta_1 & & & \\ \omega_1 & \omega_2 + \theta_2 & & \\ \beta_3 \omega_1 & \beta_3 \omega_2 & \omega_3 + \theta_3 & \\ \beta_3 \beta_4 \omega_1 & \beta_3 \beta_4 \omega_2 & \beta_4 \omega_3 & \omega_4 \end{bmatrix}$$

Then it may be easily verified that there is a one-two-one relation between the two sets of parameters, assuming, of course, that the variances ω_i, $i = 1, 2, 3, 4$, are positive:

$$\beta_3 = \lambda_3 \varrho / \lambda_2 , \ \beta_4 = \lambda_4 / \lambda_3 , \ \omega_1 = \lambda_1 \lambda_2 , \ \omega_2 = \lambda_2^2 , \ \omega_3 = \lambda_3^2 , \ \omega_4$$

$$= \lambda_4^2 + \psi_4 , \ \theta_1 = \lambda_1^2 + \psi_1 - \lambda_1 \lambda_2 , \ \theta_2 = \psi_2 , \ \theta_3 = \psi_3$$

and

$$\lambda_1 = \omega_1 / \sqrt{\omega_2} , \ \lambda_2 = \sqrt{\omega_2} , \ \lambda_3 = \sqrt{\omega_3} , \ \lambda_4 = \beta_4 \sqrt{\omega_3} ,$$

$$\varrho = \text{sgn} \ (\beta_3)(\omega_2 |\beta_3| / \omega_3)^{\frac{1}{2}} , \ \psi_1 = \omega_1 + \theta_1 - \omega_1^2 / \omega_2 , \ \psi_2 = \theta_2$$

$$\psi_3 = \theta_3 , \ \psi_4 = \omega_4 - \beta_4^2 \omega_3$$

It should be noted that in both models the single overidentifying constraint is that $\sigma_{41} / \sigma_{31} = \sigma_{42} / \sigma_{32}$.

Which of the two parametrizations is to be preferred? If one is interested in the correlation ϱ between two latent variables then model 1 is the one. On the other hand, if one is interested in the autoregression coefficients β_3 and β_4 then model 2 is the one. The reason why these two seemingly different models are equivalent in this case is that the quasi-Markov-simplex model is not really meaningful until one has at least five time points. Since there is an indeterminacy associated with the first and the last variable one must have at least three variables in between these to be able to test the first-order Markov property.

Checking the data. In addition to the structural assumptions imposed by the model specification the analysis makes the usual assumptions of linearity and additivity, independence of observations and homoskedasticity of error terms. Another crucial question is whether second-order information is sufficient. Usually one needs to assume that the observed variables are approximately multinormal. There is one important exception to this, however, namely when the x-variables are fixed variables. Then these variables may be any variables including dummy categorical variables. We are then only concerned with conditional distribution of y for given x which should be approximately multinormal. Before estimating the model, the assumption of multinormality should be assessed, including a check on outliers. If the observed distributions deviate far from the multinormal I recommend that the variables be transformed and/or that robustified estimates of variances and covariances be used. For methods of assessing multinormality and computing robust estimates see e.g. Gnanadesikan (1977).

Checking the model. The first and most obvious test of the model is whether the parameter estimates are reasonable. If some parameters fall outside the admissible range, either the model is fundamentally wrong or the data is not informative enough. An example of this is Professor Andersen's second example and I shall consider this below.

The fit of the model may be assessed by various means. One is the use of the overall χ^2-measure and its associated degrees of freedom and probability level. However this does not indicate which part of the model is wrong. A more detailed assessment of fit can be obtained by an inspection of the normalized residual covariances. The normalized residual covariance is $s_{ij} - \hat{\sigma}_{ij}$ divided by the square root of its asymptotic variance which is estimated as $(\hat{\sigma}_{ii}\hat{\sigma}_{jj} + \hat{\sigma}_{ij})/N$. A Q-plot of these residuals gives a very effective summary. Residuals which are larger than two in magnitude are indicative of a specification error in the model and the corresponding indices i and j usually give a hint as to where this error is.

Another alternative is to look at the first-order derivatives of the fitting function with respect to the fixed and constrained parameters, these being the Lagrangian multipliers corresponding to the constraints of the model, see Sörbom (1975). One problem with this approach, however, is that these derivatives depend on the magnitude of the data and the parameter values. In principle, one must assess each derivative in relation to its standard error. But the computation of these standard errors requires the inversion of a matrix of the order of all elements in all parameter matrices and this is obviously not feasible in most cases. A reasonable compromise may be to consider the ratio of the first-order derivative to the corresponding second-order derivative. The fixed parameter corresponding to the largest such index is probably the one which when relaxed will improve fit maximally. I recommend this procedure only when relaxing this parameter makes sense from a substantive point of view and when the values of this parameter can be clearly interpreted.

Multiple solutions and non-admissible solutions. It may happen that there are several local minima of the fitting function. The only way to avoid this is to have a model which is appropriate for the data and a large random sample. Experience indicates, however, that multiple solutions seldom occur, and when they do, it is usually with solutions on the boundary of or outside the admissible parameter space, as in Professor Andersen's second example. The computer program does not constrain variances to be positive, correlations to be less than one in magnitude, etc. The only constraint imposed by the program is that the matrix Σ reproduced by the model is positive definite. Apart from this there is nothing that prevents the program from going outside the admissible parameter space. Although constrained estimation would be possible, the minimization algorithm for this would be much more time-consuming even in the case when the solution is admissible. In my opinion, this would not be worth while. If the solution is in the interior of the parameter space one will just spend more computer time to find it. If, on the other hand, the solution is inadmissible, the current program will find the solution outside the parameter space whereas a program which uses constrained estimation will find the solution on the boundary of the parameter space. In both cases the conclusion will be that the model is wrong or that the sample size is too small.

Using LISREL it is often possible, though not always, to use various tricks to force the program to stay within the admissible parameter space. Let me demonstrate this using Professor Andersen's second example. This covariance structure is generated by the following simplex model

$$y_t = \eta_t + \varepsilon_t, \ t = 1, 2, 3 \tag{1}$$

$$\eta_t = \beta_t \eta_{t-1} + \zeta_t, \quad t = 2, 3 \tag{2}$$

with $\mathrm{Var}\,(\eta_t) = \omega_t$ and $\mathrm{Var}\,(\varepsilon_t) = \theta$, $t = 1, 2, 3$. Let $\psi_t = \mathrm{Var}\,(\zeta_t)$. Then there is a one-to-one correspondence between (ψ_1, ψ_2, ψ_3) and $(\omega_1, \omega_2, \omega_3)$ $(\psi_1 = \omega_1, \psi_2 = \omega_2 - \beta_2^2\omega_1, \psi_3 = \omega_3 - \beta_3^2\omega_2)$ so I will use ψ_t instead of ω_t, $t + 1$, 2, 3. The trick is to write (1) and (2) for LISREL as

$$
\begin{pmatrix} y_1 \\ y_3 \\ y_3 \end{pmatrix} = \begin{pmatrix} 1 & 0 & 0 & 1 & 0 & 0 \\ 0 & 1 & 0 & 0 & 1 & 0 \\ 0 & 0 & 1 & 0 & 0 & 1 \end{pmatrix} \begin{pmatrix} \eta_1 \\ \eta_2 \\ \eta_3 \\ \varepsilon_1 \\ \varepsilon_2 \\ \varepsilon_3 \end{pmatrix} + \begin{pmatrix} 0 \\ 0 \\ 0 \end{pmatrix}
$$

$$
\begin{pmatrix} 1 & 0 & 0 & 0 & 0 & 0 \\ -\beta_2 & 1 & 0 & 0 & 0 & 0 \\ 0 & -\beta_3 & 1 & 0 & 0 & 0 \\ 0 & 0 & 0 & 1 & 0 & 0 \\ 0 & 0 & 0 & 0 & 1 & 0 \\ 0 & 0 & 0 & 0 & 0 & 1 \end{pmatrix} \begin{pmatrix} \eta_1 \\ \eta_2 \\ \eta_3 \\ \varepsilon_1 \\ \varepsilon_2 \\ \varepsilon_3 \end{pmatrix}
$$

$$
= \begin{pmatrix} \sqrt{\psi_1} & 0 & 0 & 0 & 0 & 0 \\ 0 & \sqrt{\psi_2} & 0 & 0 & 0 & 0 \\ 0 & 0 & \sqrt{\psi_3} & 0 & 0 & 0 \\ 0 & 0 & 0 & \sqrt{\theta} & 0 & 0 \\ 0 & 0 & 0 & 0 & \sqrt{\theta} & 0 \\ 0 & 0 & 0 & 0 & 0 & \sqrt{\theta} \end{pmatrix} \xi + \begin{pmatrix} 0 \\ 0 \\ 0 \\ 0 \\ 0 \\ 0 \end{pmatrix}
$$

corresponding to (27) with $\mathrm{Cov}\,(\xi) = I$. LISREL can then be used to estimate β_2, β_3, $\sqrt{\psi_1}$, $\sqrt{\psi_2}$, $\sqrt{\psi_3}$ and $\sqrt{\theta}$. The square roots may actually come out negative or zero but, when squared, these yield maximum likelihood estimates of ψ_1, ψ_2, ψ_3 and θ. The maximum likelihood estimates of ω_t are then computed as $\hat{\omega}_1 = \hat{\omega}_1$, $\hat{\omega}_2 = \hat{\beta}_2^2\hat{\psi}_1 + \hat{\psi}_2$ and $\hat{\omega}_3 = \hat{\beta}_3^2\hat{\psi}_2 + \hat{\psi}_3$. The actual values come out as $\hat{\omega}_1 = 0.792$, $\hat{\omega}_2 = 1.26$, $\hat{\omega}_3 = 1.03$, $\hat{\beta}_2 = 0.758$, $\hat{\beta}_3 = 0.476$, $\theta = 0.00$ so that Professor Andersen's conjecture is indeed correct.

Causality. I agree with Dr. Schweder that causality cannot be inferred from just any correlational study. On the other hand, I would not go so far as to say that causal effects can only be demonstrated by means of carefully controlled experiments. In the social sciences, notably in econometrics, causal inference is obtained by means of tightly specified models. In most cases these models are dynamic and estimated from time-series or cross-sectional data. Such models can also be estimated by LISREL. For an example where both time-series and cross-sectional data are used, see Jöreskog (1978).

Other models and methods. As pointed out by Dr. Kiiveri there are alternative model formulations which are both more general and more compact in the sense of requiring fewer parameter matrices. While such an approach is mathematically elegant and makes it possible to include models for categorical variables in the same framework, I think, the computational algorithm for such an approach will be more extensive since it will have to deal with very large matrices. The E-M algorithm provides a mean for iterating in very simple steps but my experience with the E-M algorithm for factor analysis estimation is that it requires an extremely large number of iterations to converge.

References

Dunn, J. E. (1973). A note on a sufficiency condition for uniqueness of a restricted factor matrix. Psychometrika 38, 141-143.

Geraci, V. J. (1976). Identification of simultaneous equation models with measurement error. Journal of Econometrics 4, 263-283.

Gnanadesikan, R. (1977). Methods for statistical data analysis of multivariate observations. Wiley, New York.

Goldberger, A. S. (1964). Econometric theory. Wiley, New York.

Howe, H. G. (1955). Some contributions to factor analysis. Report No. ORNL-1919. Oak Ridge National Laboratory, Oak Ridge, Tenn.

Jennrich, R. I. (1978). Rotational equivalence of factor loading matrices with specified values. Psychometrika 43, 421-426.

Jöreskog, K. G. (1974). Analyzing psychological data by structural analysis of covariance matrics. In Contemporary developments in mathematical psychology, Vol. II (ed. R. C. Atkinson, D. H. Krantz, R. D. Suppes), pp. 1-56. W. H. Freeman & Co., San Francisco.

Jöreskog, K. G. (1978). An econometric model for multivariate panel data. Annales de l'INSEE, No. 30-31.

Jöreskog, K. G. & Sörbom, D. (1979). Advances in factor analysis and structural equation models. Abt Books, Cambridge, Mass.

Sörbom, D. (1975). Detection of correlated errors in longitudinal data. British Journal of Mathematical and Statistical Psychology 28, 138-151.

13

Analyzing Political
Participation Data
with a MIMIC Model

David C. Stapleton

This chapter illustrates the use of multiple-indicator
multiple-cause (MIMIC) models by an application to political
participation. In such models, observed indicators (that is,
imperfect measures) of latent variables are used to make inferences about parameters of structural equations in which the
latent variables are determined by observed exogenous causes.
The model received its name from Jöreskog and Goldberger (1975),
although it had previously been discussed by Zellner (1970),
Hauser and Goldberger (1971), and Goldberger (1972) and is
in fact a special case of the LISREL model developed by Jöreskog
(1973).

Previous discussion has centered around the single latent
variable case of the MIMIC model; Robinson (1974) and Abraham
and Ledolter (1975) have considered more general cases. Here
we work with several latent variables and illustrate a useful
approach to identification. Estimation of a MIMIC model and
tests of overidentifying restrictions are also discussed. The
results from the political participation model demonstrate that
imposing the restrictions in estimation may substantially reduce
the variance of the estimators.

Reprinted from Sociological Methodology, edited by K. F.
Schuessler (San Francisco: Jossey-Bass, 1977), pp. 52-74.
Reprinted by permission.

MODEL SPECIFICATION

The MIMIC model with multiple latent variables is given by

$$y^* = \Gamma \cdot x + \zeta \tag{1}$$

$$y = \Lambda \cdot y^* + \varepsilon \tag{2}$$

where y^* is a column vector of H latent variables, x is a column vector of K observed exogenous variables, y is a column vector of L observed indicators, ζ is a column vector of H structural disturbances, ε is a column vector of L measurement errors, and Γ and Λ are H × K and L × H parameter matrices, respectively. Without loss of generality, all variables are assumed to have zero expectations. The disturbances and measurement errors are independent of x; furthermore the structural disturbances are independent of the measurement errors but not necessarily of each other. In the standard MIMIC model, the measurement errors are assumed to be independent of each other; but this restriction may be relaxed. Here we work with two versions of the model: The first incorporates the restriction and the second does not. Letting $\Theta = V(\varepsilon)$, the first version has Θ diagonal and the second version has Θ full. Finally we let $\Phi = V(x)$ and $\Psi = V(\zeta)$.

As an empirical example, we consider a model of political participation in which propensities of individuals to participate in three modes of politics are the latent variables, actions of the individuals are the indicators of these propensities, and background characteristics of the individuals are the causes. The data, as well as the general structure of our model, are taken from a major study by Verba and Nie (1972) in which four modes of participation are defined: voting, campaigning, participating in community activities, and contacting local and national officials personally. Their analysis is based on a national survey of 2,549 individuals conducted by the National Opinion Research Center in 1967.

Although Verba and Nie do not present their model explicitly, its structure is that of Equations (1) and (2) with Θ diagonal. Our own model differs in three major respects. First, only the first three of their four propensities and only three of their five indicators of the third propensity are used. This modification is introduced because of our dissatisfaction with the underlying theory. Their indicators for the fourth propensity (personal contacting) and the two we omit for the third are based on exploration of the data and are accompanied by rather

weak ad hoc justifications. Second, our set of exogenous variables differs considerably from theirs. Because of our interest
in the effect of education on political participation, we use
educational attainment variables in place of their socioeconomic
status variable, which is based on income and occupational
status as well as educational attainment. There are other differences in the exogenous variables as well; notably neither respondent's income nor occupational status is included. These
variables were omitted so that the coefficients of the educational
variables would reflect indirect effects of education on participation, via effects on income and occupational status, as well as
more direct effects. Third, as previously noted, the case of
correlated measurement errors is considered here as well as the
case of independent errors.

There are also two important differences in statistical
approach. First, Verba and Nie use a two-step approach to
estimation whereas the method used here is full-information
maximum likelihood. Their version of Equation (2) is first
estimated by a factor-analytic technique; to estimate their
version of Equation (1) they construct factor scores for the
participation propensities and use them as dependent variables
in regressions on their set of exogenous variables. This two-
step approach is examined in Jöreskog and Goldberger (1975,
pp. 636-637) and is shown to be less efficient than maximum
likelihood. The second statistical difference is that Verba and
Nie make no attempt to test their overall model; here several
tests will be considered.

The three latent variables, eleven indicators, and fifteen
causes in the political participation model are defined as follows:

y_1^* propensity of the individual to participate in political
campaigns
y_2^* propensity of the individual to vote
y_3^* propensity of the individual to participate in community
groups attempting to solve a community problem
y_1 frequency of attempts to influence the vote of others
(0 = never; 1 = rarely; 2 = sometimes; 3 = often)
y_2 frequency of working for a political party or candidate
(same code as for y)
y_3 attending political rallies in the past 3 or 4 years
(1 = yes; 0 = no)
y_4 contributing to parties or candidates in the past 3 or
4 years (1 = yes; 0 = no)
y_5 current membership in a political club (1 = yes; 0 = no)
y_6 voting in the 1964 presidential election (1 = yes; 0 = no)

y_7 voting in the 1960 presidential election (1 = yes; 0 = no)

y_8 frequency of voting in local elections (0 = never; 1 = rarely; 2 = usually; 3 = always)

y_9 working with others to solve a community problem (1 = yes; 0 = no)

y_{10} forming a group to work on a community problem in the last 3 or 4 years (1 = yes; 0 = no)

y_{11} number of active memberships in community problem-solving organizations

x_1-x_5 dummy variables for the last five of the following six schooling levels: grade school or less, some high school, high school degree, some college, college degree, and at least some work toward an advanced degree (x_i = 1 if the individual has completed the ith schooling level and 0 otherwise; a person who attended college but did not graduate is assigned the following values: x_1 = 1; x_2 = 1; x_3 = 1; x_4 = 0; x_5 = 0)

x_6 individual's age in years

x_7 $x_6^2/1{,}000$

x_8 sex (1 = female; 0 = male)

x_9 race (1 = nonwhite; 0 = white)

x_{10} nativity (1 = foreign born; 0 = U.S. born)

x_{11} father's nativity (1 = father foreign born; 0 = father U.S. born)

x_{12} father's occupational status (Duncan scale)

x_{13} father's educational attainment (0 = no schooling; 7 = at least some grade school; 10 = some high school; 12 - finished high school; 14 = some college; 16 = finished college; 17 = at least some graduate work)

x_{14} head of household (1 = yes; 0 = no)

x_{15} family size (number of children)

In our model, the coefficient matrix Γ is assumed to be full; that is, every exogenous variable is assumed to affect all three of the unobserved propensities to participate. If the vector y^* were observed, Equation (1) would be a conventional multivariate regression model. Of course y^* is not observed but is reflected in the observed indicator vector y according to Equation (2). With Θ diagonal, Equation (2) is the basic equation of factor analysis. In our model, y_1 through y_5 are taken to be indicators of the propensity to participate in campaigns, y_6 through y_8 are taken to be indicators of the propensity to vote, and y_9 through y_{11} are taken to be indicators of the propensity to participate in community groups. Thus Λ is assumed to have the following structure:

$$\Lambda' = \begin{pmatrix} \lambda_1 & \lambda_2 & \lambda_3 & \lambda_4 & \lambda_5 & 0 & 0 & 0 & 0 & 0 & 0 \\ 0 & 0 & 0 & 0 & 0 & \lambda_6 & \lambda_7 & \lambda_8 & 0 & 0 & 0 \\ 0 & 0 & 0 & 0 & 0 & 0 & 0 & 0 & \lambda_9 & \lambda_{10} & \lambda_{11} \end{pmatrix} \quad (3)$$

Each element of y may be viewed as a measure of one element of y* with a random additive measurement error and a nonunit slope.

Several of the indicators are dummy variables or are otherwise restricted. We ignore this problem for present purposes, although some progress toward handling dichotomous variables in LISREL-type models has recently been made by Muthen (1976).

IDENTIFICATION

We now determine whether the parameters of the model can be identified in terms of the population (variance) covariance matrix of the observed variables. The latter is given by

$$\Sigma = \begin{pmatrix} \Sigma_{yy} & \Sigma_{yx} \\ \Sigma'_{yx} & \Sigma_{xx} \end{pmatrix} = \begin{pmatrix} \Lambda\Gamma\Phi\Gamma'\Lambda' + \Lambda\Psi\Lambda' + \Theta & \Lambda\Gamma\Phi \\ \Phi\Gamma'\Lambda' & \Phi \end{pmatrix} \quad (4)$$

$$= \begin{pmatrix} \Pi'\Phi\Pi + \Omega & \Pi'\Phi \\ \Phi\Pi & \Phi \end{pmatrix}$$

where Σ_{yy} is the $L \times L$ covariance matrix of y, Σ_{xx} is the $K \times K$ covariance matrix of x, and Σ_{yx} is the $L \times K$ covariance matrix of y and x; $\Pi = \Gamma'\Lambda'$ and $\Omega = \Lambda\Psi\Lambda' + \Theta$. Can we solve the $(L + K)$ $(L + K + 1)/2$ equations of Equation (4) uniquely for the elements of Λ, Γ, Φ, Ψ, and Θ in terms of the elements of Σ? If we cannot, our model cannot be distinguished from an infinity of alternatives. Suppose for instance, that Λ and Γ are unrestricted, as well as Ψ and Φ, with Θ diagonal. Let M be any nonsingular $H \times H$ matrix and define $\Lambda^* = \Lambda M$ and $\Gamma^* = M^{-1}\Gamma$. Then replacing Λ and Γ in Equations (1) and (2) with Λ^* and Γ^* will lead to the same covariance matrix for the observed variables as the original equations; that is, the two models are indistinguishable and the solutions for Λ and Γ in terms of Σ are not unique.

A necessary, but not sufficient, condition for identification is simply that there be no more parameters than equations in Equation (4). Convenient sufficient conditions for identification of MIMIC models are not yet available. (For some special cases see Robinson, 1974.) For particular models, identification can be verified by obtaining an explicit solution. Fortunately, for many models, including our participation model, this verification is quite straightforward.

To examine the identification of our participation model, we consider its reduced form. By substitution of Equation (1) into Equation (2), we get

$$y = \Lambda \Gamma x + \Lambda \zeta + \varepsilon = \Pi' x + u \tag{5}$$

where $u = \Lambda \zeta + \varepsilon$ so that $V(u) \equiv \Lambda \Psi \Lambda' + \Theta = \Omega$. Ignoring the structure of Π and Ω for the moment, Equation (5) is seen to be a conventional multivariate regression model. Clearly Π and Ω can be easily solved in terms of Σ from Equation (4) and in fact contain all the information relevant for identifying Λ, Γ, Ψ, and Θ; Φ is simply the variance of x. Therefore the question of identification can be resolved by determining whether the structural parameters can be obtained uniquely from the reduced-form parameters Π and Ω. We also ask: Do the underlying structures of Π and Ω imply constraints? That is: Is the model overidentified?

We note first that the scale of the unobserved variables is unknown. Equations (1) and (2) may be transformed via the arbitrary nonsingular diagonal matrix K to yield

$$\tilde{y}^* = Ky^* = K\Gamma x + K\zeta = \tilde{\Gamma} x + \tilde{\zeta} \tag{6}$$

$$y = \Lambda K^{-1} Ky^* + \varepsilon = \tilde{\Lambda} \tilde{y}^* + \varepsilon \tag{7}$$

where $\tilde{y}^* = Ky^*$, $\tilde{\Gamma} = K\Gamma$, $\tilde{\Lambda} = \Lambda K^{-1}$, and $\tilde{\zeta} = K\zeta$. Also we get $V(\tilde{\zeta}) = K\Psi K' = \tilde{\Psi}$ and $V(\tilde{y}^*) = K\tilde{V}(y^*)K'$. Note that $\tilde{\Lambda}$ has the same structure as Λ in Equation (3). In the absence of further restrictions on Λ and Γ, it is clear that by arbitrarily changing the scale of the unobserved variables one can generate an infinity of models that have the same reduced form as Equations (1) and (2).

To remove this indeterminacy, it is necessary to fix the scales for the unobserved variables. A natural normalization would be to assign each latent variable a unit variance. But an alternative is in fact more convenient for estimation: We fix one nonzero coefficient in each column of Λ at unity. Then the corresponding coefficients in $\tilde{\Lambda} = \Lambda K^{-1}$ will not be unity unless $K = I$; that is, changing the scale of y* would violate the normalization. In this sense, fixing the scale of Λ fixes the scale of y*.[1]

Specifically, we set $\lambda_1 = \lambda_6 = \lambda_9 = 1$. Recalling the structure of Λ we get

$$\Pi = [\pi_1, \ldots, \pi_k] \tag{8}$$

$$= [\gamma_1, \lambda_2\gamma_1, \lambda_3\gamma_1, \lambda_4\gamma_1, \lambda_5\gamma_1, \gamma_2, \lambda_7\gamma_2, \lambda_8\gamma_2, \gamma_3, \lambda_{10}\gamma_3, \lambda_{11}\gamma_3]$$

where γ_h is the hth column of Γ' and π_j is the jth column of Π. The elements of Γ are identified by

$$\gamma_1 = \pi_1 \qquad \gamma_2 = \pi_6 \qquad \gamma_3 = \pi_9 \tag{9}$$

and then the unknown λ's are readily determined. Furthermore π_2, π_3, π_4, and π_5 are proportional to π_1; π_7 and π_8 are proportional to π_6; and π_{10} and π_{11} are proportional to π_9. Hence the rank of the 15×11 matrix Π is only 3—a very restrictive specification. To determine the number of overidentifying restrictions, we note that, for $k = 1, \ldots, 15$:

$$\lambda_j = \pi_{kj}/\pi_{k1} \qquad j = 2, 3, 4, 5 \tag{10}$$

$$\lambda_j = \pi_{kj}/\pi_{k6} \qquad j = 7, 8$$

$$\lambda_j = \pi_{kj}/\pi_{k9} \qquad j = 10, 11$$

Since there are fifteen ways to solve for each of the eight λ_j's, one for each k, there are $8 \times 14 = 112$ overidentifying restrictions on Π. These will be imposed in our estimation procedure.

When Θ is assumed to be diagonal, the structure of Ω is seen to be the same as the structure of the variance matrix from a confirmatory factor-analysis model. With Λ and Γ already identified in terms of Π, only the $6 + 11 = 17$ parameters of Ψ and Θ are left to be identified in terms of the 66 distinct elements of Ω. Because of the structure of Λ (see Equation 3), a typical element of Ω is given by

$$\omega_{jk} = \lambda_j \lambda_k \psi_{hi} + \theta_{jk} \tag{11}$$

where ω_{jk} is the covariance between the reduced form disturbances for y_j and y_k, θ_{jk} is the covariance between the corresponding measurement errors, and the covariance between the structural disturbances for latent variables y_h^* and y_i^* (associated with indicators y_j and y_k respectively) is ψ_{hi}. Also $\theta_{jk} = 0$ for $j \neq k$ and $\lambda_1 = \lambda_6 = \lambda_9 = 1$. The elements of Ψ can be solved from those equations in Equation (11) for which $j \neq k$, after which the θ_{jj}'s are readily obtained. Thus when Θ is assumed to be diagonal, there are $66 - 17 = 49$ further restrictions to be imposed on the estimate of Ω.

When Θ is assumed to be full, there are a total of $66 + 6 = 72$ parameters in Θ and Ψ and identification in terms of the 66 elements of Ω is not possible. Since Θ is full, there are no restrictions to be imposed on the estimate of Ω. Note that if, in the second version, we permit $\text{cov}(\zeta, \varepsilon) \equiv \Delta \neq 0$ (3×11), we get $\Omega = \Delta\Psi\Delta' + \Lambda\Delta' + \Delta\Lambda' + \Theta$. Since there are no restrictions on Ω implied by $\Delta = 0$ when Θ is full, nothing is lost by relaxing that assumption.

In summary, once the scale of the latent variables is fixed, Λ and Γ are (over)identified in terms of Π alone and, when measurement errors are uncorrelated, the parameters of Θ as well as Ψ are (over)identified in terms of Ω, given Λ and Γ. When measurement errors are correlated, we cannot identify Θ and Ψ and only the reduced-form error variance matrix, Ω, can be estimated.

While we have treated only our own participation model explicitly, it is apparent that identification of any MIMIC model for which the nonzero elements of the columns of Λ do not overlap (see Equation 3) can be easily checked from equations like (8) and (11) under the same normalizations. Furthermore, the first step of reducing the problem to identification in terms of Π and Ω appears to be generally useful. It should also be noted that while fixing the scale of the unobserved variables arbitrarily limits quantitative interpretations of the coefficients in Λ and Γ, the choice of scale will not affect qualitative results (signs and significance) or the quantitative results for the reduced-form coefficient matrix Π or the reduced-form error variance matrix Ω.

ESTIMATION

Under the assumption that the T observations on (y', x') are independent drawings from a multinormal distribution, the maximum-likelihood estimates for the parameters of the first version of our model will minimize

$$F_1 = \ln |\Sigma| + \text{tr}(\Sigma^{-1}S) \tag{12}$$

where Σ is a function of the parameters as given in Equation (4) and S is the sample variance-covariance matrix of the observed variables; see Jöreskog (1973, p. 88).[2]

For the second version of our model (correlated measurement errors) F_1 reduces to

$$F_2 = \ln |W| \tag{13}$$

where

$$W = (1/T)(T - X\Gamma'\Lambda')'(\gamma - X\Gamma'\Lambda') \tag{14}$$

and T and X are $T \times L$ and $T \times K$ matrices of observations on y' and x'; see Jöreskog (1973, p. 94). Equation (13) is simply the maximum-likelihood criterion for a multivariate regression model.

A numerical minimization program due to Jöreskog and Sörbom (1976) can be used to minimize either F_1 or F_2. If this program is not readily available, any one of a number of general numerical minimization routines requiring user-supplied first-partial and, in some cases, second-partial derivatives of the function being minimized might be used. First derivatives of F_1 are a straightforward simplification of those found in Jöreskog and Van Thillo (1973, pp. 7-8); both first and second derivatives of F_2 are given in the Appendix to this chapter.

In using either alternative, difficulties may be encountered in obtaining convergence. Two suggestions are offered. First, convergence may be faster if the observed variance-covariance matrix is scaled so that diagonal elements are of the same order of magnitude. (Note that this implies a scaling of the parameters as well.) Second, reasonable starting values should be provided. Good starting values for Λ and Γ for either version of the MIMIC model can be obtained from an unconstrained least-squares esti-mate of Π. For instance, in the political participation model we get consistent estimates for Λ and Γ by using the coefficients from the regression of y_1 on x as starting values for γ_1, the average ratio of the coefficients from the regression of y_2 on x to those in the first regression for λ_2, and so forth. For version 1 the diagonal elements of

$$\bar{W} = (1/T)(T - XP)'(T - XP) \tag{15}$$

where $P = (X'X)^{-1}X'T$ is the least-squares estimate of Π, may be used as starting values for Θ, using starting values of zero for all elements of Ψ. For version 2, W is in fact a consistent esti-mate of Ω (starting values for Ω are not needed to minimize F_2).

The maximum-likelihood estimates of the two versions of our political participation model are presented in Tables 13.1 and 13.2. Version 1 estimates were obtained using an earlier version of LISREL III (Jöreskog and Van Thillo, 1973); version 2 estimates were obtained using a Fletcher-Powell minimization routine (Gruvaeus and Jöreskog, 1970). In both versions only 2,445 of the 2,549 observations were used due to missing data.

Clearly the assumption that y and x are drawings from a multinormal distribution is untrue: Many of our variables are either binary or truncated at zero. While this objection certainly prevents us from claiming that the estimates presented in the

TABLE 13.1

Maximum-Likelihood Estimates of Political Participation Model with Uncorrelated Measurement Errors

Λ'

	y_1	y_2	y_3	y_4	y_5	y_6	y_7	y_8	y_9	y_{10}	y_{11}
y_1^*	1.000	0.998** (0.028)	0.403** (0.014)	0.307** (0.012)	0.473** (0.016)	0	0	0	0	0	0
y_2^*	0	0	0	0	0	1.000	0.992** (0.023)	2.363** (0.056)	0	0	0
y_3^*	0	0	0	0	0	0	0	0	1.000	0.627** (0.029)	0.2091** (0.082)

Γ

	x_1	x_2	x_3	x_4	x_5	x_6	x_7	x_8	x_9	x_{10}	x_{11}	x_{12}	x_{13}	x_{14}	x_{15}
y_1^*	0.089* (0.041)	0.124** (0.039)	0.084* (0.044)	0.178* (0.067)	0.206* (0.093)	0.334** (0.046)	-0.300** (0.047)	-0.139** (0.041)	0.024 (0.044)	-0.127* (0.053)	0.117 (0.065)	0.038** (0.011)	0.011* (0.004)	-0.068 (0.041)	0.004 (0.009)
y_2^*	0.076** (0.021)	0.136** (0.020)	-0.005 (0.023)	0.065 (0.034)	0.005 (0.047)	0.472** (0.025)	-0.373** (0.025)	-0.080** (0.020)	-0.061* (0.022)	-0.095** (0.025)	0.183** (0.034)	0.008 (0.005)	-0.000 (0.001)	-0.052* (0.021)	0.020** (0.004)
y_3^*	0.110* (0.045)	0.185** (0.043)	0.083 (0.049)	0.087 (0.070)	0.320* (0.101)	0.371** (0.051)	-0.311** (0.052)	-0.128** (0.044)	0.137* (0.045)	-0.128* (0.055)	0.087 (0.068)	0.020 (0.011)	0.014* (0.005)	-0.080 (0.044)	0.042** (0.010)

ψ

	ζ_1	ζ_2	ζ_3
ζ_1	3.606** (0.074)		
ζ_2	0.887** (0.071)	2.246** (0.046)	
ζ_3	2.298** (0.087)	0.838** (0.080)	3.346** (0.077)

Diagonal of Θ

θ_1	θ_2	θ_3	θ_4	θ_5	θ_6	θ_7	θ_8	θ_9	θ_{10}	θ_{11}
0.949** (0.016)	0.486** (0.010)	0.122** (0.002)	0.097** (0.002)	0.177** (0.003)	0.113** (0.003)	0.121** (0.002)	0.731** (0.014)	0.168** (0.003)	0.107** (0.002)	0.732** (0.014)

Note: Standard errors are given in parentheses.
*At least two standard errors away from zero but less than three.
**Three or more standard errors away from zero.

TABLE 13.2

Maximum-Likelihood Estimates of Political Participation Model with Correlated Measurement Errors

Λ'

	y_1	y_2	y_3	y_4	y_5	y_6	y_7	y_8	y_9	y_{10}	y_{11}
y_1^*	1.000	0.592** (0.065)	0.293** (0.031)	0.310** (0.032)	0.287** (0.038)	0	0	0	0	0	0
y_2^*	0	0	0	0	0	1.000	1.307** (0.050)	2.567** (0.101)	0	0	0
y_3^*	0	0	0	0	0	0	0	0	1.000	0.533** (0.064)	0.2929** (0.266)

Γ

	x_1	x_2	x_3	x_4	x_5	x_6	x_7	x_8	x_9	x_{10}	x_{11}	x_{12}	x_{13}	x_{14}	x_{15}
y_1^*	0.074 (0.048)	0.133* (0.045)	0.115** (0.055)	0.285** (0.083)	0.200 (0.113)	0.392** (0.058)	-0.358** (0.056)	-0.189** (0.049)	-0.030 (0.050)	-0.188** (0.061)	0.137 (0.076)	0.045** (0.012)	0.013** (0.005)	-0.098 (0.050)	0.006 (0.010)
y_2^*	0.062** (0.018)	0.090** (0.017)	-0.005 (0.018)	0.074* (0.028)	-0.023 (0.038)	0.426** (0.024)	-0.765** (0.051)	-0.088** (0.028)	-0.070** (0.018)	-0.088** (0.022)	0.150** (0.027)	0.004 (0.004)	-0.001 (0.002)	-0.035 (0.018)	-0.019** (0.004)
y_3^*	0.033 (0.018)	0.067** (0.017)	0.096** (0.019)	0.059* (0.027)	0.099** (0.040)	0.144** (0.023)	-0.275** (0.049)	-0.054* (0.022)	0.032 (0.019)	-0.054* (0.023)	0.044 (0.029)	0.006 (0.005)	0.005* (0.002)	-0.033 (0.018)	0.018** (0.004)

Note: Standard errors are given in parentheses.
*At least two standard errors away from zero but less than three.
**Three or more standard errors away from zero.

next section have the sampling properties of maximum-likelihood estimators, our estimates nevertheless have an intuitive justification. Since when Σ is unconstrained the minimum of F_1 is given by $\hat{\Sigma} = S$, F_1 can be viewed simply as a scalar criterion by which to fit Σ to S when Σ is constrained.

The standard errors given in the tables are the square roots of the diagonal elements of the negative of the inverse of the second derivatives of the likelihood functions (for F_2 see the Appendix). Because of the nature of the dependent variables, significance tests should be interpreted cautiously.

TESTING THE MODEL

Before discussing the results of any overidentified model such as ours, its overall validity should be considered. Tests of the overidentifying restrictions are considered below. Define $\hat{L}$ as the value of the likelihood function at the maximum under the hypotheses of the model being estimated, $\hat{L}_0$ as the value of the function at the maximum for a just-identified model, and d as the number of restrictions on Σ implied by the model. Under normality, the test statistic $\omega = -2(\ln\hat{L} - \ln L_0)$ is distributed as chi square with d degrees of freedom if the restrictions are valid. This statistic is, in fact, the likelihood ratio.

For the first version, we have $\ln\hat{L}^1 = -(T/2)(F_1 + C_1)$, where C_1 is a constant; for the second version $\ln\hat{L}^2 = -(T/2)(F_2 + C_2)$. A just-identified model gives

$$\ln\hat{L}_0^1 = -(T/2)[\ln|S| + \operatorname{tr}(S^{-1}S) + C_1] \qquad (16)$$
$$= -(T/2)(\ln|S| + 11 + 15 + C_1)$$

and

$$\ln\hat{L}_0^2 = -(T/2)(\ln|\bar{W}| + C_2) \qquad (17)$$

The test statistics for the two versions of the model are presented in Table 13.3 along with critical values. Under the normality assumption on (y', x') both versions of the model must be rejected. This conclusion should be modified for two reasons, however. First, since we know that the normality assumption is violated, the distribution of ω is unknown even if restrictions are valid. Second, when T is large any over-identified model will in practice be rejected using the chi-square test.[3] Since the number of observations here (2,445) would be considered large by most standards, this possibility must be considered.

TABLE 13.3

Tests of the Models

Model	Test Statistic ω	Number of Restrictions d	Critical Value of $\chi^2(d)$ at 0.01 Significance Level
Version 1 (constraints on Π and Θ)	754	161	207
Version 2 (constraints on Π alone)	303	112	150

Nevertheless the test statistic is of some use as a measure of fit. The statistic itself, since it is a decreasing function of $\hat{L}$, will always become smaller when any binding restrictions are relaxed. If we look at ω as a loss function, then the statistic ω/d measures loss relative to the number of restrictions imposed. If ω were truly distributed as $\chi^2(d)$, the expectation of this ratio would be unity. For the first version this ratio is 4.7 while for the second it is 2.7, suggesting substantial reduction of loss relative to the number of restrictions relaxed.

Another less formal procedure for assessing the validity of a model is to look at the "residual matrix" ($\hat{\Sigma}$ - S). A useful summary statistic for this matrix is

$$r^2 = 1 - \frac{\underset{i \neq j}{\sum \sum} (\hat{\sigma}_{ij} - s_{ij})^2}{\underset{i \neq j}{\sum \sum} s_{ij}^2} \qquad (18)$$

where $\hat{\sigma}_{ij}$ and s_{ij} are typical elements of $\hat{\Sigma}$ and S respectively. This statistic, which is essentially the reliability coefficient proposed by Tucker and Lewis (1973), may be loosely called the percentage of covariation in the observed variables explained by the model.

Note that for both versions there are no restrictions on Σ_{xx}, so $\hat{\Sigma}_{xx} = S_{xx}$ is the maximum-likelihood estimate of Σ_{xx}. Therefore omitting the corresponding terms from the ratio makes the statistic a more meaningful test of the remaining restrictions. For the first version of our model r^2 is 0.96 and for the second it is 0.98. That is, over 95 percent of the covariation among the y's and between the y's and x's is explained by either model. Both versions fit well by this criterion. For the second version our estimate of Ω is given by $\hat{\Omega} = S_{yy} - \hat{\Pi}'\hat{\Phi}\hat{\Pi}$ so that $\hat{\Sigma}_{yy} = S_{yy}$. Omitting the corresponding elements from the computation of the statistic for this version gives $r^2 = 0.93$. That is, only 7 percent of the covariation between y and x is left unexplained when 112 restrictions are placed on the 165 elements of Σ_{yx} (that is, on Π).

Because of the first version's poor performance on the chi-square test and because of our a priori doubts as to the validity of uncorrelated disturbances, we are led to reject the first version of the model. The restrictions on Π, however, appear to be reasonable. A final heuristic test for the restrictions on Π alone is a comparison of the constrained estimate of Π, $\hat{\Pi} = \hat{\Gamma}'\hat{\Lambda}'$, from the second version of the model, with the unconstrained least-squares estimate of Π: P. The matrices $\hat{\Pi}$ and P along with approximate standard errors and the difference matrix $\hat{\Pi} - P$ are presented in Tables 13.4, 13.5, and 13.6. The differences between individual elements of $\hat{\Pi}$ and P are generally small relative to their magnitudes. Of the 165 estimated coefficients, only 10 have a different sign in $\hat{\Pi}$ than in P—and the unconstrained estimates of all 10 are nonsignificantly different from zero.[4]

Although there is no rigorous criterion for accepting or rejecting either version of the model, it appears that the restrictions on Π are sufficiently close to being satisfied in the population to allow us to make meaningful conclusions based on the estimate of the second version of the model. The signs of most coefficients correspond to our expectation—keeping in mind that income and occupational status have not been included in x. Further discussion of the results may be found in Stapleton (1976).

As a simpler alternative to maximum-likelihood estimation, one might use the unconstrained, least-squares estimates of Π and Ω, P and $\bar{W}$, to make inferences about the hypotheses being considered. In fact, consistent estimates of Λ, Γ, and, for version 1, Ψ and Θ, can be obtained directly by using P and $\bar{W}$ to replace Π and Ω in Equations (8) and (11), although they will not be unique. There are, however, three important reasons for obtaining restricted estimates of Π and, for version 1, Ω.

First, this gives us the opportunity to test the underlying theory by testing the implied restrictions. Second, if the restrictions are appropriate, the restricted estimates may be sub-

TABLE 13.4

Constrained Estimates of the Reduced-Form Coefficients with Uncorrelated Measurement Errors

	X₁	X₂	X₃	X₄	X₅	X₆	X₇	X₈	X₉	X₁₀	X₁₁	X₁₂	X₁₃	X₁₄	X₁₅
y_1	0.074 (0.048)	0.133* (0.045)	0.115* (0.055)	0.285** (0.083)	0.200 (0.113)	0.392** (0.058)	-0.358** (0.056)	-0.189** (0.049)	-0.030 (0.050)	-0.188** (0.061)	0.137 (0.076)	0.045** (0.012)	0.013* (0.005)	-0.098 (0.050)	0.006 (0.010)
y_2	0.045 (0.030)	0.081* (0.041)	0.069* (0.032)	0.171** (0.046)	0.123 (0.069)	0.234** (0.042)	-0.212** (0.038)	-0.099* (0.012)	-0.018 (0.031)	-0.099* (0.038)	0.084 (0.048)	0.027** (0.008)	0.008* (0.003)	-0.058 (0.031)	0.004 (0.007)
y_3	0.022 (0.015)	0.039* (0.014)	0.033* (0.016)	0.084** (0.024)	0.060 (0.035)	0.116* (0.018)	-0.105* (0.017)	-0.049** (0.012)	-0.010 (0.015)	-0.049* (0.018)	0.041 (0.022)	0.013** (0.004)	0.004* (0.002)	-0.029 (0.015)	0.002 (0.004)
y_4	0.024 (0.046)	0.042* (0.015)	0.036* (0.017)	0.089** (0.025)	0.063 (0.035)	0.122* (0.019)	-0.111** (0.018)	-0.052** (0.013)	-0.010 (0.017)	-0.052* (0.020)	0.044 (0.024)	0.014** (0.004)	0.004* (0.002)	-0.030 (0.016)	0.002 (0.003)
y_5	0.022 (0.015)	0.039* (0.014)	0.033** (0.016)	0.083** (0.024)	0.058 (0.033)	0.113** (0.022)	-0.103** (0.020)	-0.048** (0.012)	-0.009 (0.015)	-0.048** (0.018)	0.040 (0.022)	0.013** (0.004)	0.004* (0.002)	-0.029 (0.015)	0.002 (0.004)
y_6	0.062** (0.018)	0.090** (0.017)	-0.005 (0.018)	0.074* (0.028)	-0.023 (0.038)	0.426** (0.024)	-0.340** (0.023)	-0.088** (0.028)	-0.070** (0.018)	-0.088** (0.022)	0.150** (0.027)	0.004 (0.004)	-0.001 (0.002)	-0.035 (0.018)	0.019** (0.004)
y_7	0.081** (0.023)	0.117** (0.023)	-0.007 (0.027)	0.096* (0.036)	-0.029 (0.053)	0.557** (0.039)	-0.445** (0.035)	-0.115** (0.037)	-0.092** (0.024)	-0.115** (0.029)	0.196** (0.038)	0.005 (0.005)	-0.001 (0.002)	-0.046 (0.023)	0.025** (0.005)
y_8	0.159** (0.046)	0.230** (0.041)	-0.013 (0.047)	0.188 (0.071)	-0.057 (0.101)	1.093** (0.077)	-0.874** (0.069)	-0.226 (0.073)	-0.180** (0.048)	-0.226** (0.059)	0.385** (0.073)	0.010 (0.009)	-0.002 (0.004)	-0.090 (0.046)	0.049 (0.011)
y_9	0.033 (0.018)	0.067** (0.017)	0.096** (0.019)	0.059* (0.027)	0.099* (0.040)	0.144* (0.023)	-0.127** (0.013)	-0.054* (0.022)	0.032 (0.019)	-0.054* (0.023)	0.044 (0.029)	0.006 (0.005)	0.005* (0.002)	-0.033 (0.018)	0.018** (0.004)
y_{10}	0.018 (0.009)	0.037** (0.011)	0.053** (0.013)	0.032* (0.016)	0.054* (0.023)	0.079** (0.017)	-0.068** (0.015)	-0.030* (0.013)	0.018 (0.011)	-0.030* (0.013)	0.024 (0.015)	0.003 (0.002)	0.003* (0.001)	-0.018 (0.010)	0.010** (0.002)
y_{11}	0.097 (0.052)	0.200** (0.057)	0.297** (0.066)	0.173* (0.085)	0.292* (0.124)	0.427** (0.079)	-0.372** (0.074)	-0.163* (0.068)	0.097 (0.059)	-0.163* (0.068)	0.130 (0.087)	0.016 (0.011)	0.016* (0.006)	-0.097 (0.052)	0.054** (0.013)

Note: Standard errors are given in parentheses.

*At least two standard errors away from zero but less than three.

**Three or more standard errors away from zero.

TABLE 13.5

Unconstrained Estimates of the Reduced-Form Coefficients

	x_1	x_2	x_3	x_4	x_5	x_6	x_7	x_8	x_9	x_{10}	x_{11}	x_{12}	x_{13}	x_{14}	x_{15}
y_1	0.132* (0.065)	0.098 (0.061)	0.155* (0.071)	0.291* (0.103)	-0.076 (0.146)	0.337* (0.071)	-0.363** (0.073)	-0.282** (0.063)	-0.065 (0.068)	-0.188* (0.080)	0.216* (0.103)	0.056** (0.016)	0.009 (0.007)	-0.079 (0.066)	0.001 (0.014)
y_2	0.111* (0.051)	0.081 (0.048)	0.098 (0.056)	0.064 (0.082)	0.184 (0.116)	0.337** (0.056)	-0.274** (0.058)	-0.108* (0.050)	-0.018 (0.054)	-0.163* (0.064)	0.125 (0.081)	0.037** (0.012)	0.005 (0.006)	-0.045 (0.052)	0.008 (0.011)
y_3	0.033 (0.024)	0.097* (0.023)	0.026 (0.027)	0.114* (0.038)	0.096 (0.054)	0.103** (0.027)	-0.093** (0.027)	-0.053* (0.023)	0.041 (0.025)	0.022 (0.030)	0.031 (0.038)	0.009 (0.006)	0.006 (0.003)	-0.030 (0.024)	0.005 (0.005)
y_4	0.003 (0.021)	0.067** (0.019)	0.031 (0.023)	0.056 (0.033)	0.170** (0.047)	0.109** (0.023)	-0.113** (0.023)	-0.057* (0.020)	-0.017 (0.022)	-0.062* (0.026)	0.009 (0.033)	0.014* (0.005)	0.005* (0.002)	-0.048* (0.021)	-0.000 (0.004)
y_5	0.034 (0.029)	0.068* (0.027)	-0.001 (0.032)	0.081 (0.046)	0.080 (0.066)	0.164** (0.032)	-0.152** (0.033)	-0.031 (0.028)	0.050 (0.031)	-0.007 (0.036)	0.056 (0.046)	0.017* (0.007)	0.006 (0.003)	-0.014 (0.030)	-0.004 (0.006)
y_6	0.065* (0.025)	0.165** (0.024)	-0.013 (0.028)	0.034 (0.040)	0.052 (0.057)	0.412** (0.028)	-0.325** (0.029)	-0.089** (0.025)	-0.019 (0.027)	-0.067* (0.032)	0.171** (0.040)	0.010 (0.006)	-0.000 (0.003)	-0.069 (0.026)	0.021** (0.006)
y_7	0.078** (0.024)	0.098** (0.022)	-0.020 (0.026)	0.103* (0.038)	-0.035 (0.054)	0.575** (0.026)	-0.462** (0.027)	-0.052* (0.023)	-0.081** (0.025)	-0.108** (0.030)	0.179** (0.038)	0.005 (0.006)	-0.003 (0.003)	-0.041 (0.024)	0.027** (0.005)
y_8	0.205** (0.061)	0.329** (0.057)	0.048 (0.067)	0.143 (0.098)	-0.031 (0.138)	0.999** (0.067)	-0.797** (0.070)	-0.238** (0.060)	-0.233** (0.064)	-0.252** (0.076)	0.470** (0.097)	0.021 (0.015)	0.004 (0.007)	-0.111 (0.062)	0.034* (0.013)
y_9	0.051 (0.028)	0.117** (0.026)	0.043 (0.031)	-0.019 (0.045)	0.173* (0.063)	0.195** (0.031)	-0.175** (0.032)	-0.070* (0.027)	0.090** (0.029)	-0.077* (0.035)	0.051 (0.044)	0.021* (0.007)	0.005 (0.003)	-0.040 (0.028)	0.017* (0.006)
y_{10}	0.046* (0.022)	0.023 (0.020)	0.056* (0.024)	0.011 (0.035)	0.129* (0.049)	0.083** (0.024)	-0.067* (0.025)	-0.051* (0.021)	0.059* (0.023)	-0.022 (0.027)	-0.018 (0.034)	-0.001 (0.005)	0.004 (0.002)	-0.020 (0.022)	0.014* (0.005)
y_{11}	0.092 (0.057)	0.184** (0.053)	0.334** (0.063)	0.243** (0.091)	0.211 (0.128)	0.393** (0.063)	-0.350** (0.064)	-0.082* (0.055)	0.057 (0.060)	-0.121 (0.071)	0.153 (0.090)	0.010 (0.014)	0.016* (0.006)	-0.085 (0.058)	0.045** (0.012)

Note: Standard errors are given in parentheses.
*At least two standard errors away from zero but less than three
**Three or more standard errors away from zero.

282

TABLE 13.6

Differences between Constrained and Unconstrained Estimates of the Reduced-Form Coefficients

	x_1	x_2	x_3	x_4	x_5	x_6	x_7	x_8	x_9	x_{10}	x_{11}	x_{12}	x_{13}	x_{14}	x_{15}
y_1	-0.058	0.035	-0.040	-0.006	0.276	0.055	0.005	0.093	0.035	0.000	-0.079	-0.011	0.004	-0.019	0.005
y_2	-0.066	0.000	-0.029	-0.018	-0.061	-0.103	0.062	0.009	0.000	0.064	-0.041	-0.010	0.003	-0.013	-0.004
y_3	-0.011	-0.058	0.007	-0.030	-0.036	0.013	-0.012	0.004	-0.051	-0.071	0.010	0.004	-0.002	0.001	-0.003
y_4	0.021	-0.025	0.005	0.033	-0.107	-0.013	0.002	0.005	0.007	0.010	0.035	0.000	-0.001	0.018	0.002
y_5	-0.012	-0.029	0.034	0.002	-0.022	-0.051	0.049	-0.017	-0.059	-0.041	-0.016	-0.004	-0.002	-0.015	0.006
y_6	-0.003	-0.075	0.008	0.040	-0.075	0.014	-0.015	0.001	-0.051	-0.021	-0.021	-0.006	0.001	0.035	-0.002
y_7	0.003	0.019	0.013	-0.007	0.006	-0.018	-0.017	-0.063	-0.011	-0.007	0.017	0.000	0.002	-0.005	-0.002
y_8	-0.046	-0.099	-0.061	0.045	-0.026	0.094	-0.077	0.012	0.053	0.026	-0.085	-0.011	-0.006	0.021	0.015
y_9	-0.018	-0.050	0.053	0.078	-0.074	-0.051	0.048	0.016	-0.058	0.023	-0.007	-0.015	0.000	0.007	0.001
y_{10}	-0.028	0.014	-0.003	-0.021	-0.075	-0.004	-0.001	0.021	-0.041	-0.008	0.042	0.004	-0.001	0.002	-0.004
y_{11}	0.005	0.016	-0.037	-0.070	0.081	0.034	-0.022	-0.081	0.040	-0.042	-0.023	0.006	0.000	-0.014	0.009

stantially more efficient. This is demonstrated by the results for the political participation model. Comparing the reduced-form coefficient matrix from the second version of our model in Table 13.4 with the unconstrained estimate in Table 13.5, we find an average reduction in the estimated variances of the coefficients of 36 percent. As a result, 105 of the 165 coefficients in Table 13.4 are two or more standard errors away from zero, whereas this significance criterion is satisfied by only 84 of the coefficients in Table 13.5.

A final reason for imposing the restrictions, when appropriate, is that the results will, in general, be more readily interpretable. For example, the unconstrained estimate of Π for our model in Table 13.5 suggests that obtaining a college degree had a positive effect on voting in the 1964 presidential election but no significant effect in the 1960 election or on the frequency of voting in local elections. From this information alone one would not be able to draw a conclusion about the effect of obtaining a degree on general voting behavior. After imposing the restrictions (Table 13.4), the coefficients on the college degree dummy are significantly positive for all three voting indicators, leading to the conclusion that obtaining a degree affects participation by voting positively.

SUMMARY

We have examined the structure of the multiple-indicator multiple-cause model with several latent variables and have presented a model of political participation as an example. Checking the identification of a model has also been discussed. While general rules for checking identification are not available, the process may not be very difficult when zero restrictions and normalizations are used to identify the model. Transforming the problem to one of checking identification in terms of reduced-form parameters appears to be a useful first step. For an over-identified model, restrictions on the reduced-form parameters are implied.

Maximum-likelihood estimation of two versions of the model—one with measurement errors uncorrelated, the other with correlated errors—has also been treated. The results for the political participation model demonstrate that imposing the over-identifying restrictions on the reduced-form parameter estimates may substantially reduce their variance as well as make them more readily interpretable. Finally we have seen that full-information maximum-likelihood estimation allows us to test the appropriateness of the model statistically.

APPENDIX

As shown in Equation (5), the standard MIMIC model is reduced to a multivariate regression model with proportionality constraints on the coefficient matrix Π and a factor-analytic structure for the variance matrix of the disturbances. With no restrictions on the contemporaneous covariance matrix of measurement errors, there are none on the matrix Ω. For the latter case, Robinson (1974) has developed maximum-likelihood estimates for Λ and Γ when there are just sufficient restrictions on these matrices to allow identification of their parameters in terms of Π. For our model, however, Λ and Γ are overidentified in terms of Π by normalizations and zero restrictions on Λ. One method of obtaining maximum-likelihood estimates of Λ and Γ when these matrices are overidentified is to minimize F_2 numerically as in Equation (13). Using the first and, if necessary, the second partial derivatives of F_2 given below in conjunction with a general minimization program, one can easily estimate models in which fixed elements of Λ and Γ (including zero restrictions and normalizations) overidentify these matrices. These results are derived in Stapleton (1976, App. C).

The first partial derivatives of F_2 with respect to Λ and Γ are

$$\frac{\partial \ln |W|}{\partial \Lambda} = 2W^{-1}D\Gamma' \tag{19}$$

$$\frac{\partial \ln |W|}{\partial \Gamma} = 2\Lambda'W^{-1}D \tag{20}$$

where $D = \Pi'S_{xx} - S_{yx}$. After stacking the columns of Λ and Γ into a single-parameter vector, say θ, it can be shown that the second partial derivatives form an $(LH + HK) \times (LH + HK)$ symmetric matrix:

$$\frac{\partial^2 \ln W}{\partial \theta \partial \theta'} \equiv G = -2 \begin{pmatrix} A & B' \\ B & C \end{pmatrix} \tag{21}$$

where

$$A = W^{-1} \otimes \Gamma(S_{xx} - D'W^{-1}D)\Gamma' \quad (LH \times LH) \tag{22}$$

$$B = \Lambda'W^{-1} \otimes (S_{xx} - D'W^{-1}D)\Gamma' \tag{23}$$

$$+ (i_H \otimes W^{-1}D) \odot (I - {}'W^{-1}D\Gamma') \quad (HK \times LH)$$

$$C = \Lambda'W^{-1}\Lambda \otimes (S_{xx} - D'W^{-1}D) \quad (HK \times HK) \tag{24}$$

where i_H is a unit vector of length H, a Kronecker matrix product is indicated by $\otimes$, and a Khatri-Rao matrix product is indicated by $\odot$. The latter is defined by

$$
\begin{pmatrix} A_1 \\ \cdot \\ \cdot \\ \cdot \\ A_N \end{pmatrix} \odot \begin{pmatrix} B_1 \\ \cdot \\ \cdot \\ \cdot \\ B_N \end{pmatrix} \equiv \begin{pmatrix} A_1 \otimes B_1 \\ \cdot \\ \cdot \\ \cdot \\ A_N \otimes B_N \end{pmatrix} \tag{25}
$$

Since the actual log-likelihood function is $(-T/2)F_2$ (apart from an additive constant), the information matrix is given by $E = (-T/2)G$.

Note that only elements of the first derivatives corresponding to free elements of Λ and Γ, and elements of G corresponding both row-wise and column-wise to free elements of Λ and Γ, will be used. These results can also be used in estimating models with equality restrictions by following the development of Gruvaeus and Jöreskog (1970).

NOTES

1. A third alternative would be to fix the diagonal elements of ψ at unity (or any positive number) so that $\tilde{\psi} = K\psi K'$ will not have unities on the diagonal unless $K = I$. However, this alternative may only be used when Θ is diagonal.

2. Note that our Σ is a special case of the LISREL variance-covariance matrix with $\Lambda_y = \Lambda$, $\Theta_\varepsilon = \Theta$, $\Lambda_x = I$, $\Theta_\delta = 0$, and $B = I$.

3. Consider the chi-square statistic for the first likelihood function:

$$
\omega_1 = T[\ln|\hat{\Sigma}| + \text{tr}(\hat{\Sigma}^{-1}S) - \ln|S| - (11 + 15)]
$$

If the model were perfectly specified, $\hat{\Sigma}$ as well as S would approach Σ as T went to infinity, and the term in brackets would go to zero. At best our model is a simplification of reality and the term in brackets will approach a positive limit so that ω_1 will go to infinity. An analogous argument holds for ω_2.

4. Since $P = S_{xx}^{-1}(\Sigma_{xy} - S_{xy})$, this is the same in principle as comparing Σ_{xy} to S_{xy}. Since some meaning has been attached to the coefficients in Π, however, the effect of imposing the restrictions may more readily be assessed.

REFERENCES

Abraham, B., and Ledolter, J. 1975. "The multiple indicator-
multiple cause model with several latent variables." Manuscript. University of Wisconsin, Madison.

Goldberger, A. S. 1972. "Maximum-likelihood estimation of
regressions containing unobservable independent variables."
International Economic Review 12:1-15.

Gruvaeus, G. T., and Jöreskog, K. G. 1970. "A computer
program for minimizing a function of several variables."
Research Bulletin RB-70-14. Princeton, N.J.: Educational
Testing Service.

Hauser, R. M., and Goldberger, A. S. 1971. "The treatment
of unobservable variables in path analysis." Pp. 81-177 in
H. L. Costner (Ed.), Sociological Methodology 1971. San
Francisco: Jossey-Bass.

Jöreskog, K. G. 1973. "A general method for estimating a
linear structural equation system." Pp. 85-112 in A. S.
Goldberger and O. D. Duncan (Eds.), Structural Equation
Models in the Social Sciences. New York: Seminar Press.

Jöreskog, K. G., and Goldberger, A. S. 1975. "Estimation of
a model with multiple indicators and multiple causes of a
single latent variable." Journal of the American Statistical
Association 70:631-639.

Jöreskog, K. G., and Sorbom, D. 1976. "LISREL III: Estimation of linear structural equation systems by maximum likelihood methods." Computer manual. Chicago: National
Educational Resources.

Jöreskog, K. G., and Van Thillo, M. 1973. "LISREL: A general
computer program for estimating a linear structural equation
system involving multiple indicators of unmeasured variables."
Research Report 73-5. Uppsala, Sweden: University of
Uppsala, Department of Statistics.

Muthén, B. 1976. "Structural equation models with dichotomous
dependent variables." Research Report 76-17. Uppsala,
Sweden: University of Uppsala, Department of Statistics.

Robinson, P. M. 1974. "Identification, estimation, and large sample theory for regressions containing unobservable variables." International Economic Review 15:680-692.

Stapleton, D. C. 1976. "Social benefits of education: An assessment of the effect of education on political participation." Workshop Paper 7606. University of Wisconsin, Social Systems Research Institute.

Tucker, L. R., and Lewis, L. 1973. "A reliability coefficient for maximum likelihood factor analysis." Psychometrika 38: 1-10.

Verba, S., and Nie, N. 1972. Participation in America: Political Democracy and Social Equality. New York: Harper & Row.

Zellner, A. 1970. "Estimation of regression relationships containing unobservable variables." International Economic Review 11:441-454.

14

A Comparative Analysis of
Two Structural Equation Models:
LISREL and PLS Applied
to Market Data

Claes Fornell and Fred L. Bookstein

INTRODUCTION

Although they were introduced to marketing only recently,
structural equation models with unobservables are beginning to
change the conventions of marketing research methodology
(Bagozzi, 1980). For social science in general, the new structural equations approach is strongly identified with maximum
likelihood factor analysis procedures generalized by Karl Jöreskog
(1970, 1973, 1979) and the associated computer program LISREL
(Jöreskog and Sörbom, 1978, 1981). For marketing, in particular, nearly every application of structural modeling has used
LISREL for parameter estimation. But it is not realistic to
assume that all problems amenable to use of structural equation
models are also suited to LISREL. There are other protocols
of structural estimation that impose different assumptions about
data, theory, and the ties between unobservables and indicators.
Marketing data do not often satisfy the requirements of multinormality and interval scaling, or attain the sample size required
by maximum-likelihood estimation. More fundamentally, two
serious problems often interfere with meaningful modeling:
improper (that is, inadmissable) solutions and factor indeterminacy.

This chapter draws heavily from Claes Fornell and Fred L.
Bookstein, "Two Structural Equation Models: LISREL and PLS
Applied to Consumer Exit-Voice Theory," Journal of Marketing
Research, November 1982.

Herman Wold's method of Partial Least Squares (PLS)[1] avoids many of the restrictive assumptions underlying maximum-likelihood (ML) techniques and ensures against improper solutions and factor indeterminacy. Toward a comparison of PLS with LISREL, we estimate three models for a single data set. The first model compares traditional common-factor ML estimates with PLS estimates for a case in which LISREL produces improper solutions. We show that the improper estimates do not stem from sample variance or from lack of fit but can be traced, instead, to the path-analytic fitting objective behind LISREL. It is then shown that the removal of factor indeterminacy via PLS provides an effective cure. In a second model, wherein all factors are explicitly defined, LISREL avoids its improper solutions by giving estimates identical to those of PLS. The third model extends the second in a direction consistent with the consumer behavior theory it embodies. It too presents no improper solutions but illustrates a systematic difference between LISREL and PLS results.

PARTIAL LEAST SQUARES IN STRUCTURAL MODELING

In PLS the set of model parameters is divided into subsets estimated by use of ordinary multiple regressions that involve the values of parameters in other subsets. An iterative method provides successive approximations for the estimates, subset by subset, of loadings and structural parameters. Extending his theory of fixed-point estimation Herman Wold (1965) developed this method for structural models with unobservables (1974, 1975, 1980a, 1980b). As is the case with LISREL, many of the early elaborations and applications originate from Sweden (Ågren, 1972; Bergström, 1972; Bodin, 1974; Lyttkens, 1966, 1973; Noonan and Wold, 1977; Areskoug et al., 1975). In the United States, Hui (1978) has extended the model to nonrecursive systems, and Bookstein (1980, 1981) has provided a geometrical restatement of its protocols.

Recent discussions of both PLS and LISREL are available in Jöreskog and Wold (1982). Applications of PLS appear in a variety of disciplines, including economics (Apel, 1977), political science (Meissner and Uhle-Fassing, 1982), psychology of education (Noonan, 1980; Noonan and Wold, 1980), chemistry (Kowalski et al., 1982), and marketing (Jagpal, 1981).

MODEL STRUCTURE

The systematic part of the predictor relation in PLS is the conditional expectation of the predictands for given values of the predictors (see Chapter 15 by Wold in this volume). The structural relations ("inner relations") are thus specified as stochastic. We write this as:

$$E(\underset{\sim}{\eta}|\underset{\sim}{\eta}, \underset{\sim}{\xi}) = \underset{\sim}{\beta}^*\underset{\sim}{\eta} + \underset{\sim}{\Gamma} \underset{\sim}{\xi} \tag{1}$$

$\underset{\sim}{\eta} = (\eta_1, \eta_2, \ldots \eta_m)$ and $\underset{\sim}{\xi} = (\xi_1, \xi_2, \ldots \xi_n)$ are vectors of unobserved criterion and explanatory variables, respectively; $\beta^*(m \times m)$ is a matrix of coefficient parameters (with zeros in the diagonal) for $\underset{\sim}{\eta}$, and $\underset{\sim}{\Gamma}$ $(m \times n)$ is a matrix of coefficient parameters for $\underset{\sim}{\xi}$. This implies that $E(\underset{\sim}{\eta}\underset{\sim}{\zeta}') = E(\underset{\sim}{\xi}\underset{\sim}{\zeta}') = E(\underset{\sim}{\zeta}) = 0$, where $\underset{\sim}{\zeta} = \eta - \tilde{E}(\eta)$.

The measurement equations ("outer relations") are

$$\underset{\sim}{y} = \underset{\sim}{\Lambda}_y\underset{\sim}{\eta} + \underset{\sim}{\varepsilon} \qquad \text{Regress } y \text{ on } \Lambda \tag{2}$$

$$\underset{\sim}{x} = \underset{\sim}{\Lambda}_x\underset{\sim}{\xi} + \underset{\sim}{\delta} \qquad \text{Regress } x \text{ on } \xi \tag{3}$$

in which $y' = (y_1, y_2, \ldots y_p)$ and $x' = (x_1, x_2, \ldots x_q)$ are the observed criterion and explanatory variables, respectively; $\underset{\sim}{\Lambda}_y(p \times m)$ and $\underset{\sim}{\Lambda}_x(q \times n)$ are the corresponding regression matrices; and $\underset{\sim}{\varepsilon}$ and $\underset{\sim}{\delta}$ are residual vectors. Predictor specification implies $E(\underset{\sim}{\varepsilon}) = E(\underset{\sim}{\delta}) = E(\underset{\sim}{y}\varepsilon') = E(\underset{\sim}{x}\delta') = 0$. We standardize such that $E(\underset{\sim}{\eta}) = E(\underset{\sim}{\xi}) = 0$ and $\text{Var}(\underset{\sim}{\eta_i}) = \text{Var}(\xi_j) = \text{Var}(x_k) = \text{Var}(y_r) = 1$, all i, j, k, r. We define the following variance-covariance matrices $E(\underset{\sim}{\varepsilon}\underset{\sim}{\varepsilon}') = \underset{\sim}{\Theta}_\varepsilon$, $E(\underset{\sim}{\delta}\underset{\sim}{\delta}') = \underset{\sim}{\Theta}_\delta$, $E(\underset{\sim}{\xi}\underset{\sim}{\xi}') = \underset{\sim}{\phi}$ and $E(\underset{\sim}{\zeta}\underset{\sim}{\zeta}') = \psi$. For convenience we set $E(\underset{\sim}{x}) = E(\underset{\sim}{y}) = 0$.

Further, we estimate the unobservables as exact linear combinations of their empirical indicators:

$$\underset{\sim}{\eta} = \underset{\sim}{\pi}_\eta \underset{\sim}{y} \qquad \text{Regress } \eta \text{ on } y \tag{4}$$

$$\underset{\sim}{\xi} = \underset{\sim}{\pi}_\xi \underset{\sim}{x} \tag{5}$$

where $\underset{\sim}{\pi}_\eta$ $(p \times m)$ and $\underset{\sim}{\pi}_\xi$ $(p \times m)$ are regression matrices.

MODE SELECTION: THE RELATIONSHIP BETWEEN UNOBSERVED AND MEASURED VARIABLES

In estimation, the unobserved constructs (η, ξ) can be viewed as underlying factors or as indices produced by the observables. That is, the observed indicators can be treated as reflective or formative. Reflective indicators are typical of classical test theory and factor analysis models; they are invoked in an attempt to account for observed variances or covariances. Formative indicators are more directly concerned with the delineation of abstract relationships.

The choice of indicator mode, which substantially affects estimation procedures, has hitherto received only sparse attention in the literature. Figure 14.1 exemplifies the choices to be made. In deciding how unobservables and data should be related, there are three major considerations: study objective, theory, and empirical contingencies.

Should the study intend to account for observed variances, reflective indicators (Figure 14.1, Mode A) are most suitable. If the objective is explanation of abstract or "unobserved" variance, formative indicators (Figure 14.1, Mode B) would give greater explanatory power. Both formative and reflective indicators can also be used within a single model. For instance, if one intends to explain variance in the observed criterion variables by way of the unobservables, the indicators of the endogenous construct should be reflective, and those of the exogenous, formative; the result is a mixed-mode estimation (Figure 14.1, Mode C). Modes A and B represent two separate principles—mode A minimizes the trace of the residual variances in the "outer" equations, mode B minimizes the trace of the residual variances in the "inner" equations, both subject to certain systematic constraints that we will discuss shortly. Mode C is a compromise between the two principles.

Indicator mode is also shaped by an aspect of the substantive theory behind the model: the way in which the unobservable is conceptualized. Constructs such as "personality" or "attitude" are typically viewed as underlying factors that give rise to something that is observed. Their indicators tend to be realized, then, as reflective. On the other hand, when constructs are conceived as explanatory combinations of indicators (such as "population change"[2] or "marketing mix") that are determined by a combination of variables, their indicators should be formative.

Finally, there is an empirical element to the choice of indicator mode. In the formative mode, sample size and indicator multicollinearity affect the stability of indicator coefficients, which in this mode are based upon multiple regressions. In the reflective mode, indicator coefficients, based on simple regressions, are not affected by multicollinearity. If the indicators are highly collinear but one nonetheless desires optimiza-

FIGURE 14.1

**Three Different Modes of Relating Unobservables
to Empirical Indicators**

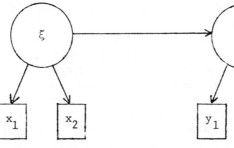

Mode A: Reflective indicators   Regress x_i on ξ

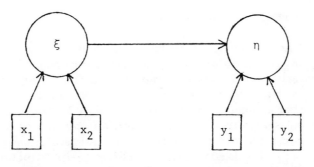

Mode B: Formative indicators :   Regress ξ on x, n any

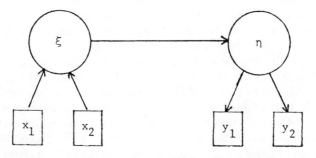

Mode C: Formative indicators for the exogenous construct,
reflective indicators for the endogenous construct

tion of explained structural model variance, one might estimate mode B but use loadings, rather than regression weights, for interpretation. This will be illustrated in our subsequent analyses.

Should the considerations involving study objective, theory or conceptualization, and empirical contingencies be contradictory, the selection of indicator mode may be difficult. For example, one may wish to minimize residual variance in the structural portion of the model, which suggests use of formative indicators, even though the constructs are conceptualized as giving rise to the observations (which suggests use of reflective indicators). In such cases the analyst might estimate twice, once in either mode. If the results correspond, there is no problem. If they differ, a compromise might be worked out using the factor structures of the blocks separately (as suggested by Bookstein, 1981); otherwise, a decision as to the overriding concern must be made.

FIXED-POINT ESTIMATION

The PLS model is estimated by determining the loadings (Λ_y, Λ_x) or weights (π_η, π_ξ), which describe how the observations relate to the unobservables, and the structural relations (β, Γ), whereby values of unobservables influence values of the other unobservables in the system. Instead of optimizing a global scalar function, PLS estimates by way of a nonlinear operator for which the vector of all estimated item loadings (Λ_y, Λ_x) is a fixed point. Following its introduction by Wold in 1963, the properties of fixed-point (FP) estimation have been discussed in Lyttkens (1966, 1973), Areskoug (1982), and in a collection of papers edited by Wold (1981). Several developments using some form of FP can be found in the recent psychometric literature (deLeeuw et al., 1976; Perreault and Young, 1980; Kroonenberg and deLeeuw, 1980; Sands and Young, 1980; Kruskal, 1980; and Carroll et al., 1980).

FP differs from ML models such as LISREL in its basic principles and assumptions. In ML estimation the probability of the observed data given the hypothesized model is maximized. Wold's PLS estimation, which is a least squares approach, minimizes residual variances under an FP constraint. ML estimators assume a parametric model, a family of joint distributions for all observables; PLS operates as a series of interdependent OLS regressions, presuming no distributional form at all. FP estimation, then, bears no resemblance to the search for zeroes of certain derivatives that characterize the estimation of ML models.

The distinction between optimizing and fixed-point methods may be compared with the two main models for solving multiple regressions. If we state as our goal the construction of a linear form $e + b_{yx_1 \cdot x_2} x_1 + b_{yx_2 \cdot x_1} x_2$, which estimates y with least error variance, we face a problem in direct minimization. But we may instead invoke the vocabulary of path analysis, referring to the <u>total effect</u> b_{yx_1} of variable x_1 on variable y and attempting to partition this into a <u>direct effect</u> $b_{yx_1 \cdot x_2}$ and an <u>indirect effect</u> $b_{x_2x_1} b_{yx_2 \cdot x_1}$ mediated by x_1's effect on x_2. The direct and indirect effects taken together must comprise the total: we must have

$$b_{yx_1 \cdot x_2} + b_{x_2x_1} b_{yx_2 \cdot x_1} = b_{yx_1} \tag{6}$$

and similarly for x_2. The result is a system of two equations with two unknowns which are, of course, identical to the normal equations of the usual approach. In this second version all explicit minimization is relegated to the bivariate analyses, the coefficients b_{yx_1}, b_{yx_2}, $b_{x_1x_2}$ being solutions of an easier optimum problem, the simple regression. Multiple regression may be thought of, then, as a revision by joint constraints of simple regression coefficients independently arrived at.

A similar distinction separates PLS from LISREL in the structural analysis of systems of latent variables. LISREL poses and solves the global optimization problem (maximization of likelihood) explicitly. PLS limits its explicit optimization computations to the now-familiar case of ordinary multiple regression where the separate analyses are jointly adjusted by nonlinear algebraic constraints.

ANALYSES

We compare LISREL and PLS using a small data set, with collinear indicators, in the context of a study of consumer dissatisfaction.

In his influential book of 1970, Albert O. Hirschman developed a theory of consumer reaction to dissatisfaction. He describes two basic modes: <u>Exit</u> and <u>Voice</u>. The exiting consumer makes use of the market by switching brands, terminating usage, or by shifting patronage—all economic actions. In contrast, Voice is a political action: a verbal protest directed at the seller, and, if remedy is not obtained, sometimes via third parties.

Hirschman posits that when the Exit option is blocked or when cross-elasticities are low, Voice will increase. By this reasoning, Exit should dominate in highly competitive markets, whereas the more a market resembles a monopoly, the more Voice would be expected. Since Seller Concentration is a measure of monopoly power, we hypothesize that Concentration is negatively related to Exit and positively related to Voice.

DATA

Exit and Voice data were obtained from a nationwide study by Best and Andreasen (1977). The 1972 Census of Manufacturers' four-digit concentration ratios were used as measures of market concentration. In total, seven variables were used: four measures of Concentration, two measures of Voice (aided and unaided recall), and one measure of Exit. The Voice and Exit measures were expressed as the proportion of respondents who recalled having taken action in each of those categories. From a total of 34 product and service categories in the Best and Andreasen study, 25 were retained for this analysis. Together they represent a majority of annual consumer purchases. The deleted categories either had no corresponding SIC code for concentration ratios or were too general to be meaningful.[3] The resulting correlation matrix is presented in Table 14.1.

MODEL 1

In LISREL applications the most common way of relating unobservables to data is by means of reflective indicators. In this mode the model attempts to explain the <u>observed correlations</u>. Reflective indicators in a PLS model imply that the primary objective is to explain the variances of the <u>observed variables</u>. As a starting point for comparative analysis we estimated Model 1 using reflective indicators.

In equation form, Model 1 is set forth as follows:

$$\begin{bmatrix} 1 & 0 \\ 0 & 1 \end{bmatrix} \begin{bmatrix} \eta_1 \\ \eta_2 \end{bmatrix} = \begin{bmatrix} \gamma_1 \\ \gamma_2 \end{bmatrix} \begin{bmatrix} \xi \end{bmatrix} + \begin{bmatrix} \zeta_1 \\ \zeta_2 \end{bmatrix} \tag{7}$$

$$\begin{bmatrix} y_1 \\ y_2 \\ y_3 \end{bmatrix} = \begin{bmatrix} 1 & 0 \\ 0 & \lambda_{y_2} \\ 0 & \lambda_{y_3} \end{bmatrix} \begin{bmatrix} \eta_1 \\ \eta_2 \end{bmatrix} + \begin{bmatrix} 0 \\ \varepsilon_2 \\ \varepsilon_3 \end{bmatrix} \tag{8}$$

TABLE 14.1

Correlation Matrix

		Exit (y_1)	Voice 1 (y_2)	Voice 2 (y_3)	CR 4 (x_1)	CR 8 (x_2)	CR 20 (x_3)	CR 50 (x_4)
Exit	(y_1)	1.0000						
Voice 1	(y_2)	-.0079	1.0000					
Voice 2	(y_3)	-.0797	.7094	1.0000				
CR 4	(x_1)	-.1024	.2881	.1827	1.0000			
CR 8	(x_2)	-.1757	.2759	.1820	.9594	1.0000		
CR 20	(x_3)	-.2000	.3057	.1959	.8797	.9561	1.0000	
CR 50	(x_4)	-.1311	.2813	.1539	.7810	.8664	.9662	1.0000

$$\begin{bmatrix} x_1 \\ x_2 \\ x_3 \\ x_4 \end{bmatrix} = \begin{bmatrix} \lambda_{x_1} \\ \lambda_{x_2} \\ \lambda_{x_3} \\ \lambda_{x_4} \end{bmatrix} [\xi] + \begin{bmatrix} \delta_1 \\ \delta_2 \\ \delta_3 \\ \delta_4 \end{bmatrix} \tag{9}$$

RESULTS: LISREL

The results in Figure 14.2 illustrate a problem that most LISREL users probably have encountered more than once (compare Jöreskog, 1979; Bentler, 1976; Areskoug, 1981; Driel, 1978): One of the variance estimates (in this case δ_3, the error of variance of x_3) is negative and the corresponding standardized loading (correlation) is greater than 1. This is an unacceptable result.

A common practice for circumventing the problem is to fix the negative variance at zero and reestimate the model, apparently on the grounds that the offending estimate is typically low and insignificant. However, this approach has both theoretical and practical flaws. The model to which it leads is based on neither the principal components nor the common-factor model (Bentler, 1976). Also, forcing one offending variance to zero will quite possibly cause the problem to reappear in other variance estimates. This is illustrated in Figure 14.3. When δ_3 is fixed at zero, the error variance ε_3 of y_3 becomes negative.

One cause of improper solutions might be failure of the model to fit the data (Driel, 1978). Given the large chi-square for the models of Figures 14.2 and 14.3, this possibility cannot be ruled out without further analysis. If the model fit is to be blamed, the improper solutions should be vitiated if $\underset{\sim}{\theta}_\delta$ is specified as symmetric rather than diagonal. But the results of Figure 14.4 show that this modification also fails. Even though the model now fits well, it does so by virtue of several negative variances, some quite large. Thus improper solutions are not necessarily circumvented and certainly not resolved by fixing the offending parameters or improving the model's fit. This is because we are dealing with an algebraic rather than a numerical or statistical problem: a matter not of "likelihood" or multivariate normality but of patterns of signs and magnitudes of the correlation matrix. Recall that LISREL's objective is to reproduce as closely as possible the observed correlation matrix $\underset{\sim}{S}$ by a matrix whose entries are explicit nonlinear functions of the parameters allowed to the model. For Model 1, with $\underset{\sim}{\theta}_\varepsilon$ and $\underset{\sim}{\theta}_\delta$ diagonal, $\underset{\sim}{\Sigma}$ is modeled as:

FIGURE 14.2

Model 1: PLS and LISREL Estimates with Reflective Indicators

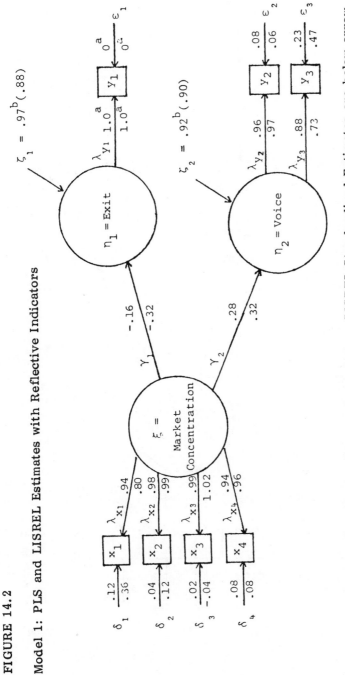

Note: Standardized Estimates are above arrow; LISREL Standardized Estimates are below arrow.
aFixed parameter.
bPLS estimate; corresponding LISREL estimate in parentheses.
$\chi^2 = 49.72$ (13 d.f.) $p \approx 0.$

FIGURE 14.3

Model 1: LISREL Estimates with Offending Variance Estimate Fixed to Zero

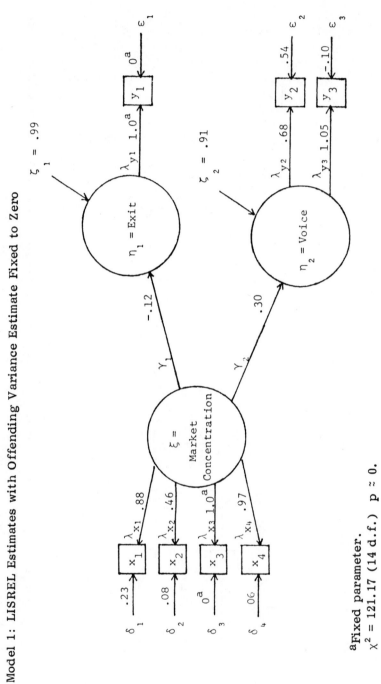

[a]Fixed parameter.

$\chi^2 = 121.17$ (14 d.f.) $p \approx 0.$

FIGURE 14.4

Model 1: LISREL Estimates with Correlated Measurement Errors for the Exogenous Construct

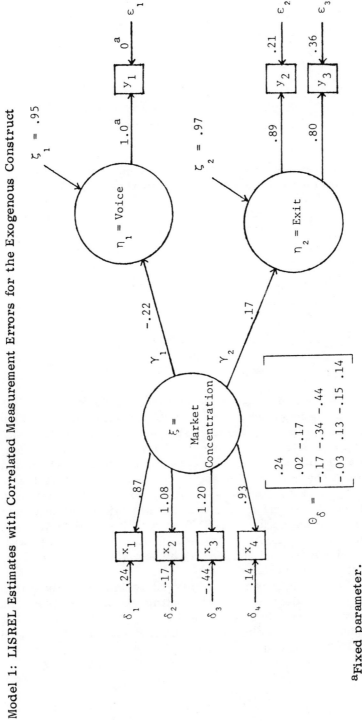

$$\Theta_\delta = \begin{bmatrix} .24 & & & \\ .02 & -.17 & & \\ -.17 & -.34 & -.44 & \\ -.03 & .13 & -.15 & .14 \end{bmatrix}$$

aFixed parameter.

$\chi^2 = 2.41$ (d.f. = 7) p ≈ .89.

$$
\underset{\sim}{\Sigma} =
\begin{bmatrix}
1.0000 & & & & & & & \\
\gamma_1\gamma_2\lambda_{y_2} & 1.0000 & & & & & & \\
\gamma_1\gamma_2\lambda_{y_3} & \lambda_{y_1}\lambda_{y_2} & 1.0000 & & & & & \\
\lambda_{x_1}\gamma_1\lambda_{y_1} & \lambda_{x_1}\gamma_2\lambda_{y_2} & \lambda_{x_1}\gamma_2\lambda_{y_3} & 1.0000 & & & & \\
\lambda_{x_2}\gamma_1\lambda_{y_1} & \lambda_{x_2}\gamma_2\lambda_{y_2} & \lambda_{x_2}\gamma_2\lambda_{y_3} & \lambda_{x_2}\lambda_{x_1} & 1.0000 & & & \\
\lambda_{x_3}\gamma_1\lambda_{y_1} & \lambda_{x_3}\gamma_2\lambda_{y_2} & \lambda_{x_3}\gamma_2\lambda_{y_3} & \lambda_{x_3}\lambda_{x_1} & \lambda_{x_3}\lambda_{x_2} & 1.0000 & & \\
\lambda_{x_4}\gamma_1\lambda_{y_1} & \lambda_{x_4}\gamma_2\lambda_{y_2} & \lambda_{x_4}\gamma_2\lambda_{y_3} & \lambda_{x_4}\lambda_{x_1} & \lambda_{x_4}\lambda_{x_2} & \lambda_{x_4}\lambda_{x_3} & 1.0000 &
\end{bmatrix}
\quad (10)
$$

For the model to be exactly true, it must reproduce the correlations among the indicators of concentration:

$$\lambda_{x_2}\lambda_{x_1} = .9594 \tag{11}$$

$$\lambda_{x_3}\lambda_{x_1} = .8797 \tag{12}$$

$$\lambda_{x_3}\lambda_{x_2} = .9561 \tag{13}$$

$$\lambda_{x_4}\lambda_{x_1} = .7810 \tag{14}$$

$$\lambda_{x_4}\lambda_{x_2} = .8664 \tag{15}$$

$$\lambda_{x_4}\lambda_{x_3} = .9662 \tag{16}$$

Notice that the correlation between these pairs of indicators is a strong negative function of the absolute difference between their subscripts—a characteristic of the familiar "Heywood" and simplex models of items in series, for which no single-factor model is adequate. For example, from equations 11-16 it is evident that $\lambda_{x_1} \geq .9594$ and $\lambda_{x_4} \geq .9662$. It follows that $\lambda_{x_4}\lambda_{x_1} \geq .9268$—but $r_{x_1x_4} = .7810$. Since no λ can exceed unity (as the standardized error variance in LISREL is equal to $1 - \lambda^2$) it is not possible for $\underset{\sim}{\Sigma}$ to fit $\underset{\sim}{S}$ without impropriety in the solutions; and the closer we come to fitting $\underset{\sim}{S}$, the smaller will be the chi-square and the larger the magnitude of the estimated negative variances.

Driel (1978) recommended that variables involved in improper solutions be deleted. By eliminating variables x_2 (CR8) and x_3 (CR20) and the Exit (y_3) variable from the data set—reducing, incidentally, the collinearity hobbling the x-block—the model now looks like Mode A in Figure 14.1:

$$\begin{array}{c} \text{Voice 1} \\ \text{Voice 2} \\ \text{CR 4} \\ \text{CR 50} \end{array} \begin{bmatrix} 1.0000 & & & \\ .7094 & 1.0000 & & \\ .2881 & .1827 & 1.0000 & \\ .2813 & .1539 & .7810 & 1.0000 \end{bmatrix} =$$

$$\begin{bmatrix} 1.0000 & & & \\ \lambda_{y_1}\lambda_{y_2} & 1.0000 & & \\ \lambda_{y_1}\gamma\lambda_{x_1} & \lambda_{y_2}\gamma\lambda_{x_1} & 1.0000 & \\ \lambda_{y_1}\gamma\lambda_{x_2} & \lambda_{y_2}\gamma\lambda_{x_2} & \lambda_{x_1}\lambda_{x_2} & 1.0000 \end{bmatrix} \qquad (17)$$

Again we consider the estimates necessary for an exact fit. Algebraically (compare Fornell and Larcker, 1981b), $\lambda_{y_1}^2$ may be estimated either by $(r_{x_2y_1}r_{y_1y_2}/r_{x_2y_2}) = 1.30$ or by $(r_{x_1y_1}r_{y_1y_2}/r_{x_1y_2}) = 1.12$. The error variances for y_1 are thus $-.30$ and $-.12$, respectively. ML estimation will conflate these two values into a single composite estimate; their pooling thus imputes a solution that is still improper. The impropriety is not a result of collinearity; one can have very high correlations and still obtain interpretable estimated models if the ratios of cross correlations $r_{x_iy_j}/r_{x_iy_k}$ are reasonably low. The improper solution is rather a consequence of either the failure of a single-factor model to explain the correlation submatrix for a particular block (unobservable), or inhomogeneities of the cross-correlation submatrices between blocks that cause them to be obviously of full rank. These are precisely the assumptions of LISREL submodels, block by block, which are seldom verified in the course of overall modeling (which is, of course, why they cause problems).

RESULTS: PLS

In Figure 14.2 the estimates via PLS for Model 1 are presented above or to the left of the corresponding LISREL estimates. Comparing them, we note lower structural parameters γ_1 and γ_2 in PLS but mostly higher loadings Λ_y, Λ_x. PLS does not produce improper estimates, as all residual variances are actual regression residuals; they are not inferred from the data. The PLS results are thus interpretable; they suggest a weak negative relationship between Concentration and Exit and a positive relationship between Concentration and Voice, along the lines suggested by Hirschman. The model is satisfactory insofar as the loadings, our primary focus in accounting for observed variances, are quite large. In general, PLS estimates of models with reflective indicators impute smaller measurement errors and weaker structural relationships than does LISREL. The algorithm by which the PLS estimates were obtained may be found in the Appendix to this chapter.

MODEL 2

The cause of improper solutions is generally the attempt to account for observed correlation matrices by patterned products of model parameters that are inadequate. The problem may be circumvented by attending to variances instead of correlations, that is, by working with components rather than factors. Components, which are exact linear combinations of their indicators, "maximize variance," while factors "explain covariance." To explore this amelioration we take advantage of the circumstance that certain component structures can be estimated by both PLS and LISREL. The MIMIC model (Jöreskog and Goldberger, 1975; Stapleton, 1978; Bagozzi, Fornell, and Larcker, 1981) is one such case.

Let this LISREL model be:

$$\eta_1 = [\gamma_1 \; \gamma_2 \; \gamma_3 \; \gamma_4] + \begin{bmatrix} \xi_1 = x_1 \\ \xi_2 = x_2 \\ \xi_3 = x_3 \\ \xi_4 = x_4 \end{bmatrix} \qquad (18)$$

should be multiplied?

$$\begin{bmatrix} y_1 \\ y_2 \\ y_3 \end{bmatrix} = \begin{bmatrix} \lambda_{y_1} \\ \lambda_{y_2} \\ \lambda_{y_3} \end{bmatrix} [\eta_1] + \begin{bmatrix} \varepsilon_1 \\ \varepsilon_2 \\ \varepsilon_3 \end{bmatrix}$$

with $\underset{\sim}{\Theta}_\varepsilon$ symmetric. The LISREL estimates are presented in Figure 14.5. For purposes of interpretation we make reference to the loadings for the x-block, computed as:

$$\underset{\sim}{\Lambda}_x = \underset{\sim}{R}_{xx} \underset{\sim}{\Gamma}, \qquad (20)$$

to yield $\lambda_{x_1} = .42$, $\lambda_{x_2} = .51$, $\lambda_{x_3} = .58$, $\lambda_{x_4} = .45$. From these loadings we compute the elements of $\underset{\sim}{\Theta}_\delta$ in Figure 14.5.

The PLS version of the above model is:

$$\eta = \gamma \xi + \zeta \qquad (21)$$

$$\eta = [\pi_{\eta_1} \; \pi_{\eta_2} \; \pi_{\eta_3}] \begin{bmatrix} y_1 \\ y_2 \\ y_3 \end{bmatrix} \qquad (22)$$

FIGURE 14.5

Model 2: LISREL Estimates

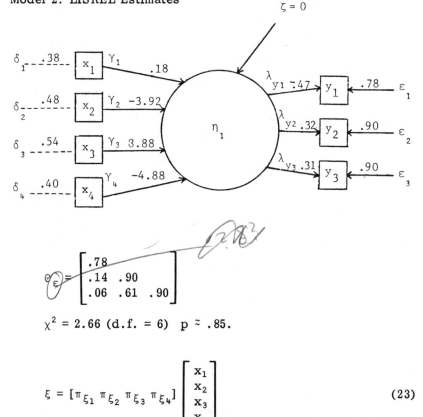

$\chi^2 = 2.66$ (d.f. = 6) p $\approx$ .85.

$$\xi = [\pi_{\xi_1} \; \pi_{\xi_2} \; \pi_{\xi_3} \; \pi_{\xi_4}] \begin{bmatrix} x_1 \\ x_2 \\ x_3 \\ x_4 \end{bmatrix} \qquad (23)$$

The results, according to a two-construct mode B estimation (as described in the Appendix), appear in Figure 14.6. Clearly, the PLS and LISREL solutions have identical x-weights (γ's in LISREL, π's in PLS). Further, the loadings for the y-side (λ_y's in LISREL) are equal to the loadings of the PLS results up to a factor of $1/\gamma_{PLS}$.[4]

Thus, if LISREL is specified in a MIMIC version with $\underset{\sim}{\theta}_\epsilon$ symmetric, it produces results identical to those from a PLS model with formative indicators. The formative specification with $\zeta = 0$—that unobservables be exact linear combinations of their indicators—is not as restrictive as it may appear. When the errors of the y-variables are correlated, the error term ζ of the unobservable is distributed instead throughout the elements of $\underset{\sim}{\theta}_\epsilon$ (see Hauser and Goldberger, 1971).

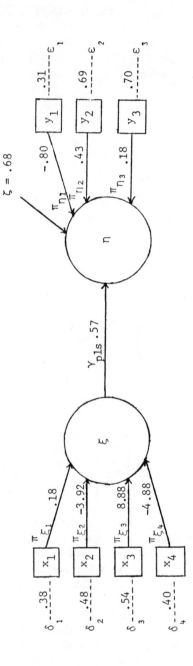

FIGURE 14.6

Model 2: PLS Estimates

306

The PLS estimate of the structural parameter γ_{PLS} in Model 2, Figure 14.6, is larger than either of the two estimates γ_1, γ_2 in Model 1, Figure 14.2. This is because the primary objective of a formative-mode model is to minimize the trace of the residual matrix ψ (the variance-covariance matrix of ζ), so that the measurement portion of Model 2 absorbs the largest possible part of the total residual, subject to the constraint of being a fixed point. The formative formulation, then, imputes a stronger relationship between Concentration and the construct combining Exit with the two Voice measures. The inverse relationship between Exit and Voice is reflected in the negative weight for Exit ($\pi_{\eta_1} = -.80$) in this construct.

RECAPITULATION OF MODELS 1 AND 2

We pause here to review certain important distinctions we have established. Model 1 presumes indicators that are reflective of the constructs. The LISREL estimation yielded negative variances and standardized loadings (correlations) greater than one. We showed that these unacceptable estimates ultimately derived from the LISREL objective of fitting a pattern of parameter products to the correlations observed. The PLS estimates, which by construction cannot yield such improprieties, exhibited smaller measurement variances but also lower estimated structural parameters. This accords with the purpose of PLS mode A, which is to explain the variance of the observed variables by minimizing mean-squared measurement residuals.

Model 2, by contrast, involved formative indicators. Estimates in this form focus on the variance in the structural portion of the model so that more of the net failure-of-fit is partitioned into measurement error. We emphasize that the choice between formative and reflective indicators is not merely a matter of empirical statistical fact. Choice of indicator mode brings conceptual, theoretical, and empirical observations to bear together on the objectives of the study; the partitioning of error variance can be manipulated only insofar as it depends on this choice. In particular, a MIMIC model specified without a disturbance term ζ but with correlated measurement errors is equivalent to a formative PLS model with two constructs. In this case PLS and LISREL produce identical estimated structures.

MODEL 3

Even though Model 2 produced consistent results from both LISREL and PLS, it is not an attractive formulation because no distinction is made, at the abstract level, between Exit and Voice. Our third model makes this distinction. Let us assume that the objective is to explain the observed y-variables (that is, in LISREL, their correlations; in PLS, their variances). In order to avoid uninterpretable results from LISREL, assume also that Concentration is formed by its indicators without any surplus variance. We arrive at a model with both formative and reflective indicators. The LISREL equations are:

$$\begin{bmatrix} 1 & 0 \\ \beta & 1 \end{bmatrix} \begin{bmatrix} \eta_1 \\ \eta_2 \end{bmatrix} = \begin{bmatrix} \gamma_1 & \gamma_2 & \gamma_3 & \gamma_4 \\ 0 & 0 & 0 & 0 \end{bmatrix} \begin{bmatrix} x_1 \\ x_2 \\ x_3 \\ x_4 \end{bmatrix} + \begin{bmatrix} 0 \\ \zeta_2 \end{bmatrix} \tag{24}$$

$$\begin{bmatrix} y_1 \\ y_2 \\ y_3 \end{bmatrix} = \begin{bmatrix} \lambda_{y_1} & 0 \\ 0 & \lambda_{y_2} \\ 0 & \lambda_{y_3} \end{bmatrix} \begin{bmatrix} \eta_1 \\ \eta_2 \end{bmatrix} + \begin{bmatrix} \varepsilon_1 \\ \varepsilon_2 \\ \varepsilon_3 \end{bmatrix} \tag{25}$$

where η_1 is Concentration, η_2 is Voice, and y_1 is Exit.
The PLS model is:

$$\begin{bmatrix} 1 & 0 \\ 0 & 1 \end{bmatrix} \begin{bmatrix} \eta_1 \\ \eta_2 \end{bmatrix} = \begin{bmatrix} \gamma_1 \\ \gamma_2 \end{bmatrix} [\xi] + \begin{bmatrix} \zeta_1 \\ \zeta_2 \end{bmatrix} \tag{26}$$

$$\xi = [\pi_{\xi_1} \ \pi_{\xi_2} \ \pi_{\xi_3} \ \pi_{\xi_4}] \begin{bmatrix} x_1 \\ x_2 \\ x_3 \\ x_4 \end{bmatrix} \tag{27}$$

$$\begin{bmatrix} y_1 \\ y_2 \\ y_3 \end{bmatrix} = \begin{bmatrix} 1 & 0 \\ 0 & \lambda_{y_2} \\ 0 & \lambda_{y_3} \end{bmatrix} \begin{bmatrix} \eta_1 \\ \eta_2 \end{bmatrix} + \begin{bmatrix} 0 \\ \varepsilon_2 \\ \varepsilon_3 \end{bmatrix} \tag{28}$$

where η_1 is Exit (y_1), η_2 is Voice, and ξ is Concentration.
The two sets of estimates presented in Figure 14.7 are similar.[5] The disturbance terms in the structural portion are slightly higher in PLS, whereas the measurement residuals are higher in LISREL. These differences derive from the different fitting objectives. We may explore them in detail by use of the descriptive statistics from the testing system of Fornell and Larcker (1981a).

FIGURE 14.7

Model 3: PLS and LISREL Estimates

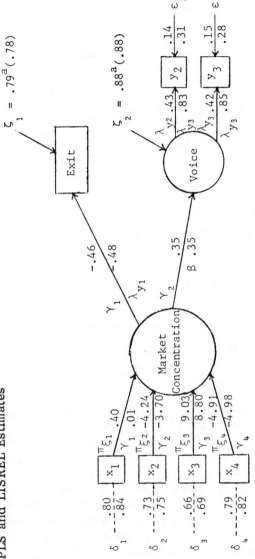

Note: PLS Standardized Estimates are above arrows; LISREL Standardized Estimates are below arrows.

[a]PLS estimate; corresponding LISREL estimate in parentheses.

$\chi^2 = 3.347$ (d.f. = 8) $p \approx .91$.

TABLE 14.2

Model 3: Descriptive Statistics

	PLS	LISREL
Average Variance Accounted (AVA) for the structural model	.165	.177
Average Variance Extracted (AVE) by the unobservables		
ρ_{vc} (market concentration)	.256	.228
ρ_{vc} (voice)	.856	.706
ρ_{vc} (exit)	1.0	1.0
Redundancy		
$\bar{R}^2_{y_1}$/market concentration	.212	.230 $= \lambda^2_{y_i}$
$\bar{R}^2_{y_2, y_3}$/market concentration	.105	.087
Operational variance (OV)		$= \beta^2 \cdot AVE$
$\hat{\bar{R}}^2_{y_1}/x_1 \ldots x_4$	.212	.230
$\hat{\bar{R}}^2_{y_2, y_3}/x_1 \ldots x_4$	.105	.072

Note: See Fornell and Larcker (1981a) for formulas and a detailed description of these statistics.

Referring to Table 14.2, the Average Variance Accounted for (AVA) is the mean R^2 of the structural model. This statistic is slightly higher in LISREL. Because of the similarity of model specifications, the difference in AVA is small. Average Variance Extracted (AVE) by the unobservables is the mean-squared loading for blocks of indicators separately. This statistic is consistently higher in PLS than in LISREL, due to the smaller imputed measurement errors. For each endogenous construct, Redundancy (which is the product of AVA by AVE) measures the power of the exogenous constructs for predicting the y-variables. As Exit (y_1) has no measurement error by construction, its AVE is 1. Thus, Redundancy for y_1 is γ^2 in PLS and $\lambda^2_{y_1}$ in LISREL. Redundancy for predicting the observed Voice indicators is calculated as the product of γ^2_2 (PLS) or β^2 (LISREL) with the corresponding AVE. Because the measurement errors are smaller, Redundancy is higher for PLS.

The sharpest difference between the PLS and LISREL estimates is seen in Operational Variance (OV), which is equal to Redundancy times the squared multiple correlation of the exogenous indicators with their unobservable construct. This statistic expresses the degree to which the variances of the y-observables are accounted for, via the model, by the observed x-variables. In the PLS solution, because all unobservables are exactly defined, it is identical to Redundancy; in LISREL it is attenuated and therefore lower. For instance, when the Redundancy of Voice is multiplied by the squared multiple correlation of Concentration with its indicators,[6] there results an OV of .072 in LISREL; the corresponding value for PLS is .105.

DISCUSSION

Under the classical assumptions of independence and normally distributed residuals, ML and OLS estimates in regression analysis are identical. In structural equation modeling this is not the case. Except as applied to certain MIMIC models, PLS and LISREL have different objectives and present systematically different results. However, as illustrated in Models 2 and 3, the more similar the model specification, the more similar the results will be.

As we have argued above, LISREL attempts to account for observed correlations, whereas PLS aims to account for variances at the observed or abstract level (depending upon indicator mode). Other major differences between the models include assumptions about factor structure, mechanisms of statistical inference, matters of identification,[7] interpretation of measurement error, as well as frequency of convergence. Unlike ML techniques, PLS makes minimal demands about measurement scales, sample size, or the distribution of residuals. Small sample sizes—sometimes fewer than the number of variables (Wold, 1980c)—can be sufficient for meaningful PLS analyses; and, in contrast to ML, PLS estimation does not involve a statistical model, thereby avoiding the need for assumptions regarding scales of measurement. Nominal, ordinal, and interval-scaled variables are permissible in PLS in the same ways as in ordinary regression.

A primary difference between PLS and LISREL concerns the structure of unobservables. LISREL specifies the residual structures, while PLS specifies the estimates of the unobservables explicitly. This difference bears important implications that have long been debated in the psychometric literature (see the review by Steiger, 1979). The main defense of the factor model is that it allows for imperfect measurement by assigning surplus variance to the unobservables. But such measurement error implies certain disturbing consequences. An infinite number of unobservables may bear the same pattern of correlations with observed variables and yet be only weakly or even negatively correlated with each other (Mulaik and McDonald, 1978). For exploratory analyses, such indeterminacy can be very problematic. In confirmatory structural equations, modeling indeterminacy has been thought to be less of a problem by reason of presumed existence of "prior knowledge" ruling out conflicting explanations. Because the chi-square statistic of fit in LISREL is identical for all possible unobservables satisfying the same structure of loadings, a priori knowledge is necessary. However, indeterminacy can create difficulties for confirmatory studies as well. There have been cases where several hypothesized models account for the same data equally well (see Mulaik, 1976). Thus, confirmatory studies are not necessarily free from the problem of having several interpretations. In this study we suggest an equally serious problem: Only indeterminate factors can have improper loadings. Since improper loadings lead to negative variances, such results do not have several interpretations, but, rather, none at all.

For the discipline of marketing, which is often concerned with prediction and control, other drawbacks flow from indeterminacy. As factor scores cannot be calculated, specific case predictions are not possible without prior estimation of the scores themselves. Similarly, testing for outliers and modeling of factor scores cannot be done either. PLS avoids factor indeterminacy by explicitly defining the unobservables. Factor scores for prediction or further modeling are then readily available.

In achieving optimal prediction, PLS estimators lack the parameter precision of ML estimation. Given multivariate normality, LISREL estimates are efficient in large samples and support analytic estimates of asymptotic standard errors. In exchanging far greater a priori assertion for statistical inference, LISREL is a model for theory testing. PLS is more applicable when prior information is wanting and theory is less developed. However, the theoretical underpinnings of PLS estimation are also less well developed. Although it appears to converge more quickly than LISREL (Areskoug, 1981), an analytic proof of PLS convergence for general models has yet to be worked out.

The example used in this chapter is not amenable to statistical inference because the sample is small and nonrandom; but

PLS modeling is not necessarily devoid of statistical inference. Although statistical tests are not integrated within the protocol of estimation, one may use the standard-error formulae of multiple regression or (perhaps more sensibly) a Stone-Geisser (Stone, 1974; Geisser, 1974) test for predictive significance (that is, redundancy). This provides jackknifed standard errors for the individual parameters, whereas LISREL calculates standard errors from the inverse of the information matrix. Both methods have limitations. In PLS there may be a problem in selecting the appropriate subgroup size for estimating pseudo-values in jackknifing. However, the larger the subsample size, the more reliable the statistic. The standard errors estimated by LISREL are also fallible. Since each unobservable requires a scale identification restriction affecting the information matrix, standard errors and critical ratios appear to vary by choice of restriction and can yield rather unpleasant paradoxes of interpretation (Pijper and Saris, 1982). By contrast, in recursive models the fixed-point estimation scheme of PLS is free of identification problems.

The tests for predictive significance and Operational Variance can be used to assess the overall "fit" of both LISREL and PLS models. A corresponding test for overidentified models in LISREL is the likelihood ratio chi-square. However, this test requires strong data assumptions and does not support any conclusions beyond those dealing specifically with accounting for observed correlations. Fornell and Larcker (1981a, 1981b) showed that the strength of variable relationships is negatively associated with goodness of fit in models otherwise isomorphic.

SUMMARY

For the marketing analyst the choice between LISREL and PLS is neither arbitrary nor straightforward. Both apply to the same class of models—structural equations with unobservables and measurement error—but they have different structures and objectives.

LISREL attempts to account for observed correlations, while PLS aims at explaining variances (of variables observed or unobserved).

LISREL offers statistical precision in the context of stringent assumptions; PLS trades parameter efficiency for prediction accuracy, simplicity, and fewer assumptions.

The factor model underlying LISREL allows more errors in measurement than the components model invoked by PLS. In

LISREL, unobservables are truly unobservable; PLS foresakes the consequent enhancements of theoretical explanation in order to avoid the ambiguities and improprieties that often ensue.

In sum, nothing less than the general research setting can determine the appropriate modeling approach. It is within this context that the LISREL benefits of statistical parameter efficiency and higher estimated relationships among unobservables should be weighed against the problems associated with indeterminacy. The analyses of the study suggest that when relevant correlation submatrices (block by block) are not of the appropriate reduced rank, factor indeterminacy is more serious than generally acknowledged. The frequent occurrences of improper and uninterpretable solutions advise against the use of LISREL unless its assumptions are verifiably true and its objectives consistent with the objectives of the study; and, when they are not, PLS presents a viable alternative.

NOTES

1. Previously known as NIPALS (Nonlinear Iterative Partial Least Squares) or NILES (Nonlinear Iterative Least Squares).

2. "Population change" is presumed to be determined by natality, mortality, and migration. This example is from Hauser's (1973) criticism of sociologists' overreliance on reflective indication.

3. No attempt was made to pool concentration ratios for several categories to obtain a better fit between the Best and Andreasen data and the SIC codes. Therefore, substantive conclusions based on the analysis reported here must be interpreted with caution. Also, given the nature of the data, the sample size, and the sampling procedure, statistical inference will play a very minor role in our analyses.

4. The y-loadings in PLS are:

$$\underset{\sim}{\Lambda}_y = \frac{1}{\gamma_{PLS}} \ [\lambda_{y_1} \ \lambda_{y_2} \ \lambda_{y_3}]$$

where γ_{PLS} is the estimated structural parameter from PLS, and $\lambda_{y_1} \dots \lambda_{y_3}$ are the estimated loadings from LISREL. Numerically, we have

$$\underset{\sim}{\Lambda}_y = \frac{1}{.57} \ [-.47 \ .31 \ .31] = [-.83 \ .56 \ .55]$$

which are the y-loadings. This can easily be verified by computing the corresponding loadings for the PLS solution as $\Lambda_y = \underset{\sim}{R}_{yy} \underset{\sim}{\Xi}_\eta$. The estimate γ_{PLS} is not available from LISREL but can be computed as:

$$(\underset{\sim}{R}_{n_1y} \; \underset{\sim}{R}_{yy}^{-1} \; \underset{\sim}{R}_{n_1y})^{\frac{1}{2}}.$$

Since γ_{PLS} is the largest eigenvalue of $\underset{\sim}{R}_{yy}^{-1} \underset{\sim}{R}_{yx} \underset{\sim}{R}_{xx}^{-1} \underset{\sim}{R}_{xy}$, both PLS and LISREL provide general models for canonical correlation.

5. In other comparisons we have found significant structural relations in a LISREL model while corresponding PLS estimates have been insignificant, even when the latter was estimated via Mode B to minimize the structural error. As shown by Bagozzi, Fornell, and Larcker (1981), this paradox is traceable to the factor score indeterminacy in LISREL.

6. This adjustment is identical to the index of factor score indeterminacy proposed by Green (1976).

7. For recursive models, PLS basically has no problem on this score—see Bookstein (1982).

REFERENCES

Ågren, A. (1972). Extensions of the Fixed-Point Method. Published doctoral dissertation, Department of Statistics, University of Uppsala, Sweden.

Apel, H. (1977). Simulation sozio-ökonomischer Zusammanhange-Kritik and Modification von Systems Analysis. Doctoral dissertation, J. W. von Goethe University, Frankfurt am Main, Germany.

Areskoug, B. (1982). "Some Asymptotic Properties of PLS Estimators and a Simulation Study for Comparisons between LISREL and PLS." In Systems under Indirect Observation: Causality, Structure, Prediction, ed. K. G. Jöreskog and H. Wold. Amsterdam: North Holland, in press.

Areskoug, B., H. Wold, and E. Lyttkens (1975). "Six Models with Two Blocks of Observables as Indicators for One or Two Latent Variables." Research Report No. 6, Department of Statistics, University of Gothenburg, Sweden.

Bagozzi, Richard P. (1980). Causal Models in Marketing. New York: John Wiley.

Bagozzi, Richard P., Claes Fornell, and David F. Larcker
(1981). "Canonical Correlation Analysis as a Special Case
of a Structural Relations Model." Multivariate Behavioral
Research 16, 437-54.

Bentler, P. M. (1976). "Multistructure Statistical Model Applied
to Factor Analysis." Multivariate Behavioral Research 11,
3-25.

Bergström, R. (1972). "An Investigation of the Reduced Fixed-
Point Method." Seminar paper, Department of Statistics,
University of Uppsala, Sweden.

Best, Arthur and Alan R. Andreasen (1977). "Consumer Re-
sponse to Unsatisfactory Purchases: A Survey of Perceiving
Defects, Voicing Complaints and Obtaining Redress." Law &
Society 11, 701-42.

Bodin, L. (1974). Recursive Fixed-Point Estimation: Theory
and Application. Published doctoral dissertation, Department
of Statistics, University of Uppsala, Sweden.

Bookstein, Fred L. (1980). "Data Analysis by Partial Least
Squares." In Evaluation of Econometric Models, ed. J. Kmenta
and J. B. Ramsey. New York: Academic Press, 75-90.

_____ (1982). "The Geometric Meaning of Soft Modeling with
Some Generalizations." In Systems under Indirect Observation:
Causality, Structure, Prediction, ed. K. G. Jöreskog and
H. Wold. Amsterdam: North Holland, in press.

Carroll, J. Douglas, Sandra Pruzansky, and Joseph Kruskal
(1980). "CANDELINC: A General Approach to Multidimensional
Analysis of Many-Way Arrays with Linear Constraints on
Parameters." Psychometrika 45, 3-24.

deLeeuw, J., F. W. Young, and Y. Takane (1976). "Additive
Structure in Qualitative Data: An Alternating Least Squares
Method with Optimal Scaling Features." Psychometrika 41,
471-503.

Driel, Otto van (1978). "On Various Causes of Improper Solutions in Maximum Likelihood Factor Analysis." Psychometrika
43, 225-43.

Fornell, Claes, and David F. Larcker (1981a). "Evaluating Structural Equation Models with Unobservable Variables and Measurement Error." Journal of Marketing Research 18 (February), 39-50.

_____ (1981b). "Structural Equation Models with Unobservable Variables and Measurement Error: Algebra and Statistics," Journal of Marketing Research 18 (August), 382-88.

Geisser, S. (1974). "A Predictive Approach to the Random Effect Model." Biometrika 61, 101-07.

Green, Bert F. (1976). "On the Factor Score Controversy." Psychometrika 41, 263-66.

Hauser, Robert M. (1973). "Disaggregating a Social-Psychological Model of Educational Attainment." In Structural Equation Models in the Social Sciences, A. S. Goldberger and O. D. Duncan, eds. New York: Seminar Press, 255-84.

Hauser, Robert M. and Arthur S. Goldberger (1971). "The Treatment of Unobservable Variables in Path Analysis." In Sociological Methodology, ed. H. L. Costner. San Francisco: Jossey-Bass, 81-117.

Hirschman, Albert O. (1970). Exit, Voice, and Loyalty—Responses to Decline in Firms, Organizations, and States. Cambridge, Mass.: Harvard University Press.

Hui, B. S. (1978). "The Partial Least Squares Approach to Path Models of Indirectly Observed Variables with Multiple Indicators." Unpublished doctoral thesis, University of Pennsylvania.

Jagpal, Harsharanjeet S. (1981). "Measuring Joint Advertising Effects in Multiproduct Firms." Journal of Advertising Research 21, no. 1, 65-69.

Jöreskog, Karl G. (1970). "A General Method for Analysis of Covariance Structures." Biometrika, 57, 239-51.

_____ (1973). "A General Method for Estimating a Linear Structural Equation System." In Structural Equation Models in the Social Sciences, ed. A. S. Goldberger and O. D. Duncan. New York: Seminar Press, 85-112.

_____ (1979). "Structural Equation Models in the Social Sciences: Specification, Estimation and Testing." In Advances in Factor-Analysis and Structural Equation Models, ed. Karl G. Jöreskog and Dag Sörbom. Cambridge, Mass.: ABT Books, 105-27.

Jöreskog, Karl G. and Arthur S. Goldberger (1975). "Estimation of a Model with Multiple Indicators and Multiple Causes of a Single Latent Variable." Journal of the American Statistical Association 70, 631-39.

Jöreskog, Karl G. and Dag Sörbom (1978). LISREL IV: Analysis of Linear Structural Relationships by the Method of Maximum Likelihood. Chicago: National Educational Resources.

_____ (1981). LISREL V: Analysis of Linear Structural Relationships by Maximum Likelihood and Least Squares Methods. Chicago: National Education Resources.

Jöreskog, Karl G. and Herman Wold (1982), eds. Systems under Indirect Observation: Causality, Structure, Prediction. Amsterdam: North Holland, in press.

Kowalski, B. R., R. W. Gergerlach, and H. Wold (1982). "Chemical Systems under Indirect Observation." In Systems under Indirect Observation: Causality, Structure, Prediction, ed. K. G. Jöreskog and H. Wold. Amsterdam: North Holland, in press.

Kroonenberg, Pieter M. and Jan deLeeuw (1980). "Principal Component Analysis of Three-Mode Data by Means of Alternating Least Squares Algorithms." Psychometrika 45, 69-97.

Kruskal, Joseph (1980). "An Elegant New/Old Approach to Estimating Path Models (Structural Equation Models) with Unobserved Variables." Technical Report. Murray Hill, N.J.: Bell Laboratories.

Lyttkens, E. (1966). "On the Fixed-Point Property of Wold's Iterative Estimation Method for Principal Components." In Multivariate Analysis, ed. P. R. Krishnaiah. New York: Academic Press, 335-50.

_____ (1973). "The Fixed-Point Method for Estimating Interdependent Systems with the Underlying Model Specification." Journal of the Royal Statistical Society A136, 353-94.

Meissner, W. and M. Uhle-Fassing (1982). "PLS—Modeling and Estimation of Politimetric Models." In Systems under Indirect Observation: Causality, Structure, Prediction, ed. K. G. Jöreskog and H. Wold. Amsterdam: North Holland, in press.

Mulaik, Stanley A. (1976). "Comments on the Measurement of Factorial Indeterminacy." Psychometrika 41, 249-62.

Mulaik, Stanley A. and Roderick P. McDonald (1978). "The Effect of Additional Variables on Factor Indeterminacy in Models with a Single Common Factor." Psychometrika 43, 177-92.

Noonan, Richard (1980). "School Environments and School Outcomes: An Empirical Comparative Study Using IEA Data." Working paper series No. 26, Institute of International Education, University of Stockholm, Sweden.

Noonan, Richard and Herman Wold (1977). "NIPALS Path Modelling with Latent Variables: Analyzing School Survey Data Using Nonlinear Iterative Partial Least Squares." Scandinavian Journal of Educational Research 21, 33-61.

_____ (1980). "PLS Path Modelling with Latent Variables: Analyzing School Survey Data Using Partial Least Squares— Part II." Scandinavian Journal of Educational Research 24, 1-24.

Perreault, William D. and Forrest W. Young (1980). "Alternating Least Squares Optimal Scaling: Analysis of Nonmetric Data in Marketing Research." Journal of Marketing Research 17 (February), 1-13.

Pijper, de W. M. and W. E. Saris (1982). "The Effect of Identification Restriction on the Test Statistic in Latent Variable Models." In Systems under Indirect Observation: Causality, Structure, Prediction, ed. K. G. Jöreskog and H. Wold. Amsterdam: North Holland, in press.

Sands, R. and F. W. Young (1980). "Component Models for Three-Way Data: An Alternating Least Squares Algorithm with Optimal Scaling Features." Psychometrika 45, 39-87.

Stapleton, D. C. (1978). "Analyzing Political Participation Data with a MIMIC Model." In Sociological Methodology. San Francisco: Jossey-Bass, 52-74.

Steiger, James H. (1979). "Factor Indeterminacy in the 1930's and the 1970's—Some Interesting Parallels." Psychometrika 44, 157-67.

Stone, M. (1974). "Cross-Validatory Choice and Assessment of Statistical Predictions." Journal of the Royal Statistical Society B36, 111-33.

Wold, Herman (1963). "Toward a Verdict on Macroeconomic Simultaneous Equations." In Semaine d'étude sur le rôle de l'analyse économétrique dans la formulation des plans de développement, ed. P. Salviucci. Scripta Varia 28 (Pontifical Academy of Science, Vatican City). Cited in Wold, H. ed. (1981). The Fixed Point Approach in Interdependent Systems. Amsterdam: North Holland.

_____ (1965). "A Fixed-Point Theorem with Econometric Background, I-II." Arkiv för Matematik 6, 209-40.

_____ (1974). "Causal Flows with Latent Variables: Partings of the Ways in the Light of NIPALS Modeling." European Economic Review 5, 67-86.

_____ (1975). "Path Models with Latent Variables: The NIPALS Approach." In Quantitative Sociology: International Perspectives on Mathematical and Statistical Model Building, ed. H. M. Blalock et al. New York: Academic Press, 307-57.

_____ (1980a). "Model Construction and Evaluation When Theoretical Knowledge Is Scarce—Theory and Application of Partial Least Squares." In Evaluation of Econometric Models, ed. J. Kmenta and J. G. Ramsey. New York: Academic Press, 47-74.

_____ (1980b). "Soft Modelling: Intermediate between Traditional Model Building and Data Analysis." Mathematical Statistics 6, 333-46.

_____ (1980c). "Factors Influencing the Outcome of Economic Sanctions: An Application of Soft Modeling." Paper presented at the Fourth World Congress of the Econometric Society, Aix-en-Provence, France.

_____ (1981) ed. The Fixed Point Approach to Interdependent Systems. Amsterdam: North Holland.

APPENDIX: PLS ALGORITHMS FOR MODELS 1, 2,
AND 3 OF THE TEXT

The computations of PLS estimation are performed by iterations of explicit simple and multiple regressions. This can easily
be accomplished within such computer packages as MIDAS, SAS,
and TROLL. Specialized PLS programs are also available. Interested readers may contact Claes Fornell at the Graduate School
of Business Administration, The University of Michigan, Ann
Arbor, Michigan, 48109, concerning these programs.

Model 1 (Figure 14.2)

Initialize

Set $\eta_1 = y_1 = $ Exit, $\eta_2 = y_2 + y_3 = $ Voice, $\xi = x_1 + x_2 + x_3 + x_4 = $ Concentration.

Loop

Normalize η_1, η_2, and ξ to variance unity. Regress η_1,
η_2 on x_1, x_2, x_3, x_4 separately:

$$\eta_1 = \lambda_{1i}x_i + \varepsilon_{1i}$$
$$\eta_2 = \lambda_{2i}x_i + \varepsilon_{2i}, \quad i = 1, 2, 3, 4.$$

Construct

$$\hat{\xi} = \sum_{i=1}^{4} (\lambda_{1i} - \lambda_{2i})x_i,$$

the minus sign because η_1 and η_2 have covariances of opposite
sign with ξ.

Regress $\hat{\xi}$ on y_2 and y_3 separately:

$$\hat{\xi} = \lambda_{3i}y_i + \varepsilon_{3i}, \quad i = 2, 3.$$

Compute

$$\eta_2 = \sum_{i=2}^{3} \lambda_{3i}y_i.$$

There is no η_1, since this block has only one indicator.

Test

If $\hat{\xi}$ is not equal to ξ or $\hat{\eta}_2$ to η_2, Loop again. Otherwise,

Finish

Regress η_1 and η_2 separately on ξ for the structural parameters γ_1, γ_2.

Model 2 (Figure 14.6)

Initialize

Set $\eta = y_1 + y_2 + y_3$, $\xi = x_1 + x_2 + x_3 + x_4$.

Loop

Normalize η and ξ to variance unity. Regress η on x_1, x_2, x_3, x_4 jointly:

$$\eta = \sum_{i=1}^{4} \pi_{\eta i} x_i + \varepsilon_\eta.$$

Construct

$$\hat{\xi} = \sum_{i=1}^{4} \pi_{\eta i} x_i.$$

Regress $\hat{\xi}$ on y_1, y_2, y_3 jointly:

$$\hat{\xi} = \sum_{i=1}^{3} \pi_{\xi i} y_i + \varepsilon_\xi.$$

Construct

$$\hat{\eta} = \sum_{i=1}^{3} \pi_{\xi i} y_i.$$

Test

If $\hat{\xi}$ is not equal to ξ or $\hat{\eta}$ to η, Loop again. Otherwise,

Finish

Regress η on ξ for the structural parameter γ.

Model 3 (Figure 14.7)

Initialize

Set $\eta_1 = y_1 = $ Exit, $\eta_2 = y_2 + y_3 = $ Voice, $\xi = x_1 + x_2 x_3 + x_4 = $ Concentration.

Loop

Normalize η_1, η_2, and ξ to variance unity. Regress η_1, η_2 on x_1, x_2, x_3, x_4 jointly:

$$\eta_1 = \sum_{j=1}^{4} \pi_{1j} x_j + \varepsilon_1,$$

$$\eta_2 = \sum_{j=1}^{4} \pi_{2j} x_j + \varepsilon_2.$$

Compute

$$\hat{\xi} = \sum_{j=1}^{4} (\pi_{1j} - \pi_{2j}) \, x_j.$$

Regress $\hat{\xi}$ on y_2, y_3 separately:

$$\hat{\xi} = \lambda_{y_2} y_2 + \varepsilon_{y_2},$$
$$\hat{\xi} = \lambda_{y_2} y_2 + \varepsilon_{y_2}.$$

Compute

$$\hat{\eta}_2 = \sum_{i=2}^{3} \lambda_{y_i} y_i.$$

Test

If $\hat{\xi}$ is not equal to ξ or $\hat{\eta}_2$ to η_2, Loop again. Otherwise,

Finish

Regress η_1 and η_2 separately on ξ for the structural parameters γ_1, γ_2.

15

Systems Under Indirect
Observation Using PLS

Herman Wold

INTRODUCTION

"Soft modeling" has become the widely used name for the methodology for PLS estimation of path models with latent variables indirectly observed by multiple indicators. Several introductory expositions of PLS soft modeling are now available. In this chapter I shall first recall the recent origin of path modeling with latent variables, then restate the elements of PLS soft modeling with comments on its flexibility and broad scope, adduce some comparative remarks on LISREL and PLS addressed to the potential user in the field of empirical economic research, and discuss some aspects of PLS that come to the fore in applications to quantitative analysis.

Sociology in the early 1960s opened a new era in quantitative systems analysis by O. T. Duncan's merger of the latent (indirectly observed) variables in psychology and the path modeling with manifest (directly observed) variables in econometrics. As illustrated in Figure 16.1, the econometric models included multiple OLS regression, causal chain systems, and interdependent (ID) systems, whereas the psychological pedigree included

This chapter was originally prepared for the transactions of the Berlin Conference on "Modeling of Innovation Processes and Structural Change," G. O. Mensch, editor. Support of research on PLS Soft Modeling by the Stiftung Volkswagenwerk is gratefully acknowledged.

FIGURE 15.1

Path Models with Latent Variables: A Merger of Econometrics and Psychometrics

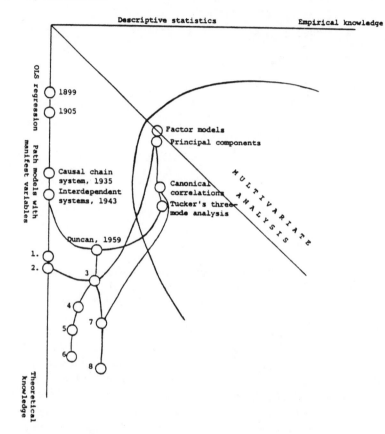

1. -2. The Fix-Point Method, H. Wold, 1963, 1980c
   1. REID (Reformulated ID) systems
   2. GEID (General ID) systems
   3. PLS estimation of principal components and canonical correlations, H. Wold, 1966
   4. ML algorithm for estimation of Factor models, K. G. Jöreskog, 1967
5. LISREL algorithm for path models with latent variables, K. G. Jöreskog, 1970
6. LISREL, further versions, K. G. Jöreskog, 1973 and Jöreskog and Sörbom, 1981
7. PLS algorithms for path models with latent variables, H. Wold, 1977, 1979-
8. PLS algorithm for path models with latent variables indirectly observed by observations over time and space, J.-B. Lohmöller, 1981.

factor analysis, principal components, and canonical correlations. On the econometrics side was the new twist I had given ID systems by the REID (Reformulated ID) and GEID (General ID) systems and my Fix-Point (FP) iterative method for the estimation of REID and GEID systems (Wold, 1965). On the psychometric side were my iterative methods for assessing principal components and canonical correlations (Wold, 1966).

Path modeling with latent variables posed entirely new statistical problems, through the 1960s, dealt with by ad hoc devices in the sociological literature.

Karl G. Jöreskog in 1967 was the first to bring the computer to operative use for ML (Maximum Likelihood) estimation of factor models, and in 1970 Jöreskog launched his LISREL algorithm for ML estimation of path models with latent variables. When seeing the LISREL algorithm I realized that principal components and canonical correlations can be interpreted as path models with one and two latent variables, respectively, and with each latent variable explicitly estimated as a weighted aggregate of its indicators; this gave me the incentive to extend the FP algorithm to general path models with latent variables indirectly observed by multiple indicators. There were two stumbling blocks in this endeavor: the step from two to three latent variables, and the step from one to two "inner" relations between the latent variables. Once these stumbling blocks were overcome it was easy to design general algorithms for iterative estimation of path models with latent variables. The ensuing algorithms I first called NIPALS (Nonlinear Iterative Partial Least Squares), later PLS. An array of somewhat different PLS versions was designed; in late 1977 I arrived at what I regard as the end station: the basic design for PLS estimation of path models with latent variables, or briefly PLS soft modeling.

EXPOSITION OF SOFT MODELING

The know-how of PLS soft modeling will now be set forth beginning with the notion of the arrow scheme. Figure 15.2 shows the arrow scheme of a specific economic example. The arrow scheme of a soft model specifies the theoretical-conceptual design of the model. The formal definition of the model, and the PLS algorithm for its estimation, can be written down directly from the arrow scheme.

FIGURE 15.2

Arrow Scheme for a Causal Model of Total Factor Productivity
Growth at Various Levels of Aggregation

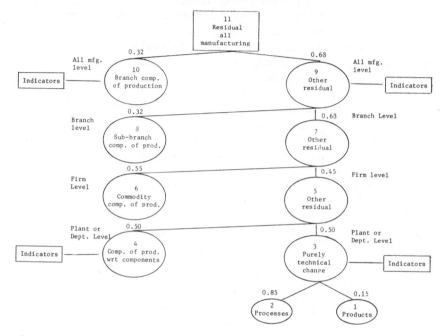

Arrow Scheme

As illustrated in Figure 15.2, a soft model has inner and
outer variables. The core of the model is a path of causal-
predictive inner relations between the inner variables. The
inner variables may be latent (indirectly observed) or manifest
(directly observed). Each latent variable is indirectly observed
by a block of manifest outer variables, called indicators of the
latent variable. The manifest variables (MVs) are illustrated
by squares or rectangles, the latent variables (LVs) by circles
or ovals. In Figure 15.2 the indicators are shown only for
some of the latent variables.

From the arrow scheme illustrated in Figure 15.2 we see
that this specific model has eleven inner variables, ten of which
are latent and the eleventh is manifest. The model has a path
of five inner relations: These form a causal chain system (no
feedbacks or interdependencies). The LVs have multiple indicators. Each indicator is connected with its LV by an outer
relation. In econometric usage the model has five endogenous
and six exogenous inner variables.

Model Design

A soft model can be interpreted as a give-and-take of information, information conveyed by inner relations between the inner variables, and by outer relations between the latent variables and their indicators. It is characteristic of soft modeling that all causal-predictive information of the model, and, in particular, all information between the different blocks of indicators, is assumed to be conveyed by the inner relations via the inner variables.

As long as he adheres to this fundamental principle, the investigator (model builder) is free to design the arrow scheme and thereby his model in accordance with his intentions, intuition, and knowledge. In particular, he is free to design which LVs and other inner variables the model should contain, which relations there are between the inner variables, and which indicators should be used for indirect observation of the various LVs.

Expected Signs

At the outset, the investigator should specify in the arrow scheme: (i) the expected signs of the correlations of the outer variables and their LVs; (ii) the expected signs of the correlations of any two inner variables that are adjacent in the sense that they are directly connected by an arrow in the arrow scheme.

Formal Specification

A soft model is formally defined by its inner and outer relations. The formal specification further includes the relations for substitutive prediction. Inner and outer relations are subject to predictor specification, and to zero residual intercorrelations in accordance with the "Model Design" section.

Notation

The following notations of variables and data are of general scope in soft modeling; however, the specifications refer to Figure 15.2.

Variables:

Inner variables:	latent or manifest variables
Latent variables:	ξ_k; $k = 1, \ldots, 10$
Exogenous variables:	$k = 1, 2, 4, 6, 8, 10$
Endogenous variables:	$k = 3, 5, 7, 9, 11$
Manifest variables:	x_{11}

Outer variables: The kth LV is indirectly observed by H_k manifest

indicators, denoted x_{kh} $k = 1, \ldots, 10; h = 1, \ldots, H_k$

The indicators of an endogenous (exogenous) LV are called endogenous (exogenous) indicators.

Data: The manifest variables are observed over time or a cross section, say,

T cases (observations): $t = 1, \ldots, T$

Inner variable: $x_{11,t}$ $t = 1, \ldots, T$

Outer variables: x_{kht} $k = 1, \ldots, 10; h = 1, \ldots, H_k; t = 1, \ldots, T$

The ranges of the subscripts k, h, t will often be tacitly understood.

latent manifest cases

Outer Relations

As always in soft modeling, the outer relations take the form:

$$x_{kh} = \pi_{kho} + \pi_{kh}\xi_k + \varepsilon_{kh} \qquad k = 1, \ldots, 10; \qquad (1)$$
$$h = 1, \ldots, H_k$$

and are subject to predictor specification; that is, the systematic part of (i) is assumed to be the expected values of the x_{kh}:

$$E(x_{kh}|\xi_k) = \pi_{kho} + \pi_{kh}\xi_k. \qquad (2)$$

This implies

$$E(\varepsilon_k) = 0; \; r(\varepsilon_{kh'}\xi_k) = 0 \qquad k = 1, \ldots, 10; \qquad (3)$$
$$h = 1, \ldots, H_k$$

In virtue of the "Model Design" principle we furthermore assume:

$$r(\varepsilon_{kh'}\varepsilon_{k'}) = 0; \qquad k, k' = 1, \ldots, 10; \qquad (4a)$$
$$h = 1, \ldots, H_k$$

$$r(\varepsilon_{kh'}\varepsilon_{k'g}) = 0; \qquad k \neq k'; \; g = 1, \ldots, H_{k'} \qquad (4b)$$

To avoid ambiguity in the product term in (1) some SSU (Standardization for Scale Unambiguity) is necessary. To

achieve SSU in soft modeling the LVs are always standardized
to unit variance:

$$\text{var}(\xi_k) = E[\{\xi_k - E(\xi_k)\}^2] = 1 \quad k = 1, \ldots, 10 \qquad (5)$$

It is convenient to extend the standardization (5) to the
manifest inner variables, in this case, to x_{11}, defining a corresponding LV:

$$x_{11} = x_{11,1} = \pi_{11,1}\xi_{11}; \quad \pi_{11,1} = \text{s.d.}(x_{11}) \qquad (6)$$

Inner Relations

In the basic design of PLS soft modeling, the inner relations
are linear and form a causal chain system. The model in Figure
15.2 has five inner relations, five endogenous LVs, and six
exogenous LVs:

$$\xi_k = \beta_{ko} + \beta_{k,k-2}\xi_{k-2} + \beta_{k,k-1}\xi_{k-1} + u_k$$
$$k = 3, 5, 7, 9, 11 \qquad (7)$$

Predictor specification:

$$E(\xi_k|\xi_{k-2}, \xi_{k-1}) = \beta_{ko} + \beta_{k,k-2}\xi_{k-2} + \beta_{k,k-1}\xi_{k-1} \qquad (8)$$

which implies

$$E(\xi_k) = 0; \quad r(\varepsilon_{k'}\xi_{k-2}) = r(\varepsilon_{k'}\xi_{k-1}) = 0. \qquad (9)$$

The inner and outer relations provide the basis for generating data for simulation experiments on the model. Observations
ξ_{kt} on a specified joint distribution for the LVs are generated
in the first place, and then observations on the residuals ε_{kht}
in accordance with the zero correlation assumptions (3)-(4a,b).

Substitutive Prediction

For the indicators of any endogenous LV, soft modeling
gives relations for substitutive prediction, obtained by substitutive elimination of the LV from the outer relations, using the
corresponding inner relation. In the present case:

$$x_{kht} = \mu_{kh} + \pi_{kh}(\beta_{k,k-2}\xi_{k-2,t} + \beta_{k,k-1}\xi_{k-1,t}) + \nu_{kht};$$
$$k = 3, 5, 7, 9, 11$$
$$(10)$$

with location parameters and residuals given by

$$\mu_{kh} = \pi_{kho} + \pi_{kh}\beta_{ko}; \quad \nu_{kht} = \varepsilon_{kht} + \pi_{kh}\nu_{kt} \tag{11}$$

Inner relations and substitutive predictions are two key tools for prediction in PLS soft modeling. An inner relation predicts an endogenous inner variable in terms of explanatory inner variables. Substitutive prediction provides prediction of an endogenous indicator in terms of the explanatory variables in the corresponding inner relation. As we shall see in the next section, a third mode of prediction is available after PLS estimation of the model.

Note that the outer relations (1) are not designed for prediction of an exogenous indicator in terms of its LV; more efficient predictions were yielded by (1) if ξ_k were the first principal component of the kth block of indicators. In PLS soft modeling an exogenous LV serves the purpose of conveying information through the inner relations, not to predict its indicators. Later we shall return to the interpretation of the outer relations.

Estimation by PLS

The key feature of PLS soft modeling is the explicit estimation of each LV, namely by a weighted aggregate of its indicators. PLS estimation proceeds in three stages. The two first stages measure the indicators as deviations from their means, giving:

$$\bar{x}_{kh} = 0 \quad k = 1, \ldots, 11; \ h = 1, \ldots, H_k \tag{12}$$

The first stage estimates the LVs; the second the multiplicative parameters of inner and outer relations, and the ensuing relations of substitutive prediction; the third stage estimates the location parameters of LVs, inner and outer relations, and substitutive predictions.

The data input of PLS estimation is either raw data (13a) or cross product data (13b):

$$x_{kht}; \ \bar{x}_{kh'} \quad c_{khk'h'} = \frac{1}{T} \Sigma_t (x_{kht} - \bar{x}_{kh})(x_{k'h} - \bar{x}_{k'h'})$$

$$\tag{13a-b}$$

Using raw data or cross product data, the ensuing parameter estimates are the same, apart from rounding errors. To estimate case values for LVs and residual variables raw data must be used.

The subsequent exposition uses raw data input. Theoretical and estimated parameters are denoted by corresponding Greek and Roman letters. PLS estimates of the LV ξ_k and its case values ξ_{kt} are denoted:

$$X_k, X_{kt} \text{ or alternatively } \hat{\xi}_k, \hat{\xi}_{kt} \tag{14}$$

PLS Estimation, First Stage

The first stage is an iterative procedure, say with steps

$$a = 1, 2, \ldots. \tag{15}$$

leading to the following form for the estimated LVs:

$$X_{kt} = \sum_{h=1}^{H_k} (w_{kh}x_{kht}) \tag{16}$$

The standardization (12) to zero mean implies (17a). In accordance with (5) the estimates (16) are standardized to unit variance, yielding (17b):

$$\bar{X}_k = 0; \quad \text{var} (X_k) = T^{-1} \sum_t (X_{kt}^2) = 1 \tag{17a-b}$$

The weights w_{kh} are determined by <u>weight relations</u>. For each k the weight relations involve estimates X_k for ξ_k and for LVs that are <u>adjacent</u> to ξ_k in the sense of being directly connected with ξ_k by an arrow in the arrow scheme. We see from Figure 15.2 that in the present model:

Adjacent to ξ_1 is ξ_4; adjacent to ξ_2 is ξ_4;

    "   " ξ_3 is ξ_6;   "   " ξ_4 are ξ_1, ξ_2, ξ_6;

. . . .

    "   " ξ_9 " ξ_{11} "   " ξ_{10} " ξ_7, ξ_8, ξ_{11}

    "   " ξ_{11} " ξ_9 ξ_{10}.

The weight relations for ξ_k involve a sign weighted sum of the estimates for the adjacent LVs. The sign weighted sum is called <u>adjacent impact</u> and is denoted U_k; to spell it out for k = 6:

$$U_6 = s_{63}X_3 + s_{64}X_4 + s_{68}X_{8t} \tag{18a}$$

where

$$s_{6c} = \text{sign of } r(X_6, X_c), \qquad c = 3, 4, 8 \qquad (18b)$$

For each LV the investigator has the option to choose between two designs for the weight relations, called Mode A and Mode B. The question about guiding rules for the choice of PLS estimation mode will be taken up presently. The weight relations of Mode A and B involve the adjacent impact and are simple and multiple OLS, respectively.

Mode A and Mode B in general give different weight w_{kh}. In (19a-b), linking up with (18), we write down the weight relations Mode A and Mode B for PLS estimation of ξ_6:

$$x_{6ht} = \underset{\bullet}{w}_{6h} U_{6t} + d_{6ht}; \quad U_{6t} = \sum_{h=1}^{H_6} (\underset{\bullet}{w}_{6h} x_{6ht}) + d_{6t} \qquad (19a\text{-}b)$$

REFLECTIVE.   FORMATIVE

where the weights $\underset{\bullet}{w}_{6h}$ are provisional and are transformed by the scalar

$$f_6 = \pm T^{1/2} \left\{ \sum_t \left[\sum_h (\underset{\bullet}{w}_{kh} x_{6ht}) \right]^2 \right\}^{-1/2} \qquad (20)$$

Then we obtain by way of

$$w_{6h} = f_6 \underset{\bullet}{w}_{6h} \qquad h = 1, \ldots, H_6 \qquad (21)$$

the weights w_{6h} for the estimation of ξ_6 by PLS Mode A and Mode B, respectively.

The Iterative PLS Algorithm for Estimation of the LVs. In each step $a = 1, 2, \ldots$ the PLS algorithm computes new proxies $\underset{\bullet}{w}_{kh}^{(a)}, w_{kh}^{(a)}, X_{kt}^{(a)}, U_{kt}^{(a)}$ for the weights, the LVs and the sign-weighted LVs.

Start, $a = 1$: Set all $\underset{\bullet}{w}_{kh}$ equal to unit,

$$\underset{\bullet}{w}_{kh}^{(1)} = 1 \qquad k = 1, \ldots, 10; \ h = 1, \ldots, H_k \qquad (22)$$

and compute

$$X_{kt}^{(1)} = \sum_h x_{kht} \quad \text{and} \quad U_{kt}^{(1)} = \sum_c (s_{kc}^{(1)} x_{ct}^{(1)}) \qquad (23a\text{-}b)$$

with

$$w_{kh}^{(1)} = f_k^{(1)} \cdot w_{kh}^{(1)} \quad \text{and} \quad s_{kc}^{(1)} = \text{sign } r(x_k^{(1)}, x_c^{(1)}) \qquad (24a\text{-}b)$$

where equation (20) for the standardizing scalar now reduces to

$$f_k^{(1)} = \pm T^{-1/2} \{ \sum_t [\sum_h x_{kht}]^2 \}^{-1/2}. \qquad (25)$$

The General Step from a to a + 1: Using the sign-weighted sums $U_{kt}^{(a)}$ obtained in step a, compute the provisional weights $w_{\cdot kh}^{(a+1)}$ either from (19a) or (19b) according as Mode A or Mode B has been chosen for the estimation of ξ_k:

$$x_{kht} = w_{\cdot kh}^{(a+1)} U_{kt}^{(a)} + d_{kht}^{(a+1)}; \ U_{kt}^{(a)} = \sum_h (w_{\cdot kh}^{(a+1)} x_{kht}) +$$

$$d_{kt}^{(a+1)} \qquad (26a\text{-}b)$$

Then use equations (16), (20), (23b), (24) and (25) to compute $X_{kt}^{(a+1)}$ and $U_{kt}^{(a+1)}$.

The iterative procedure is subject to a conventional stopping rule, say:

$$(w_{kh}^{(a+1)} - w_{kh}^{(a)}) / w_{kh}^{(a+1)} \le 10^{-5} \qquad \begin{array}{l} k = 1, \ldots, 10; \\ h = 1, \ldots, H_k \end{array} \qquad (27)$$

In the limit the superscripts are deleted, giving the same notations for the estimates w_{kh} and X_{kt} as in (16).

As is often and rightly said: A system is more than the sum of its parts. Hence systems analysis requires a holistic approach, not the reductionist approach of building separate models for the component parts of the system. At first sight it might seem that the PLS algorithm (16)-(27) is reductionist, inasmuch as in the step $a \rightarrow a + 1$ the estimate of $X_{kt}^{(a+1)}$ of ξ_k is dependent only on the estimates $X_{ct}^{(a)}$ of those LV's ξ_c which are adjacent to ξ_k. In the next step, however, the estimated interdependence of ξ_k is extended to LVs that are adjacent to the LVs adjacent to ξ_k, and so on in further steps. Hence PLS estimation of a path model with LVs is holistic in the sense that already at the first stage of the PLS algorithm all estimated LVs are involved in a system of joint interdependence.

PLS Estimation, Second Stage

The second stage is noniterative. The outer and inner relations (1), (7) are estimated by corresponding OLS regressions, without location parameters, using the LVs estimated in the first stage. The elimination of the endogenous LVs that give the relations (10) of substitutive prediction carries over to the estimated relations.

(i) Outer relations:

$$x_{kht} = p_{kh} X_{kt} + e_{kht} \qquad \begin{aligned} &k = 1, 2, \ldots, 10; \\ &h = 1, \ldots, H_k \end{aligned} \qquad (28)$$

The ambiguity of sign in (25) is part of the ambiguity of scales in (1). Now if the sign of X_k is shifted, all estimated loadings $p_{hk}(h = 1, \ldots, H_k)$ shift the sign. Hence the ambiguity of the sign in (28) resolves by so choosing the sign that for the majority of the estimated loadings p_{kh} the sign coincides with the expected signs (i) prespecified in the arrow scheme. The ensuing degree of coincidence provides a first partial test of the validity of the model.

Then when the signs of all X_k are determined, the ensuing estimated correlations $r(X_k X_{k'})$ between any two adjacent ξ_k, $\xi_{k'}$ should coincide with the expected signs (ii); this checking provides a second partial test of the validity of the model.

(ii) Inner relations:

$$X_{kt} = b_{k,k-2} X_{k-2,t} + b_{k,k-1,t} X_{k-1,t} + u_{kt};$$
$$k = 3, 5, 7, 9, 11 \qquad (29)$$

(iii) Substitutive prediction, I: Endogenous indicators predicted in terms of inner variables:

$$x_{kht} = p_{kh}(b_{k,k-2} X_{k-2,t} + b_{k,k-1} X_{k-1,t}) + v_{kht} \qquad (30a)$$

with

$$v_{kht} = e_{kht} + p_{kh} u_{kt} \qquad \begin{aligned} &k = 3, 5, 7, 9, 11; \\ &h = 1, \ldots, H_k \end{aligned} \qquad (30b)$$

In (30a) we can substitute the estimated LVs by the equivalent of their indicators. This gives:

Substitutive prediction, II: Endogenous indicators predicted in terms of outer variables:

$$x_{kht} = p_{kh} b_{k,k-2} \sum_{g=1}^{H_{k-2}} (w_{k-2,g} x_{k-2,g}) +$$

$$p_{kh} b_{k,k-1} \sum_{g=1}^{H_{k-1}} (w_{k-1,g} x_{k-1,g}) \qquad (31)$$

PLS Estimation, Third Stage: Location Parameters

Dropping the standardization (17a) of the indicators to zero means it is an immediate matter to write down the estimated location parameters for estimated LVs, outer and inner relations, and relations of substitutive prediction.

Estimated location parameters for LVs and outer relations:

$$\bar{X}_k = \sum_h (w_{kh} \bar{x}_{kh}); \quad p_{kho} = \bar{x}_{kh} - p_{kh} \bar{X}_k \qquad (32a\text{-}b)$$

Estimated location parameters for inner relations and substitutive predictions:

$$b_{ko} = \bar{X}_k - p_{kh}(b_{k,k-2} \bar{X}_{k-2} - b_{k,k-1} \bar{X}_{k-1});$$

$$m_{kh} = p_{kho} + p_{kh} b_{ko} \qquad (33a\text{-}b)$$

Special Cases: Principal Components and Canonical Correlations

Many multiple indicator problems of structure and performance in the social sciences are being treated by the well-known principal component method or by canonical correlation analysis. It is instructive to see how they are special cases of PLS soft modeling.

The First Principal Component

The one-block PLS soft model formed with indicators x_{ht} and estimated PLS Mode A is numerically equivalent to the first principal component as evaluated algebraically by the first eigenvalue and eigenvector.[1] The weight relations coincide with the estimated outer relations:

$$x_{ht} = p_{ho} + p_h X_t + e_t; \quad p_h = \overset{\bullet}{w}_h \tag{34}$$

The First Canonical Correlation

The two-block soft model formed with indicators x_{gt}, y_{ht} and estimated PLS Mode B is numerically equivalent to Hotelling's first canonical correlation.[1] For the x-block and its LV the weight relations and the estimated outer relations read:

$$Y_t = \sum_{g=1}^{G} (w_{1g} x_{gt}) + d_{1t}; \quad x_{gt} = p_{1go} + p_{1g} X_t + u_{1gt} \tag{35a-b}$$

and similarly for the y-block.

Second and Higher Dimensions (Orders) of Principal Components and Canonical Correlations

The second principal component and the second canonical correlation can be obtained by PLS Mode A and Mode B, respectively, using the same weight relations as for the first dimension, and as data input the residuals of the previous outer relations. In the same way, principal components and canonical correlations can be estimated consecutively.

In PLS terminology, principal components and canonical correlations of consecutive orders define LVs of consecutive dimensions.

Multiple OLS Regression

It is instructive to see the multiple OLS regression:

$$y_t = b_0 + b_1 X_{1t} + \ldots + b_G X_{Gt} + w_t \tag{36}$$

as a special case of PLS soft modeling, and at the same time of canonical correlation. Clearly, the multiple OLS regression is numerically equivalent to the weight relation in the first dimension of PLS estimation Mode B. In obvious notation, disregarding location parameters:

$$Y_t = (s.d. \ y)^{-1} y_t; \quad Y_t = \sum_g (\overset{\bullet}{w}_{1g} x_{gt}) + d_{1t} \tag{37a-b}$$

Thus in the multiple OLS regression of y on $x_1, \ldots, x_G$ the regression coefficient of x_{gt} is (s.d. y) $\dot{w}_{1g}$, and the regression residual is (s.d. y) d_{1t}.

In the PLS Mode B interpretation of multiple OLS regression the y-block has just one dimension. Furthermore, the inner relation coincides with the substitutive prediction relation (30a), giving (38a), where the superscript (1) is attached to estimates belonging to the first dimension of the x-block.

$$Y_t = b_{21}^{(1)} X_t^{(1)} + v_t^{(1)}; \quad Y_t = b_{21}^{(1)} X_t^{(1)} + b_{21}^{(2)} X_t^{(2)} + v_t^{(2)}$$

$$(38a\text{-}b)$$

The canonical correlation of second dimension (order) gives the expansion (38b), where $X_t^{(2)}$ denotes the estimated second dimension of the x-block. By construction, both $X_t^{(1)}$ and $X_t^{(2)}$ are weighted aggregates of the indicators x_{gt}, and $X_t^{(2)}$ and $X_t^{(2)}$ are uncorrelated over the T cases.

Using PLS estimation Mode B there is no second dimension $X_t^{(2)}$, since the residual v_t reaches minimum variance already by the first dimension $X_t^{(1)}$.

Using PLS estimation Mode A there will, in general, be two or more nonvanishing dimensions $X_t^{(d)}$. The situation is illustrated in Figure 15.3. As shown by the full-drawn line, each new dimension $X_t^{(d)}$ contributes to the explanation of Y_t, with result that the residual $v_t^{(d)}$ diminishes in variance and sooner or later coincides with the residual d_{1t} of the multiple OLS regression of y on $x_1, \ldots, x_G$. To put it otherwise, the consecutive dimensions $X_t^{(d)}$ determine an array of decreasing eigenvalues which are equal to the shared coefficients $b_{21}^{(d)}$ of the substitutive predictions (38b).

HYPOTHESIS TESTING: STONE-GEISSER'S TEST FOR PREDICTIVE RELEVANCE

Speaking generally, predictive relations are designed to yield predictions beyond the data used for estimating the relations. When subjected to Stone-Geisser's test (1974) the predictive relations are "blindfolded" and applied to the data used for estimation; an adaptation of Ball's Q^2 serves as test criterion.

FIGURE 15.3

PLS Interpretation of Multiple OLS Regression

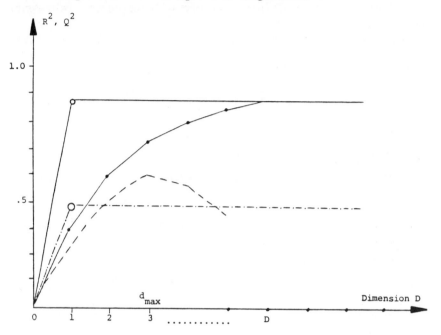

Legend: Full-drawn lines: R^2 with loss of degrees of free-dom. Broken lines: Ball's Q^2; that is, R^2 without loss of degrees of freedom.

In the special case of multiple OLS regression (36) the test criterion is

$$Q^2 = \sum_{t=1}^{T} (b_0^{(t)} + b_1^{(t)} x_{1t} + \ldots + b_G^{(t)} x_{Gt} -$$

$$y_t)^2 / \sum_{t=1}^{T} (\bar{y}^{(t)} - y_t)^2 \qquad (39)$$

where $b_g^{(t)}$ (g = 0, 1, ..., G) are the coefficients of the multiple OLS regression (36) when reestimated after omitting y_t from the given data, and $\bar{y}^{(t)}$ is the trivial prediction of y_t given by the mean $\bar{y}$ after omitting y_t. In words, Q^2 is nothing else than R^2 evaluated without loss of degrees of freedom. The predictive relation under test does or does not have predictive relevance according as $Q^2 > 0$ or $Q^2 < 0$.

The Stone-Geisser test (39) for multiple OLS regression allows straightforward extension to models that involve simultaneous estimation of several endogenous variables $y_{1t}, \ldots, y_{Kt}$, and in particular to PLS soft models; compare S. Wold (1976), H. Wold (1979).

STANDARD ERRORS: TUKEY'S JACKKNIFE

Tukey's jackknife (1958) is a general device of "blindfolding" to assess the standard errors of parameter estimates. As applied to a model estimated on the basis of T cases (observations over time) or a cross section, the jackknife omits one case after another, each time reestimating the parameters using the remaining T-1 cases; this gives T estimates for each parameter of the model; the standard deviation of these estimates is the jackknife standard error of the parameter.

Clearly, when applying Stone-Geisser's test for predictive relevance, jackknife standard errors of the parameters of the predictive relation are obtained as a by-product of Stone-Geisser's test.

THE BASIC PLS DESIGN: APPLICATIONS AND GENERALIZATIONS

In its published applications PLS soft modeling emerges as a general tool for quantitative systems analysis in terms of structure versus performance, with structure and performance indirectly observed by multiple indicators observed over time or a cross section. The applications range from natural science and medicine to social, economic, and political sciences; for references see Wold (1979, 1981).

The basic design of PLS soft modeling set forth in the previous sections has been generalized in several directions, including (i) latent variables in two or more dimensions (special cases: principal components and canonical correlations of second or higher order); (ii) hierarchic structure of the latent variables; (iii) path models with feedbacks or interdependencies in the inner relations; (iv) soft modeling of multidimensional contingency tables; (v) multiway observation of the indicators, for example, observation both over time and a cross section. In each of (i), (ii), and (iv) the generalization is immediate matter, using an adaptation of the basic PLS algorithm; see Wold (1979, 1980) and Bertholet and Wold (1981). The generalization (iii) is performed by combining the PLS and Fix-Point algorithms, see Wold (1977) and Hui (1978, 1982). The generalization (v) is a brilliant innovation due to Lohmöller (1982); see also Lohmöller and Wold (1980).

SOME COMPARATIVE COMMENTS ON PLS AND LISREL

LISREL and PLS are two estimation methods for path models with latent variables indirectly observed by multiple indicators. The conceptual-theoretical design of the model given by its arrow scheme is the same for LISREL and PLS and gives no indication in favor of one approach or the other. However, for researchers with a growing research agenda, the greater scope and flexibility of PLS should be noted.

LISREL and PLS are complementary rather than competitive. LISREL is parameter-oriented and aims at high accuracy of the parameters. PLS is prediction oriented and aims at high prediction accuracy.

There is a choice between optimal prediction and optimal parameter accuracy. You cannot have both, except in special cases where the two optimization principles give numerically the same parameters and predictions.

Data input for the LISREL algorithm is the covariance matrix of the observables (cross-product data). The theoretical covariance matrix is an algebraic function of the parameters, and LISREL is a Maximum Likelihood approach to retrieve the parameters from the covariance matrix. The estimation of parameters from the covariance matrix is the source of identification problems. LISREL estimates the distribution of each LV but cannot estimate its case values, nor its mean value.

To predict case values of the observables, case values of the latent variables must be estimated. Using raw data input, PLS makes explicit, albeit deliberately approximate, estimation of each latent variable as a weighted aggregate (16) of its indicators. The requisite weights, and later on other PLS parameters, are estimated using cross-product data input.

Apart from prediction, the explicit PLS estimation of the case values of the latent variables brings further advantages. No identification problems arise. Furthermore, the case values of the latent variables are needed for the Stone-Geisser test and for the jackknife standard errors. In addition, the case values of latent variables and of the residuals in inner and outer relations and in relations of substitutive prediction give information for improvement and further evolution of the model.

LISREL is at a premium in small models, where the separate parameters have operative use. PLS comes to the fore in large systems with many parameters, where the emphasis is on prediction, be it in terms of latent variables, or by some other strategic package of variables and parameters.

In the passage from small to large systems as needed in real-world problems, LISREL sooner or later breaks down when there are too many parameters to estimate. Current LISREL programs can estimate models with some 40 or 50 parameters. In PLS estimation the size of the model and the number of parameters is practically unrestricted. For example, a recent PLS model on alcohol and suicide problems (Bartl et al., 1981) has 102 directly observed and 23 latent variables. Nonetheless, defending its reputation of "instant estimation," the PLS algorithm converges in 200 seconds of computer time. (LISREL completed only four iterations in 30 minutes, after which the researchers decided not to attempt another run.)

In the LISREL approach each latent variable has just one dimension. The PLS approach allows consecutive estimation (i) of multidimensional latent variables, and each dimension can be tested for predictive relevance by the Stone-Geisser test.

The basic PLS algorithm requires no modification for the estimation of latent variables with hierarchic structure (ii), nor for soft modeling of multidimensional contingency tables (iv). In the ML approach the generalizations (ii) and (iv) are not covered by the LISREL algorithm.

In ML estimation it is assumed that the observables under analysis are ruled by a joint multivariate distribution, usually a normal distribution subject to independent observations. On this basis the ML methodology provides a general framework for: (a) ML parameter estimation; (b) hypothesis testing, leading to other rejection or nonrejection; or (c) assessment of standard errors for the parameters.

LS (Least Squares) estimation, including PLS, is distribution free. For path modeling with latent variables an LS counterpart to (a)-(c) is now emerging, namely: (d) PLS estimation; (e) Stone-Geisser testing for predictive relevance, leading to either nonrelevance or some degree of relevance; or (f) jackknife estimation of standard errors for the parameters.

CONCLUSION

A wide realm of real-world problems can be posed as system analysis in terms of structure and performance. Economics is an example par excellence for this characterization. In this broad perspective, then, it is needless to say that PLS soft modeling is at an early stage of methodological evolution in statistical theory and econometric application. A growing number of econometric applications of soft modeling attest to its broad scope and flexibility. At the present stage I am confident that soft modeling will prove quite powerful where standard econometric methods have performed unsatisfactorily: in the analysis of structural change and change in economic performance. This finally answers the question posed in the beginning.

In conclusion, reference is made to some current lines of research on PLS soft modeling; methodological advances and applied work:

Systematic comparison of the ML and PLS approaches (a)-(c) and (d)-(f), with respect to hypothesis testing and standard errors; compare Lohmöller and Wold (1982).

Revision of the PLS algorithm for estimation of latent variables in higher dimensions. In the new PLS algorithm the PLS estimate of a latent variable is uncorrelated not only with the lower dimensions of that same variable but also with the lower dimensions of the adjacent latent variables.

REFERENCES

Bartl, P., K. Unverdorben, and J. B. Lohmöller (1981). Sociale Probleme im Grundwehrdienst: Eine Pfadanalyse zu Alkohol—und Suicid Problemen. Forschungsbericht 81.01, Fachbereich Pädagogik, Hochschule der Bundeswehr, Munich.

Geisser, S. 1974. A predictive approach to the random effect model. Biometrika 61:101-107.

Hui, B. S. (1978). The partial least squares approach to path models of indirectly observed variables with multiple indicators. Unpublished doctoral thesis, University of Pennsylvania.

Hui, B. S. (1982). On building partial least squares models with interdependent inner relations. In K. G. Jöreskog and Wold, H. Systems Under Indirect Observation: Causality, Structure, Prediction. Amsterdam: North Holland, in press.

Jöreskog, Karl G. (1967). Some contributions to maximum likeli-hood factor analysis. Psychometrika 32, 443-482.

____. (1970). A general method for analysis of covariance structures. Biometrika 57, 239-251.

____. (1973). A general method for estimating a linear struc-tural equation system. In ed. A. S. Goldberger and Duncan, O. D. Structural Equation Models in the Social Sciences. New York: Academic Press, 85-112.

Jöreskog, Karl G. and Dag Sörbom (1981). LISREL V: Analysis of Linear Structural Relationships by Maximum Likelihood and Least Squares Methods. Chicago: National Education Re-sources, Inc.

Jöreskog, K. G. and H. Wold, eds. 1982. Systems under Indirect Observation: Causality, Structure, Prediction, Parts I-II. Amsterdam: North-Holland.

____. Chapter 12 in Part I of Jöreskog and Wold, eds., 1982.

Lohmöller, Jan-Bernd (1981). LVPLS 1.6: Latent Variables Path Analysis with Partial Least-Squares Estimation. Forschungsbericht 81.04 Hochschule der Bundeswehr, Munich.

____. 1982. Lernen, Entwicklung, Veränderung: Mathematischer Modell fur multivariate Längsschnittsdaten. Doctoral thesis, Hochschule der Bundeswehr, Munich.

Lohmöller, J.-B. and H. Wold. 1980. Three-mode path models with latent variables and PLS parameter estimation. Forsch-ungsbericht 80:3, Fachbereich Pädagogik, Hochschule der Bundeswehr, Munich.

Noonan, R. and H. Wold. 1977 and 1980. NIPALS path modeling with latent variables. Analyzing school survey data using nonlinear iterative partial least squares, I-II. Scandinavian Journal of Educational Research 21:33-61, and 1-24.

Stone, M. 1974. Cross-validatory choice and assessment of statistical predictions. Journal of the Royal Statistical Society, Series B 38:111-133.

Tukey, J. W. (1958). Bias and confidence in not-quite large samples (abstract). Annals of Mathematical Statistics, 29, 614.

Wold, H. 1963. On the consistency of least squares regression. Sankhyā A 25, Part 2:211-215.

_____. 1965. A fixed-point theorem with econometric background, I-II. Arkiv för Matematik 6, 209-240.

_____. 1966. Nonlinear estimation by iterative least squares procedures. In Research Papers in Statistics, Festschrift for J. Neyman, ed. F. N. David. New York: John Wiley, 411-444.

_____. 1977a. On the transition from pattern recognition to model building. In Mathematical Economics and Game Theory, Essays in Honor of Oskar Morgenstern, ed. R. Henn and O. Moeschlin. Berlin: Springer, 536-549.

_____. 1977b. Open path models with latent variables. In Festschrift Withclen Kreble, ed. H. Albach, E. Helmstedter, and R. Henn. Tübingen: Mohr.

_____. 1979. Model construction when theoretical knowledge is scarce. An example of the use of Partial Least Squares. Cahier 79:06, Department of Econometrics, University of Geneva.

_____. (1980). Soft modelling: Intermediate between traditional model building and data analysis. Mathematical Statistics 6, 333-346.

_____. 1980b. Model construction and evaluation when theoretical knowledge is scarce. On the theory and application of Partial Least Squares. In Model Evaluation in Econometrics, ed. J. Kmenta and J. Ramsey, New York: Academic Press.

_____, ed. 1980c. The Fix-Point Approach to Interdependent Systems, Amsterdam: North-Holland.

_____. 1982. Soft modeling: the basic design, and some generalizations, Chapter 1 in Part II of Jöreskog and Wold, eds. 1982.

Wold, Herman and Jean-Luc Bertholet (1981). The PLS approach to multidimensional contingency tables. Paper presented at the International Meetings, Institute of Statistics and Social Research, University of Rome, June 25-26.

Wold, S. 1976. Pattern recognition by means of disjoint principal components models. Pattern Recognition 8: 127-139.

16

Data Analysis by
Partial Least Squares

Fred L. Bookstein

INTRODUCTION: LATENT VARIABLES
AND SOFT MODELS

For the phenomena social scientists study, causation generally operates at the level of events: decisions, outcomes, and
opportunities. But were each event measured separately, there
would be no such activity as prediction, only prophecy. In
accounting for patterns of events in any predictive way, causal
theories must depend on measurable attributes—of individuals,
institutions, interactions—presumed stable over events. Then
the variables we measure are usually misspecified for the causal
schemes we believe to govern the process of their interrelations;
they are all proxies, all at the wrong level of aggregation.

In modern quantitative practice the method of latent variables emerges as a general response to this perplexity. By way
of compensating for the misspecification of any causally relevant
empirical attribute, we measure it variously and repeatedly.
Each "variable" becomes, in practice, a block of many items.
A latent variable (Lv) is a scale score which combines the
items of a block into a single quantity for arithmetic use later
in a particular causal model. The Lv is formed by three considerations: the items whose causal force it embodies; the other

Reprinted from Evaluation of Econometric Models, ed. J.
Kmenta and J. B. Ramsey (New York: Academic Press, 1980),
pp. 75-90. Reprinted by permission.

variables, observed or latent, in the causal scheme; and the details of the algorithm by which we generate the scale scores to combine the items of a block in a single aggregate. Latent variables may be linear expressions in the items of a block, integers, switches, or any other sort of statistical artifact required by a particular style of prediction.

Over several years of exposition and collaboration Herman Wold has developed for latent variable analyses a set of conventions which he calls soft modeling. By "soft" here Wold means weak in assumptions, undemanding. His soft models presume several blocks, each an assemblage of items that are all proxies of one Lv. The blocks are related by an arrow diagram, a causal chain specifying that certain Lvs must have an expected value depending linearly on the values of others. Each Lv is to be an explicit linear combination of the items of its block. No other information is supplied, no other assumptions are made.

To estimate such a model is to fix coefficients of two sorts: the regression weights ("inner relations"), whereby values of an Lv "affect" values of others further down the causal chain, and the item weights ("outer relations"), which describe the manner in which the items of a block severally determine the LV which represents them. Whereas conventional multiequation estimators require a stringent parametric model, namely, a family of joint distributions for all the items in all the blocks, the soft specification is translated into a collection of partial linear models separately, almost trivially, analyzed by ordinary least squares regressions. This unexpected tactic is directed by two formal themes, suboptimality and simplicity.

1. Suboptimality refers to the formal disaggregation of the computations. Each block in the model, as items together with its eventual Lv, can be interpreted as the basis of a submodel consisting just of that block and the other blocks with which it communicates, that is, to which it is linked by explicit arrows of the causal scheme. In Wold's prescriptions, each Lv is related only to the other Lvs of its submodel together with the items of its own block. Nevertheless, since the submodels overlap, the result is an interdependent system of equations between latent variables.

2. For the sake of simplicity any Lv is characterized in terms of projections of the Lvs of its submodel upon the items of its own block. The minimum-distance property of projection onto a block is the only optimization principle invoked in the course of soft modeling.

The Lvs of a soft model may be jointly characterized using
a complex, nonlinear operator for which the vector of all estimated
item weights (outer relations) serves as a fixed-point. Soft
estimation, then, does not resemble at all the search for zeroes
of certain derivatives which characterizes the estimation of
"harder" models. In its stead, in order to find the fixed-point
of a soft model, Wold proposes a succession of regressions and
linear combinations within submodels—a cycle of replacement
of "old" Lvs (earlier estimates) by new—which seems always to
converge.

In this chapter I will explore the role of these two formal
themes in Wold's exposition. Regarding his prescriptions for
submodel regressions, I will indicate which are dictated by the
needs of suboptimality or simplicity and which are left for determination by content of the theory, model, or data at hand. The
procedures I derive from the principles are somewhat more flexible than Wold's own.

THE COMMAND DIAGRAM

We begin with an arrow diagram. The specimen in Figure
16.1 sets out three blocks of items with corresponding Lvs. The
Lvs are to capture X-ness, Y-ness, and Z-ness in the context
of a joint linear determination of Z-ness by X-ness and Y-ness.
(Latent variables here are denoted by the letter L subscripted
by the block name, e.g., L_X, L_Y, L_Z.)

Under Wold's rules the arrow diagram is transformed into
a series of <u>operations</u> replacing each tentative Lv by a new

FIGURE 16.1

Typical Arrow Diagram

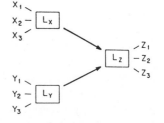

A latent variable L_X, proxied by indicators X_1, X_2, X_3, and
and Lv L_Y proxied by Y_1, Y_2, Y_3, jointly and linearly determine
the expected value of an Lv L_Z proxied by items Z_1, Z_2, Z_3.

FIGURE 16.2

Command Diagram Transcribing Figure 16.1

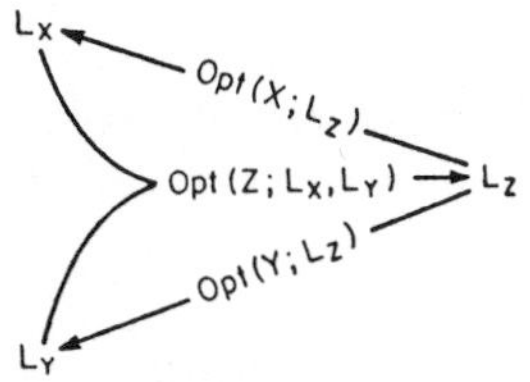

Each operation is an optimization for a new latent variable as a function of all the Lv's communicating with it. The arrows express association of inputs and outputs, not causation.

linear combination of the items in its block. The new Lv is not a function of the old Lv at all, but is rather a function of the Lvs of the other blocks of its submodel as they relate to the items of its own block. The new Lv is always some sort of orthogonal projection; we may disregard details by using the general operator Opt, without specifying its precise functional form. The operator Opt has two arguments. The first is the block for which the operator's output is the new Lv; the second is a list of all Lvs linked to the output block by direct paths in the arrow diagram—a roster of the submodel. In this manner the arrow diagram in Figure 16.1 is replaced by the command diagram in Figure 16.2, which makes explicit the arguments and outputs of the various operations, that is, the flow of partial regressions over the diagram. The various Opt commands have been written out over paths connecting their arguments to their outputs, the Lvs for the next round of iteration. For example, the path Opt $(X; L_Z)$ is a regression, of form not yet specified, which computes a new Lv L_X by relating the current Lv L_Z, a linear combination of items $Z_1 \ldots Z_3$, to the items of the X-block. The direction of the arrows in the command diagram expresses the order of operations and outputs, not the flow of causation, which will be embodied in the detailed formulation of the functions Opt.

The command diagram directly implies the algorithm to be used for estimating both sorts of coefficients, regression weights and item coefficients, as follows.

1. Begin by defining each latent variable as an arbitrary linear combination of the items in its block. Reasonable starting values might be $L_X = \Sigma \, X_i$, etc.

2. Determine the exact form of each Opt command in terms of some short sequence of regressions and linear combinations, utilizing the considerations set forth later in this exposition.

3. Execute all the commands of step 2 just once, without updating the Lv formulas.

4. Have we arrived at the fixed-point?—are the Lvs consistent with their joint characterization to some preset tolerance? If so, we are done; otherwise,

5. Replace each Lv by the output of the appropriate Opt operator, scaled to unit variance, and return to step 3.

In practice these algorithms always converge. At convergence, each Lv is a projection onto its block of the Lvs from the other blocks of its submodel. Only in these projections does optimality lie; otherwise the estimates are characterized by their self-consistency as tested in step 4.

POLYGON DIAGRAMS

The intrinsic geometric content of soft techniques stems from the equivalence of regression and orthogonal projections from point to hyperplane. The fixed-point of a soft model can be expressed in terms of the mutual orientations among hyperplanes in a high-dimensional space. Their soft modeling is mainly geometry, so we should be able to visualize what we are doing.

I shall try to demonstrate how to distinguish between alternative forms of the Opt operator with diagrams on flat paper, but to do so I shall need to establish some visual conventions. Figures 16.3a-g illustrate the standard constructions to which I shall be referring. Until one notices the apparent inconsistency of the right-angle (⌐) symbols scattered throughout the diagrams, the figures appear to be of points, lines, and planes from ordinary three-dimensional space. But, in fact, all of the objects depicted lie in the space dual to that of the items in an analysis, namely the vector space of their linear combinations. Points in these diagrams are particular scores, such as the Lvs themselves; lines and planes depict subspaces spanned by the items of a block or by sets of Lvs. The diagrams are drawn in no particular coordinate system at all, though the representation of regression as explicit orthogonal projection suggests a set of orthonormal axes as the appropriate basis. In terms of the items themselves, the variables of the raw data matrix, the inner product here is defined not in terms

FIGURE 16.3

Elements of the Polygon Diagram

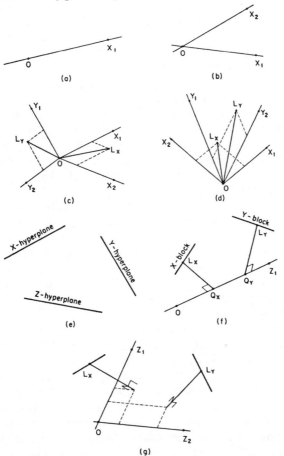

(a) The line of multiples of an item X_1. 0 is the zero vector.
(b) The hyperplane corresponding to the X-block, viewed plane-
on. (c) Two hyperplanes, each with a latent variable, viewed
from in between. For any vector of a block the dotted lines
connect it to a basis for its block. (d) The same, viewed through
the X-hyperplane to the Y-hyperplane. (e) Three hyperplanes
all edge-on. (f) Two projections from hyperplanes onto a line.
Orthogonality of projection onto a rank-1 subspace is shown by
the symbol $\ulcorner$ drawn in perspective. (g) Two projections from
hyperplanes onto a hyperplane viewed plane-on. Orthogonality
of projection onto a higher-rank subsapce is shown by a combination of symbols $\ulcorner$ for each basis vector separately.

353

of the Euclidean cosine, but in terms of the familiar covariance matrix Σ. Submatrices Σ_{XX}, Σ_{XY}, etc. of Σ incorporate the covariances within and between the separate blocks of the model.

Let us proceed through the frames of Figure 16.3.

Figure 16.3a shows a typical line of our space. Here it is the set of all multiples of an item, X_1. Note the zero vector 0, which is represented by a point on this and all other lines and higher-rank subspaces.

Figure 16.3b shows a hyperplane of this space, viewed plane-on so that we can see two of the items making it up. Note that these axes are oblique, as items X_1 and X_2 are, in general, correlated. Any block of the soft model can be drawn as a hyperplane whose points are all the possible Lvs for that block. The algorithms of soft modeling determine one Lv per block—one point per hyperplane which is in the proper relation to the Lvs of the other blocks of its submodel.

Figure 16.3c shows two hyperplanes, after the fashion of Figure 16.3b, with an Lv in each one. We indicate the (hyper) plane in which a point resides by dropping dashed lines to the lines determined by the items spanning that (hyper)plane. Note that these planes can intersect only at the point 0; elsewhere they avoid each other like two lines in a plane. We are looking at them "from in between."

Figure 16.3d shows the same two hyperplanes, from a viewing position now "through the X-block to the Y-block." Certain constructions will appear more legible from this perspective.

Figure 16.3e shows three hyperplanes, all edge-on to the viewing eye, in some inconceivable high-dimensional rotation. These three entities intersect in a single point only, the origin of coordinates, and do not intersect, even in pairs, anywhere else at all. One must imagine them curving around to the point of concurrence through dimensions unrelated to flat paper or even physical space. It should be clear why the point 0 is not drawn; there is no logical place to put it.

Figure 16.3f shows two regressions, from hyperplanes representing two blocks of a model, onto the line of multiples of an item Z_1 from the Z-block. An Lv L_X, residing within the hyperplane corresponding to the X-block of the model (here seen edge-on), is regressed on a single item Z_1, that is to say, projected orthogonally onto the line from 0 through Z_1. (Recall that 0 is included in both blocks as well as in the line.) The predicted value of L_X is the vector Q_X, some multiple of Z_1; the residual of that regression is the vector difference $L_X - Q_X$, orthogonal to the lines of multiples of Z_1, as shown. Note that the symbol for orthogonality of two lines is drawn in perspective.

Similarly, in this figure the latent variable Ly is regressed onto the same single item Z . The perception that these projections come from different directions is to be encouraged; that they seem to come from different distances is tempered by the difficulty of representing all Lvs at the same distance from 0 (as they are all normalized to the same variance).

Figure 16.3g shows two regressions onto a hyperplane viewed plane-on. The symbol for orthogonality of a projection onto a hyperplane is a combination of $\sqcap$'s, in perspective, indicating orthogonality to the basis vectors of the plane of view. The "normals" for these two projections (residuals of the regressions) come from different directions in hyperspace, resulting in a paradoxical perspective for the diagram as a whole.

A soft algorithm terminates when a loop of Opt operations returns to the very latent variables, hyperplane by hyperplane, with which it began. The fixed-point itself, though really a vector, can then be drawn as a whole closed polygon in the diagram, a path whose edges bear Lvs at one end and right angles at the other, as in Figure 16.9b and 16.10c. In terms of its symbols of perpendicularity the polygon diagram indicates the precise form of each Opt command to be programmed for calculation through regression analysis. For instance, the model of Figure 16.9a-b is executed by the following command sequences:

0. Set $L_Z = \Sigma\ Z_i$, normalized to variance 1.

1. Compute L_X by determining the predicted value of L_Z given the X-items.

2. Compute L_Y by determining the predicted value of L_Z given the Y-items.

3. Compute L_W by determining the predicted value of L_Z given L_X and L_Y.

4. Compute L_Z^* by determining the predicted value of L_W given the Z-items.

5. Set $L_Z^! = L_Z^*/\sigma_{L_Z^*}$ where the denominator is the observed variance of L_Z^*.

6. TEST. If $L_Z^!$ is sufficiently close to L_Z, exit; otherwise,

7. Set $L_Z = L_Z^!$ and go to step 1 again.

The flow of computation is drawn in Figure 16.9c by associating the step numbers from the preceding list to the appropriate edges in the polygon diagram, paths from dependent variables to predicted value which link the hyperplanes of the model.

PRECISE FORMS OF THE OPERATOR "OPT"
FOR THE RELATIONSHIP BETWEEN
TWO LATENT VARIABLES

Using the elements introduced in Figure 16.3 we can explore a diversity of Opt commands by combining projections in various ways.

Consider first the simplest submodels, those containing only two blocks of items. These correspond to Opts whose second argument is a single Lv only. In the absence of formal asymmetries among the items of the output block, there seem to be only two reasonable ways to proceed, shown in Figure 16.4 for the case of the command $L_X = Opt(X; L_Z)$ from Figure 16.2. As the notation indicates, either construction is intended to provide a tentative Lv in the X-hyperplane. That is, a linear combination of the X-items is to determined with respect to a tentative Lv L_Z in the Z-hyperplane.

Mode A projects L_Z upon the items X_i of the X-block separately, then constructs the sum of these projections. When all items are of unit variance, as will be assumed throughout the remainder of this exposition, the result is $L_X = \Sigma_{XZ} L_Z$. The geometry of this $Opt_A(X; L_Z)$ is diagrammed in Figure 16.4a.

Mode B projects L_Z onto the X-hyperplane directly, in one operation. The result, shown in Figure 16.4b, is $Opt_B(X; L_Z) = \Sigma_{XX}^{-1} \Sigma_{XZ} L_Z$, which differs from Opt_A whenever the Xs show any intercorrelation, i.e., when Σ_{XX} is nondiagonal.

To each of these modes, A and B, corresponds a familiar statistical model. Opt_B is the multiple regression of L_Z on the X-items, that linear combination L_X which minimizes the root-mean-square of ε in the regression equation $L_Z = L_X + \varepsilon$. Opt_A is factor estimation. If each indicator X_i is representable by $a_i L_Z + \varepsilon_i$, where the ε_i are mutually independent random deviates with mean zero, then $Opt_A(X; L_Z)$ is the estimator of L_Z with the smallest summed mean squared ε_i.

Though Wold requires us to choose one mode or the other, A or B, for each block, there seems to be no way to decide between them on the basis of observation of distributions or to estimate their relative import for an estimate of the "true situation" by any simple technique. This ambiguity is unfortunate, since the factor Σ_{XX}^{-1} which distinguishes the modes can have a great effect—as the indicators of any block are almost certainly correlated, this inverse will have some large eigenvalues with associated eigenvectors not necessarily uncorrelated with $\Sigma_{XZ} L_Z$. To guard against ill-conditioning of Σ, one might, for instance, "deflate" Opt_B by an additional multiplication by

FIGURE 16.4

The Two Modes of the Opt Command for a Block X
Communicating with Only One Other Block Z

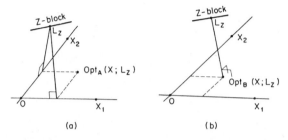

(a)   (b)

(a) Factor estimation mode. L_X is the sum of the simple regression functions of L_Z upon the X-items separately. (b) Linear estimation mode, orthogonal projection of L_Z upon the X-hyperplane.

$\Sigma_{XX.Z}$, suppressing dimensions of the X-hyperplane according to their alienation from the Z-items. Perplexities of the single-block Opt operation are a prime site for further research into soft modeling.

PRECISE FORMS OF THE OPERATOR "OPT"
FOR THE RELATIONSHIPS AMONG SEVERAL
LATENT VARIABLES: APPROACHES
TO EIGENANALYSIS

In the previous section we considered whether or not to use the covariance structure Σ_{XX} of the indicators of a single block. When we attend to larger submodels, three blocks or more, there is an analogous choice to be made: we must examine the consequences of ignoring or recognizing the covariances among the connected Lvs themselves.

Wold favors the most straightforward of approaches, simply disregarding the covariances of Lvs. Mode C will combine this strategy with mode A within blocks,* mode D with mode B within blocks. Figures 16.5a and b show the constructions which result for the operation $L_Z = \text{Opt}(Z; L_X, L_Y)$ from Figure 16.2. In

---

*This is not the mode C of Wold's exposition.

FIGURE 16.5

Opt Operators That Ignore the Covariance Structure of the Lvs
within a Submodel

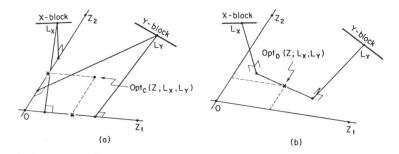

(a) Factor estimation mode Opt_C, mean of the separate Opt_{AS}
for each communicating Lv: sum of the means of all the simple
regression functions item by item. (b) Linear estimation mode
Opt_D, mean of the separate Opt_{BS} block by block.

either case the output L_Z is the simple mean of the appropriate
Opt_{AS} or Opt_{BS} computed in block Z with respect to the communicating Lvs separately. By linearity of projection, the Lv
Opt_C predicts the Zs from $L_X + L_Y$—minimizes mean squared
prediction error over all the Zs separately—while Opt_D finds
a Lv L_Z to predict $L_X + L_Y$ with minimal mean-squared error
in the units of $L_X + L_Y$.

Simple contradiction of this direct route suggests other
possibilities in which the communicating Lvs are considered
jointly. Mode E will maximize the sum of the squared <u>simple</u>
correlations of L_Z with L_X and L_Y; mode F will maximize the
squared <u>multiple</u> correlation of L_Z with L_X and L_Y. These
new optima emerge from computations involving a new hybrid
block W spanned by L_X and L_Y themselves. W is a plane, not
a hyperplane—the collection of all linear combinations of L_X and
L_Y only. It does not lie wholly in either the X- or the Y-
hyperplane, but rather straddles the "space" between them,
as in Figure 16.6.

In terms of the matrices Σ_{WW} and Σ_{WZ} of correlations among
the Lvs of the W-plane and between them and the indicators of
the Z-block, the optima of modes E and F are expressible as
eigenvectors. The vector L_Z which maximizes $R^2(L_Z | L_X, L_Y)$
is the first canonical variable of the Z-block with respect to
the W-block, the dominant eigenvector of the matrix $\Sigma_{ZW} \Sigma_{WW}^{-1} \Sigma_{WZ}$

FIGURE 16.6

The Plane W Spanned by the Lvs L_X, L_Y from Two Separate
Blocks X and Y

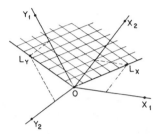

with respect to Σ_{ZZ}; the vector L_Z which maximizes $R^2(L_Z|L_X) +$
$R^2(L_Z|L_Y)$ is easily shown to be a closely related quantity,
the dominant eigenvector of the matrix $\Sigma_{ZW}\Sigma_{WZ}$ with respect to
Σ_{ZZ}.

Now, as eigenvector extractions, these would seem to be
excluded from the repertoire of soft modeling. But, in addition
to regression, we have another privilege, the passage to the
limit, which contributes the needed extension to this class of
algorithms. Figure 16.7 shows, for instance, a two-block model,
estimated according to mode B, at convergence. The normalizations after each round of optimization are shown explicitly. In

FIGURE 16.7

The Polygon Diagram for Classic Two-block Canonical Analysis

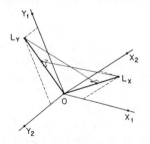

Vector L_X is in the (X_1, X_2)-plane, vector L_Y in the (Y_1, Y_2)-
plane. The expected value of L_X, when regressed on the items
of the Y-block, is proportional to L_Y; and the expected value
of L_Y regressed on the X-block is proportional to L_X.

FIGURE 16.8

Polygon Diagrams for Optimization Modes E,F

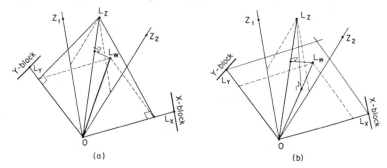

(a)   (b)

Vector L_Z is in the (Z_1,Z_2)-plane, L_X in the X-block, L_Y in the Y-block. Vector L_W is in the (L_X,L_Y)-plane. At convergence, L_Z is the predicted value of L_W given the items of the Z-block. (a) Computation of L_W by optimization mode A. The eigenanalysis shows the maximization of the mean-square of the simple correlations of L_X,L_Y with L_Z. (b) Computations of L_W by optimization mode B. The eigenanalysis is that of canonical correlation analysis, showing maximization of the multiple correlation of L_Z with L_X and L_Y jointly.

view of the algebraic characterization of mode-B optimization, we may write $L_X = \lambda\Sigma_{XX}^{-1}\Sigma_{XY}L_Y$, $L_Y = \mu\Sigma_{YY}^{-1}\Sigma_{YX}L_X$, where λ and μ are normalizing factors. Substituting either equation in the other, we see that each of L_X and L_Y is an eigenvalue of the appropriate matrix for the problem of canonical variates of the X-block with respect to the Y-block; and although they are eigenvectors, we have computed them by iteration of regressions only.

At convergence, then, the mode-F L_Z drawn out in Figure 16.8b will have maximized $R^2(L_Z|L_X,L_Y)$. Note that we cannot write this as Opt_F on the command diagram, for it is not a finite sequence of regressions—it emerges only at convergence from the succession $L_W = Opt_B$, $L_Z = Opt_B$.

Similarly, when the estimation of a model according to the polygon in Figure 16.8a has converged, it is to a variable L_Z which is an eigenvector maximizing $R^2(L_Z|L_X) + R^2(L_Z|L_Y)$. This mode-E optimization emerges at convergence from the sequence L_W - Opt_A, $L_Z = Opt_B$.

The inclusion of the ancillary plane W in the polygon diagram (or the ancillary block W in the command diagram) makes it possible to reflect in the operator Opt the differences in position of blocks within the causal structure of the model. When determination is joint, as in Figure 16.1, optimization should proceed according to mode F (command diagram, Figure 16.9a, and polygon diagram, Figure 16.9b), in which the maximand for computation of L_Z is the <u>multiple</u> correlation with L_X and L_Y. Corresponding to the postulation of a single linear dependency in this submodel, there is a single scalar to be maximized. In a submodel like that of Figure 16.10a, where two separate dependencies are being described, I believe optimization should proceed according to mode E instead, with maximand the summed squared <u>simple</u> correlations of L_Z with L_X and L_Y separately. There are two dependencies postulated here, corresponding to the two arrows out of the Z-block; in the absence of any instructions to the contrary, their strengths can only be summated.

FIGURE 16.9

Recommended Algorithm, Mode F, for the Causal Model of Figure 16.1

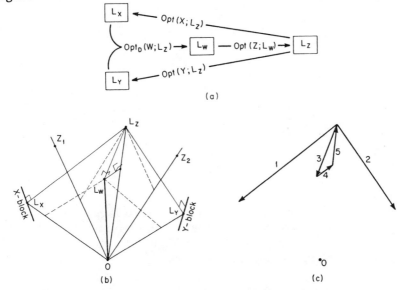

(a) The command diagram displays the ancillary L_V L_W. (b) The polygon diagram displays the ancillary plane W. (Compare Figure 16.8b.) (c) The sequence of regressions 1-4 along edges of Figure 16.9b is followed by a normalization, step 5.

FIGURE 16.10

Adjustment of Figure 16.9 When the Postulated Direction of
Causation is Reversed

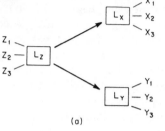

(a)

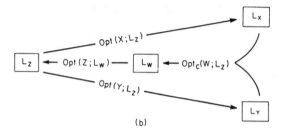

(b)

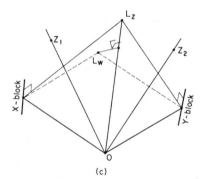

(c)

(a) The revised arrow diagram. (Compare Figure 16.1.) (b)
The command diagram. Note that L_W is computed by optimization
mode C, rather than mode D as in Figure 16.9b. (c) The polygon
diagram, incorporating Figure 16.8a.

FIGURE 16.11

Command Diagram for the Adelman-Morriss Model

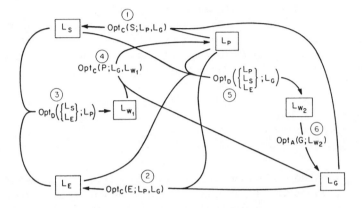

See Wold, "Model Construction and Evaluation When Theoretical
Knowledge Is Scarce—Theory and Application of Partial Least
Squares," in J. Kmenta and J. G. Ramsey, eds., <u>Evaluation of
Econometric Models</u> (New York: Academic Press, 1980), pp. 47-74
for an explanation of the blocks E, S, P, G. There are two
ancillary blocks. W_1 is spanned by Lvs L_S, L_E, and W_2 by Lvs
L_E, L_S, L_P. The numbering of the Opt commands corresponds
to the order of their computation, as described in the text.

    In a four-stage model, such as that of Adelman and Morriss,
a suitable command diagram will combine several of these modes
as in Figure 16.11. Two ancillary structures are invoked here:
a plane W_1 spanned by Lvs L_E and L_S of the two exogenous
blocks "economic levels" and "social conditions"; and a hyperplane W_2 spanned by those two Lvs and also L_P ("political conditions") as well—all the Lvs causally prior to the criterion
variable "growth rates." Latent variables L_S and L_E may be
computed (1 and 2 in the figure) from their communicating Lvs
L_P and L_G by simply averaging of the projections, mode-C
optimization, as in Figure 16.5. For the operation (4) whose
output is L_P, the instrumental variable in this little model, we
additively combine $Opt_B(W_1; L_P)$, the net projection (3) of
the exogenous variables onto the P-block (as in Figure 16.8a),
with the projection of L_G from the "other direction," causally
speaking. (The role of this intermediate block P is the same as
that of block Z in Figure 16.10a, even though one of the arrows

is reversed. The Lv Lp enters into two different linear determinations the efficacies of which—one a multiple R^2, one a simple r^2—should be strictly summed.) Finally, the operation (6) whose output is L_G invokes the ancillary Lv L_{W_2} (5) in a construction which extends that of Figure 16.8a by one further block. Without information about the matrices Σ_{EE}, Σ_{SS}, Σ_{GG}, Σ_{PP} one cannot guess at their condition numbers or prescribe mode-B projection with respect to any but the blocks W_1, W_2 of ancillaries; then operations 1, 2, 4, and 6 are all performed item by item.

HOW SOFT MODELING RELATES TO CANONICAL ANALYSIS AND TO MAXIMUM-LIKELIHOOD ESTIMATION

Expansion of so modest an assumption as the mere block structure into these detailed sequences of regressions can be justified by recourse to the rationale of canonical analysis. Were we presented (well out of context of soft modeling) with the problem of predicting Z-ness by the X- and Y-items in Figure 16.1, we would no doubt settle upon a global figure-of-merit, the proportion of variance in L_Z explained. By this criterion the optimum L_Z is the first canonical variate of the Zs with respect to the pool of the Xs and the Ys together; L_X would then be the collection of X-terms in the predictor, L_Y the Y-terms. Such an analysis neglects the existence of the X- or Y-blocks as separate pools of variables presumed to have some coherence among themselves.

I find the easiest way of declaring the block structure to be by way of side conditions upon the optimization. An insistence that L_X, the X-component in the prediction function of L_Z, be itself an optimal predictor of L_Z seems to imply the existence of the X-block satisfactorily. We shall insist, too, that L_Y be optimal for prediction of L_Z from the Y-block alone. That L_Z which is optimally predicted by a linear combination of L_X and L_Y obeying these side conditions is just the fixed-point drawn out in Figure 16.9b. In other words, at convergence of the command sequence of Figure 16.9a, L_Z is in canonical relation to the plane of its own best predictors. The soft model, by restating this optimum in terms of partial least squares, has provided a simple iterative algorithm for the analytically unwieldy solution. In all these determinations the polygon diagram lets us see what we are doing.

The preceding paragraph may be thought of as adumbrating a "model" which the partial least squares (PLS) procedure is "estimating." Soft modeling is in effect the estimation of a recursive system of simultaneous linear equations under very specialized nonlinear constraints. Nevertheless, most writers on the subject, in particular Wold and myself, choose not to make much of this interpretation, for two reasons. First, in the context of any family of distributions likely to have generated the data in practice, the constraints make no sense at all. They are not counterfactual, merely irrelevant. In specifying that the analysis be consistent with the only prior knowledge we have, the block-causal structure, they speak not about the data but about the explanatory use we propose to make of them, explanations whose strength is embodied in the sums-of-squares we are optimizing. Second, from such a model it would be pointless to pass to distribution-based estimates of standard errors in the parameters. These errors conform only to sampling variation in the units of analysis, but PLS is applied rather in the attempt to smooth out into Lvs those long lists of proxies. The sampling variation most crucial for reasoned applications is of those lists of items themselves, and as the items are not drawn randomly from any universe, no sampling theory is of any avail here at all.

Axiomatic simplicity aside, the main advantage of soft modeling by the partial least squares approach over likelihood-based estimators such as LISREL is the proximity of maximands to our intuitive statistical experience, the possibility of viewing the fixed-points in diagrams of perpendiculars and linear combinations. Only during a single regression is likelihood to be seen in the dual space, manifested as ellipses about the foot of a perpendicular. Otherwise, likelihood relates to our vector geometry not by these visible constructions but through hyper-volumes and determinants, far too subtly for any simple depiction at all.

In short, likelihood is invisible. Its maxima have no tangible attributes in terms of the subsystems of a model, the arenas wherein our explanatory theories stand or fall. At the conclusion of a partial least-squares analysis, by contrast, we have a collection of true sentences, embodying geometric facts, about precisely those local optima which we know how to interpret: little two- and three-block configurations. In "complex situations with scarce prior information," least-squares optima are treacherous enough, what with outliers, multicollinearity, item selection bias, that ubiquitous misspecification of levels, and all the other familiar difficulties. It seems gratuitous to augment these very real frustrations by the introduction of multivariate normality assumptions, various hypotheses of zero residual correlation, and the like, just for the sake of arriving at asymptotic estimates of the covariance matrix of error of estimate. The decentralization of soft modeling corresponds to a lack of faith, in interdisciplinary problems of any complexity, that the data will bear the imposition of a single function, however general, relating the observed covariance structure to a theoretical distribution.

17

A New Approach to
Nonlinear Structural Modeling
by Use of Confirmatory
Multidimensional Scaling

Claes Fornell and Daniel R. Denison

INTRODUCTION

The decade following the introduction of nonmetric multidimensional scaling (NMDS) by Shepard (1962) saw numerous
marketing applications utilizing a variety of algorithms (Green
and Rao 1972). Major advantages of the nonmetric scaling
methods over traditional metric models of analysis were that it
was not necessary to specify a specific functional form and that
ordinal input data were sufficient. Another advantage of the
nonmetric methods over previously fully nonmetric methods was
that the nonmetric input produced metric output.

Yet NMDS has several shortcomings that have limited its
use in recent years, the most serious of which is that the analysis
cannot be effectively guided by substantive theory. There are
also problems of local optima and the metric determinacy of the
solution (which depends on the number of points to be scaled
relative to the number of dimensions necessary to represent
the data). Further, it has not been possible to use statistical
theory in the evaluation of results. Hence, even in the case
where a global optimum has been reached, one is still faced
with a subjective evaluation with respect to the extent to which
the obtained results are consistent with some a priori theory
(Lingoes, 1980).

As a result of these shortcomings, which have been discussed in the marketing literature by Green and Carmone (1970),
Green and Rao (1972, 1977), and Green and Wind (1973), NMDS
has been used essentially as a descriptive tool. For purposes
of theory testing and statistical inference, other models, such

as structural equations, are more useful. Because these models can analyze systems of relationships, address errors in measurement, handle abstract or unobserved variables, and confront as well as combine empirical data with theoretical knowledge, they are very powerful tools of analysis.

Before the development of structural equation models such as Jöreskog's LISREL and Wold's PLS, researchers had to settle for analysis techniques that incorporated some but not all of these capabilities. For example, multiple regression includes a theory for confronting data with a priori hypotheses via statistical testing but assumes perfect measurement of predictor variables; traditional principal components can model unobservables but it is essentially a descriptive tool without inferential powers.

This chapter presents a new analysis methodology that not only overcomes the shortcomings of traditional NMDS but also offers an alternative or complement to existing structural equation models. Recent advances by Carroll, Purzansky, and Kruskal (1980), Bloxom (1978), Lee and Bentler (1980), Borg and Lingoes (1980), and Lingoes and Borg (1978) now make it possible to impute theoretical knowledge in the empirical analysis and to use statistical theory for hypothesis testing in a confirmatory MDS analysis. Fornell and Denison (1981) have shown that confirmatory NMDS can be used to assess convergent, discriminant, and nomological validity in a manner that is similar to the LISREL approach. In both cases constraints are imposed on a model according to certain validity criteria. The resulting and constrained model is then compared to a model without validity constraints. If the fit to the data of the constrained model is significantly worse than the fit of the unconstrained model, it is concluded that the difference in fit cannot be attributed to chance and the model is rejected.

The purpose of this chapter is to develop the NMDS methodology further and demonstrate not only that it can handle systems of relationships, impute a priori (theoretical) knowledge, and subject the result to statistical testing, but that it can also deal with abstract variables and measurement error. In addition to incorporating a priori knowledge by imposing external constraints and making inferential statements about the results, a new approach to estimating measurement errors and structural parameters will be suggested.

Consequently, it will be argued that NMDS now offers a viable alternative to better-known structural equation models. Its basic appeal is that it is virtually free of assumptions regarding level of measurement, specification of functional form, independence of observations, sample size, and variable distributions.

It also offers a flexible method for determining the degree to which validity considerations are met by the data.

Following a brief and nontechnical introduction of confirmatory nonmetric multidimensional scaling, we present our approach to the analysis of unobservables with measurement error in the context of MDS. Subsequently, we illustrate the approach with empirical data.

CONFIRMATORY NONMETRIC MULTIDIMENSIONAL SCALING

The NMDS approach to the analysis of data structure rests on two basic assumptions: relationships between variables may be interpreted as distances between points in a multidimensional space; and proximity data can be taken as a basis for establishing a set of order relations on interpoint distances. The overall objective is to find a configuration whose rank order of ratio-scaled distances best reproduces the rank order of the input data in the fewest number of dimensions.

The interpretational difficulties posed by local optima, indeterminacy, and lack of statistical inference are overcome by confirmatory MDS, which makes it possible to impose external constraints on the solution. These external constraints may be derived from measurement requirements and/or substantive theory. In the case of multiple local minima that are not significantly different in terms of their ability to represent the data, the "preferred" configuration would be one that best satisfies both external constraints and data fit. Hotelling's t-test has been adapted (Lingoes and Borg, 1980) as a means of comparing two configurations of points as representations of the same proximity matrix. Thus, one can express the difference between a configuration with and without external constraints in terms of the probability that both represent the same set of proximities presented in the original data.

There are several computer algorithms available for confirmatory MDS. In this study we will use the Constrained/Confirmatory Montone Distance Analysis (CMDA) developed by Borg and Lingoes (1980). This algorithm constrains distances in an NMDS configuration to satisfy ordinal specifications between points by use of a penalty function. It differs from traditional MDS in that the minimization of the loss function is conditional upon target matrices (for a detailed discussion, see Borg and Lingoes 1980). However, the methodology we propose and the analyses we perform are not specific to this algorithm. MDSCAL-5 and

KYST should produce equivalent results (see Carroll, Pruzansky, and Kruskal, 1980). In addition, other algorithms are available for different types of constraints. For example, CANDELINC (Carroll, Pruzansky, and Kruskal, 1980) and the technique of Bentler and Weeks (1978) impose linear constraints. An approach by Bloxom (1978) considers equality constraints. Finally, Lee and Bentler (1980) have developed a method that obtains solutions subject to nonlinear relations.

STRUCTURAL MODELING WITH CONFIRMATORY MULTIDIMENSIONAL SCALING

Much of the logic behind the analysis covariance structures derives from the multitrait-multimethod matrix of Campbell and Fiske (1959). Most analysis procedures can be seen as operationalizations of two fundamental principles: Multiple measures of the same construct must converge within as well as discriminate between the different constructs under consideration.

In addition, approaches to structural modeling include a system for imposing constraints implied not only by measurement theory but by substantive theory as well. That is, structural modeling specifies a methodology for determining which properties of the constraints constitute convergent, discriminant, and nomological validity, along with estimation procedures for assigning a numerical value to the relationships between indicators and constructs, and to the relationships between constructs. These two steps (imposing constraints that operationalize the three types of validity and estimating the indicator-construct and construct-construct links) have yet to be explored in connection with multidimensional scaling. The remainder of this section of the chapter develops an approach to accomplish this.

Convergent and Discriminant Validity

One schema for imposing a set of order relations on an MDS solution can be derived within the framework of facet theory (Guttman 1959; Borg 1977). Consider first the requirements for convergent and discriminant validity, where the former is the degree to which two or more attempts to measure the same construct are in agreement and the latter is the degree to which a construct differs from other constructs. For the sake of simplicity, assume that there are two constructs, each represented by a number of indicators. In the terminology of

MDS, a construct is a region and its indicators are points in that region. If R_1 and R_2 represent two separate regions, a definitional mapping schema for clustering points according to convergent-discriminant requirements can be expressed as:

$$\left[\begin{array}{l} \text{Each point of } R_1 \text{ is closer to } \left\{ \begin{array}{l} \text{all} \\ \text{some} \end{array} \right\} \begin{array}{l} a_1 \\ a_2 \end{array} \text{ points in} \\[3mm] R_1 \text{ than it is to } \left\{ \begin{array}{l} \text{all} \\ \text{some} \end{array} \right\} \begin{array}{l} b_1 \\ b_2 \end{array} \text{ points of } R_2. \end{array} \right] \qquad (1)$$

Constraint conditions $a_1 b_1$, for example, require each point within region 1 (or, stated differently, each indicator of construct 1) to be closer to all other points within that region than to any other point within region 2. Similarly, each point within region 2 should be closer to all other points within region 2 than to any point within region 1. $a_1 b_1$ represents the most stringent form of convergent-discriminant validity, while the other possible constraint conditions ($a_1 b_2$, $a_2 b_1$, and $a_2 b_2$) are operationalizations of weaker forms.[1]

Nomological Validity

Facet theory can also be used to derive a schema for nomological validity criteria. Nomological validity is concerned with the extent to which predictions from the theory embodied by the model are confirmed. In other words, the constraint conditions relative to convergent-discriminant validity are measurement criteria, while the constraint conditions for nomological validity are more concerned with the underlying substantive theory. For three regions, we operationalize the nomological criteria as:

$$\begin{array}{l} \text{Each point of } R_1 \text{ is closer to } \left\{ \begin{array}{l} \text{all} \\ \text{some} \end{array} \right\} \begin{array}{l} c_1 \\ c_2 \end{array} \text{ points of} \\[3mm] R_2 \text{ than to } \left\{ \begin{array}{l} \text{all} \\ \text{some} \end{array} \right\} \begin{array}{l} d_1 \\ d_2 \end{array} \text{ points of } R_3, \text{ where } R_2 \text{ is a} \\[3mm] \text{region that is adjacent to } R_1, \text{ and } R_3 \text{ is a region} \\ \text{that is distant from } R_1. \end{array} \qquad (2)$$

For example, the constraint condition $c_1 d_1$ requires that each point of region 1 be closer to all points within region 2 than to any point within region 3. This operationalizes testing criteria for a theory postulating that construct 1 is more closely related

FIGURE 17.1

Combined Validity Constraints

Measurement validity *(handwritten)*

Theory validity *(handwritten)*

	Convergent–Discriminant Validity			
	$a_1\,b_1$	$a_2\,b_1$	$a_1\,b_2$	$a_2\,b_2$
$c_1\,d_1$	Strong Convergent–Discriminant Validity. Strong Nomological Validity.			Weak Convergent–Discriminant Validity. Strong Nomological Validity.
$c_2\,d_1$				
$c_1\,d_2$				
$c_2\,d_2$	Strong Convergent–Discriminant Validity. Weak Nomological Validity.			Weak Convergent–Discriminant Validity. Weak Nomological Validity.

(left axis label: Nomological Validity)

to construct 2 than it is to construct 3. As in the convergent-discriminant case, other possible conditions ($c_1\,d_2$, $c_2\,d_1$, and $c_2\,d_2$) provide operationalizations of weaker forms of nomological validity.[2]

Combining the mapping schemas for clustering (convergent-discriminant validity) and for proximity between regions (nomological validity) generates the matrix of constraints presented in Figure 17.1.

Parameter Estimation

Given the mapping of items in a geometrical space of the smallest dimensionality, there are many ways of assigning a numerical value to the relationships between items. Not only are there several possible distance functions (for example,

Euclidian, city block, Minkowski's p-metric), there are also different ways of conflating several distances into a single estimate. The approach we will describe addresses the problem of random measurement error and is able to handle abstract unobserved variables. It draws upon classical measurement theory, is simple to use, and is not statistically advanced.

If measurement errors are random, the expected value of the indicators of a construct will approximate the true value of the construct. This is calculated by using the Euclidian distance function to find the mean value on each dimension for the points in the relevant region, or

$$C_{ik} = \frac{\sum_{j=1}^{n} x_{ijk}}{n_i}$$

Assumes each vble adds equally to the construct?

(3)

where:

C_{ik} is the projection of the ith centroid in the kth dimension
x_{ijk} is the projection of the jth indicator of the ith construct in the kth dimension, and
n_i is the number of indicators in the ith construct.

The relationships between constructs are estimated at the Euclidian distances between centroids:

$$d_{ij} = \left[\sum_{k=1}^{R} (C_{ik} - C_{jk})^2 \right]^{1/2},$$

(4)

where:

d_{ij} is the distance between the ith and the jth constructs,
R is the number of dimensions,
C_{ik} is projection of the ith centroid in the kth dimension, and
C_{jk} is projection of the jth centroid in the kth dimension.

According to classical measurement theory, measurement error is equal to the difference between observed and true values. Having defined the true value as the centroid in equation 3, we may express this error as

$$e_{iq} = \left[\sum_{k=1}^{R} (C_{ik} - X_{ikq})^2 \right]^{1/2},$$

(5)

where:

e_{iq} is the distance between the ith construct and its qth indicator, and

X_{ikq} is the projection of the qth indicator of the ith construct in the kth dimension.

Figure 17.2 illustrates indicator-construct and construct-construct relationships in a two-dimensional space. The distances between the centroids ($X_A - X_B$, $X_A - X_C$, $X_B - X_C$) are estimates of structural parameters. The distance between a centroid and its corresponding indicators (for example, $X_A - A_1$, $X_A - A_2$, $X_A - A_3$) is an estimate of the degree to which an indicator represents the construct.

Since the interpoint distances in MDS are an inverse monotone function of the corresponding correlation coefficients, the interpretation of the distance between a regional centroid and a point in that region as relative measurement error is similar to the conceptualization of measurement error in LISREL and PLS.

A PROPOSED ANALYSIS STRATEGY

Having defined the criteria for evaluating measurement quality and testing theory, and the procedure for parameter estimation, we now turn to the analysis strategy. At the outset, it may be helpful to establish two fundamental principles.

First, in simultaneous analysis procedures where parameter estimates may change throughout the model given a change in a part of the model (deletion or addition of a variable, a new constraint, and so on), it makes little sense to use a sequential assessment of validity. For example, it is entirely possible that convergent-discriminant validity may be satisfactory in a confirmatory factor analysis model but would fail in the context of nomological validity where the model is extended to structural relationships among the factors. A similar contention can be made for validity testing within CMDA. Thus we propose that the validity testing should be conducted simultaneously rather than sequentially, that is, both measurement and theory should be evaluated within the same context.

Second, there are no fixed or absolute criteria for determining whether or not a theory is false. The same is true for validity assessment of measurement (compare Campbell and Fiske, 1959). What is acceptable in one study may be unacceptable in another. Moreover, what is acceptable according to one method

FIGURE 17.2

Euclidian Estimation Procedure

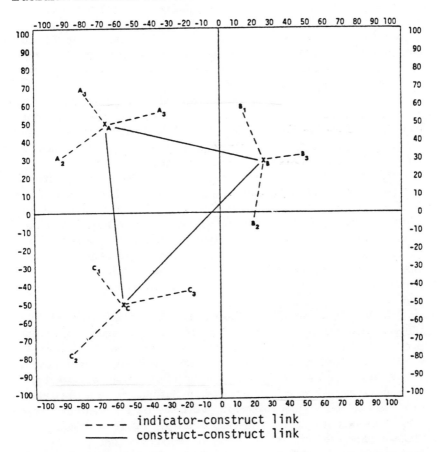

- - - - indicator-construct link
——— construct-construct link

may be unacceptable according to another. In the final analysis, evaluation of theory and measurement quality is determined by subjective judgment. What we will propose is an approach to simultaneously assess theory and measurement and to provide objective statistics upon which the subjective judgments can be made. Our clustering schemas provide a means for assessing the degree of validity.

As with traditional theory testing, the analyst can have predetermined significance criteria. Referring to Figure 17.1, the validity requirement may, for example, be set to $a_1 b_1 c_2 d_1$ at a significance level of .10. Thus, if the difference between a solution with the external constraints $a_1 b_1 c_2 d_1$ and an unconstrained configuration is insignificant ($p > .10$), it is concluded that the validity is acceptable.

An alternative to use of predetermined criteria is use of the strongest possible test. Again referring to Figure 17.1, this analysis strategy begins by subjecting the data to the constraint condition specified by a_1 b_1 c_1 d_1 (the upper left-hand cell in Figure 17.1). If the probability of a solution with these constraints is unlikely, the analysis proceeds to less stringent measurement requirements (by moving horizontally in Figure 17.1) and/or to less stringent theoretical requirements (by moving vertically in Figure 17.1). The analysis continues until a probable solution has been found. The constraints associated with this solution measure the degree of convergent, discriminant, and nomological validity that is met by the data.

In determining whether or not a model with constraints fits the data, several statistics are available. The most important of these is a t-test based on Hotelling's (1940) model for determining regression predictor effectiveness. The null hypothesis is that there is no difference between the constrained and unconstrained solutions with respect to their representations of the raw order relations in the data. Thus, similar to LISREL solutions, CMDA solutions with different constraints can be evaluated statistically.

In applying Hotelling's model to CMDA, the estimated distance is the predictor variable and the original ordering of the proximity data is the criterion variable. The test considers whether the distances obtained under a particular constraint are as effective as predictors as are the distances obtained in an unconstrained solution. Although Hotelling's test assumes interval scaling, this problem is overcome by an isotonic transformation that linearizes the relationship. Further, simulation studies (Lingoes and Borg, 1980) demonstrate that the test is insensitive to departures from normality in the dependent variable as well as to violations of the independence assumptions.

Other statistics that help distinguish the difference between CMDA solutions are the coefficient of alienation (Guttman, 1964), the rank order correlation between the order of distances in a constrained solution and the order of distances in an unconstrained solution, and the rank order correlation between the order of distances in a constrained solution and the order of distances in the original data. These statistics are essentially descriptive, although differences in the size of the rank order correlations might conceivably be evaluated through Fisher's Z transformation, which is robust against departures from normality and homoscedasticity.

ILLUSTRATION

We now present an illustration of structural modeling via multidimensional scaling. The CMDA method is used in tests of both convergent-discriminant and nomological validity. For comparison purposes, a parallel set of analyses using LISREL is also presented. Both models produce similar conclusions regarding overall structure but lead to differences in individual parameter estimation. We discuss these differences in connection with the strengths and weaknesses of each method.

Nonmetric confirmatory multidimensional scaling can be used for many different types of studies. It is particularly useful in analyses where the researcher can impute a priori knowledge not only according to indicator-construct relationships (measurement), but also according to the relative strength of relationships between constructs (substantive theory). There are many theories/models in marketing that specify such relationships. One example is the model presented by Hauser and Urban (1977) in the area of consumer choice, which postulates a sequence of related constructs: perception → preference → choice. A test of this structure would require the distance between preference and perception to be less than the distance between perception and choice. The latter distance should also be greater than the distance between preference and choice.

In addition to hierarchy-of-effects models, hypotheses involving intervening variables are also well suited to confirmatory multidimensional scaling. It is also possible to evaluate theory when a priori knowledge about construct-construct relationships is lacking. This would, of course, be a weaker test since the constraint conditions derive solely from measurement considerations (that is, convergent-discriminant validity).

The example used in this illustration is one in which relative distances between constructs are suggested by theory. The theory concerns organizational participation and is applied to consumer affairs influence in marketing decisions (Fornell, 1976). Let us briefly describe the data and the hypotheses.[3]

One way for consumers to affect the behavior of a firm is to voice their concerns to an organizational boundary unit (Adams, 1976) such as a corporate department for consumer affairs. This department may have access to different types of organizational communication channels that it can utilize to represent the consumer in marketing decision making. The theory postulates a link between a consumer affairs department's channel utilization and its active participation in marketing decisions. A second link concerns the veto power held by the consumer affairs department. Veto power is a strong form of influence and is hypothesized to operate through active participation. Consequently, there are three constructs: channel utilization, active participation, and veto power. Nomological validity implies that the distance between channel utilization and veto power is greater than the distance between active participation and veto power.

Possession and utilization of communication channels were measured by three indicators: (1) the number of planning committee memberships held by consumer affairs, (2) the extent to which consumer affairs was involved in coordinating the activities of other departments, and (3) the extent to which consumer affairs was involved in educating company personnel in consumer matters (both of the last two items were measured on a five-point scale). Participation was measured by four composite indexes representing active involvement in (1) new product development, (2) product management, (3) communication/advertising, and (4) pricing. Veto power was measured in a similar manner for the same decision areas. Thus there are three indicators of channel utilization (x_1, x_2, x_3), four indicators of active participation (y_1, y_2, y_3, y_4), and four indicators of veto power (y_5, y_6, y_7, y_8). The correlation matrix and Cronbach's alpha for the composite variables are presented in Table 17.1.

The correlation matrix was transformed to a rank-order matrix of order relations and used as input into a multidimensional scaling algorithm whose solution provides a baseline model and a determination of the minimum dimensionality needed to represent the data. This part of the analysis used the MINISSA-1 algorithm (Lingoes, 1973), which produced the three-dimensional configuration shown in Figure 17.3.

The three dimensional solution fits well. As shown in Figure 17.3, the coefficient of alienation (K) is .08, indicating that the order of the correlations can be represented by distances in three orthogonal dimensions with very few errors. In contrast to factor analysis, the dimensions are not interpreted; rather, regions or clusters of points represent constructs.4 Thus, there may be more constructs than dimensions.

Inspection of the configuration in Figure 17.3 reveals no immediate pattern of points; certainly there is no evidence of the existence of three distinct constructs. The question now is: Can we alter the solution to conform with validity criteria without significantly sacrificing goodness of fit (that is, the order of interpoint distances)?

The most stringent test of convergent-discriminant validity implies that each indicator-point of channel utilization is closer to all other indicator-points of channel utilization than to any

TABLE 17.1

Correlation Matrix (n = 138)
(Cronbach's alpha in diagonal for composites)

Variable Name	1	2	3	4	5	6	7	8	9	10	11
1. Planning committees (x_1)	.44										
2. Coordinate departments (x_2)	.32	.59									
3. Educate personnel (x_3)	.54	.31	.27								
4. New product development (y_1)	.53	.37	.20	.66							
5. Product management (y_2)	.45	.22	.11	.61	.57						
6. Communication/advertising (y_3)	.41	.20	.12	.38	.39	.70					
7. Pricing (y_4)	.25	.17	.11	.36	.31	.30	.59				
8. Veto/new product (y_5)	.48	.22	.11	.39	.23	.11	.18	.82			
9. Veto/product management (y_6)				.54	.50	.22	.33	.60	.66		
10. Veto/communication/ advertising (y_7)	.24	.14	.11	.16	.14	.49	.42	.36	.38	.67	
11. Veto/pricing (y_8)	.20	.06	.10	.14	.12	.15	.57	.38	.40	.41	.75

379

FIGURE 17.3

Unconstrained Solution

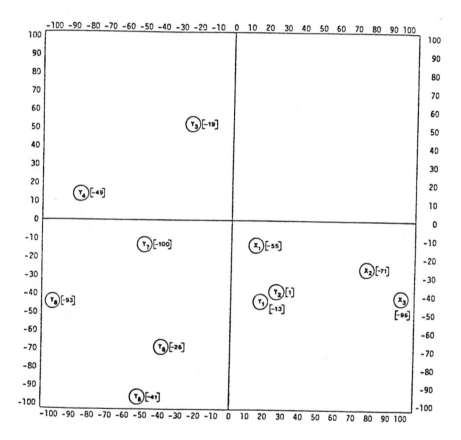

Coefficient of Alienation (K) = .08

Bracketed numbers are projections of each point on the third dimension

indicator-points of active participation and veto power. In addition, the indicator-points of active participation and veto power must conform to corresponding requirements. The most stringent test of nomological validity implies that each indicator-point of channel utilization must be closer to each indicator-point of active participation than to any indicator-point of veto power. Indicator-points of active participation must lie at a distance intermediate between channel utilization and veto power, and indicator-points of veto power must be closer to all indicator-points of active participation than to any indicator-points of channel utilization. The simultaneous test of all validity criteria according to the most stringent requirements is defined by a_1 b_1 c_1 d_1 in Figure 17.1. The results are presented in Figure 17.4.

The four statistics described earlier can be used to evaluate how well the data fit the model. p is the probability associated with Hotelling's t-test. In this case, $p < .000$ and the null hypothesis that there is no difference between the constrained solution and the unconstrained solution (Figure 17.3) is rejected. That is, the constrained model has a significantly inferior fit. K, the coefficient of alienation, is also larger. R_1 is the rank-order correlation between the order of the distances in the constrained solution and the order of the original data. R_2 is the rank-order correlation between distances in the constrained solution and distance in the unconstrained solution. These statistics indicate that the constraints have significantly altered the interpoint distances of the original input matrix as well as those of the unconstrained solution.

In sum, the four statistics suggest that the stringent validity constraints alter the original configuration substantially. Not only is the constrained solution highly improbable; it also introduces a good deal of alienation for a three-dimensional solution. A similar conclusion follows from the validity assessment via LISREL. The results of this analysis are presented in Figure 17.5. The chi-square is high, indicating a poor fit and an improbable model.

Since it is apparent that the stringent validity constraints cannot be satisfied, let us examine weaker criteria. Table 17.2 presents the statistics for a subset of the cells in Figure 17.1. As can be seen, all configurations resulting from constraints stronger than a_1 b_2 c_2 d_2 are highly improbable, whereas other constraints are easily satisfied. Given the criterion that one should select the model that best satisfies measurement and theory considerations without significantly worsening data fit, it is clear that the a_1 b_2 c_2 d_2 configuration should be selected

FIGURE 17.4

Combination of Strongest Form of Both Convergent-Discriminant
and Nomological Constraints

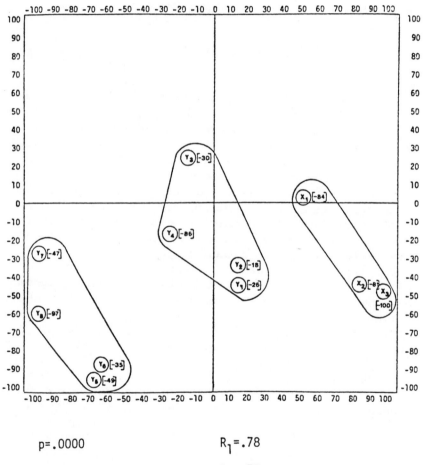

p=.0000   R$_1$=.78

K=.195   R$_2$=.79

FIGURE 17.5

Two-Stage Model
(combining convergent-discriminant and nomological validity)

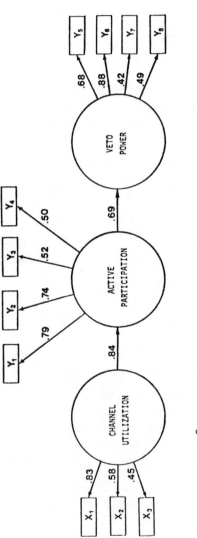

$$\frac{\chi^2}{df} = \frac{184.2328}{42} = 4.39$$

p=.0000

TABLE 17.2

Summary of CMDA Constraint Conditions

Nomological Constraints	Convergent-Discriminant Constraints		
	$a_1 \, b_1$	$a_1 \, b_2$	$a_2 \, b_2$
$c_1 \, d_1$	P $= .0000$	P $= .0000$	P $= .0000$
	K $= .195$	K $= .195$	K $= .164$
	$R_1 = .78$	$R_1 = .78$	$R_1 = .84$
	$R_2 = .79$	$R_2 = .79$	$R_2 = .87$
$c_2 \, d_2$	P $= .0000$	P $= .5761$	P $= .6723$
	K $= .217$	K $= .140$	K $= .103$
	$R_1 = .76$	$R_1 = .93$	$R_1 = .94$
	$R_2 = .77$	$R_2 = .93$	$R_2 = .97$
None	P $= .0000$	P $= .5549$	P $= .7463$
	K $= .181$	K $= .139$	K $= .105$
	$R_1 = .81$	$R_1 = .93$	$R_1 = .94$
	$R_2 = .82$	$R_2 = .93$	$R_2 = .97$

P = Probability that the difference in the ability of the constrained and unconstrained solutions to predict the order
in the original data would occur by chance.
K = Guttman-Lingoes coefficient of allienation.
R_1 = Correlation of the order of the distances in the constrained
solution and the order of the original data.
R_2 = Correlation of the order of the distances in the constrained
solution and the order of the distances in the unconstrained
solution.

FIGURE 17.6

$a_1 b_2 c_2 d_2$ Configuration

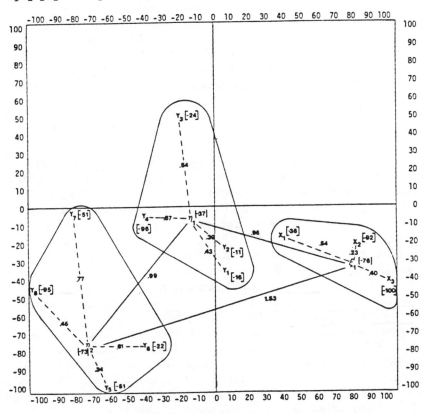

for further analysis. The results, including parameter estimates, are given in Figure 17.6.

From the clustering schemas presented earlier we can determine the degree of validity met by the data. Specifically, we have met the constraints that each indicator-point channel utilization is closer to <u>all</u> other indicator-points of channel utilization than it is to <u>some</u> indicator-points of active participation and veto power, and that each indicator-point of channel utilization is closer to <u>some</u> indicator points of active participation than to <u>some</u> indicator-points of veto power.

Thus, the CMDA analysis suggests that there is some degree of validity for our model. A nested LISREL model supports this conclusion. Since the two-stage model with three constructs (Figure 17.5) exhibited a very large chi-square, it was compared to a model postulating that all indicators belong

to a single factor. If this model is a probable representation of
the data, it means that discriminant validity (and, by implication,
nomological validity) must be rejected. The results are shown
in Figure 17.7, which depicts perfect correlations among the
constructs (which is, of course, the same as constructing a
single-factor model). The chi-square is higher than for the
two-stage, three-construct model. According to the likelihood
ratio test, it is clear that the two-stage, three-construct model
represents a significant improvement over the one-factor model.

CONCLUSION

Both LISREL and CMDA suggest a similar overall interpretation. CMDA specifies exactly what constraints are met by
the data, while LISREL indicates that the model in Figure 17.5
has a significantly better fit than the model in Figure 17.7.
When all elements are specified according to a priori theory,
both models suggest that validity has not been met. Only when
we test by weaker operationalizations of validity measurement
is the theory supported by the data.

However, if we attempt to isolate the location of errors in
the model, LISREL and CMDA do, not surprisingly, provide
different conclusions. Aside from pinpointing y_7 as having the
largest measurement error (.77 in CMDA, see Figure 17.6 and
$1-.42^2 = .82$ in LISREL, see Figure 17.5) the distribution of
relative error size does not correspond across the two models.
There are several reasons for this divergence. First, the models
are not really compatible: the LISREL model is highly improbable
($p = .000$), whereas the CMDA model is able to recover the
ordering of the original data quite well ($p = .58$). Second, the
models have different objectives and different assumptions.
LISREL attempts to account for input data in the form of a
variance-covariance matrix and assumes multinormality, large
samples, independent observations, and linear relationships.
CMDA attempts to account for input data in the form of a
proximity matrix and makes no assumptions except the requirement of monotonicity.

Further, the indicator-construct links are estimated via
very different procedures. An unobserved variable in LISREL
is assigned empirical meaning by all indicators in the model—
not merely by the indicators to which it bears epistemic relationships. As a result, it may be difficult to separate the various
sources of validity criteria that determine the construct-indicator
links. In effect, estimated epistemic relationships will be stronger

FIGURE 17.7

One-Factor Model

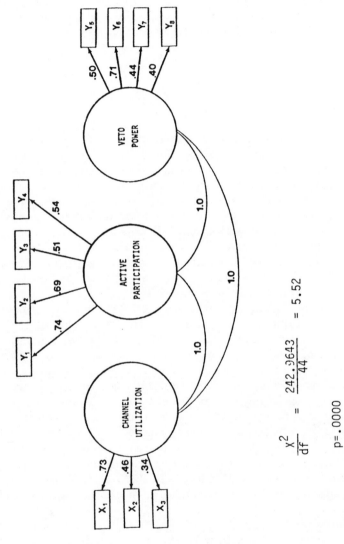

$$\frac{\chi^2}{df} = \frac{242.3643}{44} = 5.52$$

$$p = .0000$$

and measurement error smaller for indicators that covary with indicators of other, structurally related, constructs. This may lead to confounding problems (see Burt, 1976). For example, lack of discriminant validity may be interpreted as a significant structural relationship. In the language of MDS, it would be as if the centroid were not the expected value of the position of each of the indicator-points on all dimensions, but instead was estimated in a manner that optimized the relationships between constructs.

IMPLICATIONS

When the sample size is small and/or when the analyst is not able to specify the functional form, and when there is a system of relationships and many parameters to be estimated, confirmatory NMDS is an attractive alternative to structural equation models. For marketing, it should have a great deal of appeal—especially in studies of industrial markets where the number of observations across subjects often is small and the number of observations within subjects is large.

The price of liberating analysis from most of the assumptions of structural equation models is that confirmatory NMDS cannot provide all the evaluative (inferential and descriptive) statistics associated with, say, LISREL or PLS. Since the Euclidian distance is symmetric, the structural estimate does not in itself indicate directionality, let alone causality. This does not mean, however, that the testing of causal theory is irrelevant to confirmatory NMDS. Conceptualizing causal theories in terms of systems of relationships implies that there are certain structures among the components of the system. These structures can be specified via facet theory and subjected to rigorous tests of refutation within the methodology we described.

Compared to structural equation models, confirmatory NMDS is probably less able to evaluate individual parameter estimates. As with LISREL and PLS, an inferential statistic is available for the assessment of overall fit. Unlike these models, however, NMDS is not yet able to evaluate the significance of individual parameter estimates. Borg and Lingoes (1980) have found that confirmatory NMDS is not particularly powerful in discriminating among competing theories. Therefore, as is the case with LISREL (Fornell and Larcker, 1981) the usefulness of an overall goodness-of-fit measure may be limited. Therefore, it may be useful to apply a condition that constrains in a more continuous manner all but one or all but two of the points in a particular category, rather than a condition that changes from an all-point constraint to one that constrains some points. Notwithstanding, confirmatory NMDS is a promising method with much to offer, especially given the few assumptions that are required. It should find frequent application in market research as well as in other social sciences. The methodology presented here is an attempt to demonstrate one way in which it can be used.

NOTES

1. Even more stringent forms of convergent-discriminant validity could be imposed by qualifying the term "closer." One might, for example, require that the distance between any two points within the same region be one-half (or any other value) the distance between points in different regions.

2. Again, there are theoretically many different constraint conditions that could be imposed. With three points per region, for example, separate conditions could constrain three, two, and then one of those points. The same could be true in the case where there are n points per region.

3. For the substantive theory behind this model, see Fornell (1976).

4. For an early illustration, see Guttman (1965).

REFERENCES

Adams, J. S. (1976). "The Structure and Dynamics of Behavior in Organizational Boundary Roles." In Handbook of Industrial and Organizational Psychology, ed. M. Dunnette. Chicago: Rand McNally, 1175-1199.

Bentler, P. M. and D. G. Weeks (1978). "Restricted Multidimensional Scaling Models." Journal of Mathematical Psychology 17, 138-51.

Bloxom, B. (1978). "Constrained Multidimensional Scaling in N Spaces." Psychometrika 43, 397-408.

Borg, I. (1977). "Some Basic Concepts in Facet Theory." In Geometric Representations of Relational Data, ed. J. C. Lingoes, J. Roskan, and I. Borg. Ann Arbor, Mich.: Mathesis Press, 65-102.

Borg, I. and J. C. Lingoes (1980). "A Model and Algorithm for Multidimensional Scaling with External Constraints on the Distances." Psychometrika 45, 25-38.

Burt, R. S. (1976). "Interpretational Confounding of Unobserved Variables in Structural Equation Models." Sociological Methods and Research 5, 3-52.

Campbell, D. T. and D. W. Fiske (1959). "Convergent and Discriminant Validation by the Multitrait-Multimethod Matrix." Psychological Bulletin 56, 81-105.

Carroll, J. D., S. Pruzansky, and J. B. Kruskal (1980). "CANDELINC: A General Approach to Multidimensional Analysis of Many-Way Arrays with Linear Constraints on Parameters." Psychometrika 45 (1), 3-24.

Fornell, C. (1976). Consumer Input for Marketing Decisions: A Study of Corporate Departments for Consumer Affairs. New York: Praeger.

Fornell, C. and D. Denison (1981). "Validity Assessment via Confirmatory Multidimensional Scaling." In The Changing Marketing Environment: New Theories and Applications, 1981 Educators' Conference Proceedings, ed. K. Bernhardt et al., Chicago: American Marketing Association, 334-37.

Fornell, C. and David F. Larcker (1981). "Evaluating Structural Equation Models with Unobservable Variables and Measurement Error." Journal of Marketing Research 18 (February), 39-50.

Green, P. E. and F. J. Carmone (1970). Multidimensional Scaling and Related Techniques in Marketing Analysis. Boston: Allyn and Bacon.

Green, P. E. and V. R. Rao (1972). Applied Multidimensional Scaling: A Comparison of Approaches and Algorithms. New York: Holt, Rinehart and Winston.

____ (1977). "Nonmetric Approaches to Multivariate Analysis in Marketing." In Multivariate Methods for Market and Survey Research, ed. J. N. Sheth. Chicago: American Marketing Association, 237-53.

Green, P. E. and Y. Wind (1973). Multiattribute Decisions in Marketing: A Measurement Approach. Hinsdale, Ill.: The Dryden Press.

Guttman, L. (1959). "Introduction to Facet Design and Analysis." In Proceedings of the 15th International Congress of Psychology—Brussels. Amsterdam: North-Holland, 130-32.

_____ (1964). "The Structure of Interrelations among Intelligence Tests." In Proceedings of the 1964 Invitational Conference on Testing Problems, ed. C. W. Harris. Princeton, N.J.: Educational Testing Services, 25-37.

_____ (1965). "A Faceted Definition of Intelligence." Scripta Hierosolymitana 14, 166-81.

Hauser, J. R. and G. L. Urban (1977). "A Normative Methodology for Modeling Consumer Response to Innovation." Operations Research 25 (July-August), 579-619.

Hotelling, H. (1940). "The Selection of Variates for Use in Prediction with Some Comments on the General Problem of Nuisance Parameters." Annals of Mathematical Statistics 11, 271-83.

Jöreskog, K. G. and D. Sörbom (1981). LISREL V: Analysis of Linear Structural Relationships by Maximum Likelihood and Least Squares Methods. Chicago: National Education Resources.

Lee, S. Y. and P. M. Bentler (1980). "Functional Relations in Multidimensional Scaling." British Journal of Mathematical and Statistical Psychology 33, 142-50.

Lingoes, J. C. (1973). The Guttman-Lingoes Nonmetric Program Series. Ann Arbor, Mich.: Mathesis Press.

_____ (1980). "Testing Regional Hypotheses in Multidimensional Scaling." In Data Analysis and Informatics, ed. E. Diday et al. Amsterdam: North Holland, 191-207.

Lingoes, J. C. and I. Borg (1978). "CMDA-U: Confirmatory Monotone Distance Analysis-Unconditional." Journal of Marketing Research 15, 610-11.

____ (1980). "An Exact Significance Test for Dimensionality and for Choosing between Rival Hypotheses in Confirmatory Multidimensional Scaling." University of Michigan Computing Technical Report, No. 1., 1-22.

Shepard, R. N. (1962). "The Analysis of Proximities: Multi-dimensional Scaling with an Unknown Distance Function," Parts I and II. Psychometrika 27, 125-39; 219-46.